THERMODYNAMICS

Thermodynamics
A complete undergraduate course

Andrew M. Steane

OXFORD
UNIVERSITY PRESS

Great Clarendon Street, Oxford, OX2 6DP,
United Kingdom

Oxford University Press is a department of the University of Oxford.
It furthers the University's objective of excellence in research, scholarship,
and education by publishing worldwide. Oxford is a registered trade mark of
Oxford University Press in the UK and in certain other countries

© Andrew M. Steane 2017

The moral rights of the author have been asserted

First Edition published in 2017

Impression: 1

All rights reserved. No part of this publication may be reproduced, stored in
a retrieval system, or transmitted, in any form or by any means, without the
prior permission in writing of Oxford University Press, or as expressly permitted
by law, by licence or under terms agreed with the appropriate reprographics
rights organization. Enquiries concerning reproduction outside the scope of the
above should be sent to the Rights Department, Oxford University Press, at the
address above

You must not circulate this work in any other form
and you must impose this same condition on any acquirer

Published in the United States of America by Oxford University Press
198 Madison Avenue, New York, NY 10016, United States of America

British Library Cataloguing in Publication Data

Data available

Library of Congress Control Number: 2016937255

ISBN 978–0–19–878856–0 (hbk.)
ISBN 978–0–19–878857–7 (pbk.)

Printed and bound by
CPI Group (UK) Ltd, Croydon, CR0 4YY

Links to third party websites are provided by Oxford in good faith and
for information only. Oxford disclaims any responsibility for the materials
contained in any third party website referenced in this work.

To all the students with whom I have learned, and in memory of Jonathan Payne.

Acknowledgements

The impact of a textbook on the development of science is gradual and hard to measure, but if it smooths the way for teachers and students, it will liberate their time and focus their energies more productively. I would like to thank first of all Oxford University, and in particular its Department of Physics, for maintaining an academic atmosphere in which books like this can be written.

I learned this area of physics in the first instance by following the textbook by Adkins, and I would like to acknowledge teachers such as W.S.C. Williams and G. Brooker, and also other textbook writers. In alphabetical order, I thank C.J. Adkins, S.J. Blundell and K.M. Blundell, G. Carrington, R. Feynman, L. Landau, J.C. Lee, A.B. Pippard, J.A. Ramsay, J.R. Waldram, and M.W. Zemansky. Both Feynman and Lee were especially helpful in promoting physical understanding; Carrington provided a wealth of expert knowledge and detailed examples, Waldram was useful in getting quickly to the point in many areas, and clarifying some technical issues.

I thank Sonke Adlung and others at OUP for their part in bringing the book to print. Finally, I thank my family for their loving support, especially their willingness to forego whatever equally valuable activities have been squeezed out by this book project.

Contents

1	**How to use this book (†)**	1
	1.1 For the student	1
	1.2 For the teacher	1
2	**Introducing thermodynamics**	3
3	**A survey of thermodynamic ideas**	7
	3.1 Energy and entropy	7
	3.2 Concepts and terminology	12
	3.2.1 System	12
	3.2.2 State	14
	3.2.3 Extensive, intensive	18
	3.2.4 Thermodynamic equilibrium	20
	3.2.5 Temperature	22
	3.2.6 Quasistatic	22
	3.2.7 Reversible and irreversible	23
	3.2.8 Adiathermal, isentropic, adiabatic, isothermal	25
	3.2.9 Expansion coefficients, heat capacities	26
	3.2.10 Thermal reservoir	29
	3.3 The laws of thermodynamics	29
	3.4 Where we are heading	32
	Exercises	33
4	**Some general knowledge**	34
	4.1 Density, heat capacity	34
	4.2 Moles	35
	4.3 Boltzmann constant, gas constant	36
	4.4 Pressure and STP	37
	4.5 Latent heat	37
	4.6 Magnetic properties	38
5	**Mathematical tools**	40
	5.1 Working with partial derivatives	40
	5.1.1 Reciprocal and reciprocity theorems	42
	5.1.2 Integrating	45
	5.1.3 Mixed derivatives	46

†Sections marked with a dagger below are optional reading. They can be omitted without the loss of information required later in the book.

viii   *Contents*

		5.2	Proper and improper differentials, function of state	46
			5.2.1 Integrating factor	49
		5.3	Some further observations	49
			5.3.1 Alternative derivation of reciprocal and reciprocity theorems	49
			5.3.2 Integration in general	50
		Exercises		51

6 Zeroth law, equation of state 52

    6.1   Empirical temperature 54
        6.1.1   Equation of state 55
        6.1.2   Algebraic argument (†) 57
    6.2   Some example equations of state 59
        6.2.1   Ideal gas 59
        6.2.2   Thermal radiation 61
        6.2.3   Solids and wires 62
        6.2.4   Paramagnetic material 63
        6.2.5   Equations of state for other properties 65
    6.3   Thermometry 66
    Exercises 68

7 First law, internal energy 70

    7.1   Defining internal energy 70
        7.1.1   Heat and work 73
    7.2   Work by compression 74
    7.3   Heat capacities 77
        7.3.1   Energy equation 80
        7.3.2   Relation of compressibilities and heat capacities 82
    7.4   Solving thermodynamic problems 83
    7.5   Expansion 85
        7.5.1   Free expansion of ideal gas 85
        7.5.2   Adiabatic expansion of ideal gas 86
        7.5.3   Adiabatic atmosphere 87
        7.5.4   Fast and yet adiabatic? 88
    Exercises 89

8 The second law and entropy 93

    8.1   Heat engines and the Carnot cycle 93
        8.1.1   Heat pumps and refrigerators 95
        8.1.2   Two impossible things (equivalence of Kelvin and Clausius statements) 96
    8.2   Carnot's theorem and absolute temperature 97
        8.2.1   Carnot's theorem: reversible engines are equally, and the most, efficient 97
        8.2.2   Existence of an absolute temperature measure 98
        8.2.3   Hot heat is more valuable than cold heat 101

8.3	Clausius' theorem and entropy	102
8.4	The first and second laws together	105
8.5	Summary	106
	Exercises	106

9 Understanding entropy 108

9.1	Examples	109
	9.1.1 Entropy content	111
	9.1.2 Entropy production and entropy flow	112
9.2	But what is it?	113
	9.2.1 Entropy increase in a free expansion	115
9.3	Gibbs' paradox	116
	9.3.1 Entropy of mixing	118
	9.3.2 Reversible mixing	119
9.4	Specific heat anomalies	120
9.5	Maxwell's daemon	122
	9.5.1 Szilard engine	123
	9.5.2 The Feynman–Smoluchowski ratchet	125
9.6	The principle of detailed balance	127
9.7	Adiabatic surfaces (†)	128
9.8	Irreversibility in the universe	131
	Exercises	133

10 Heat flow and thermal relaxation 136

10.1	Thermal conduction; diffusion equation	136
	10.1.1 Steady state	138
	10.1.2 Time-dependent	139
10.2	Relaxation time	145
10.3	Speed of sound (†)	146
	10.3.1 Ultra-relativistic gas	148
	Exercises	148

11 Practical heat engines 150

11.1	The maximum work theorem	152
	11.1.1 Imperfections	152
11.2	Otto cycle	153
	Exercises	155

12 Introducing chemical potential 157

12.1	Chemical potential of an ideal gas	161
	12.1.1 Example: the isothermal atmosphere	163
12.2	Saha equation (†)	165
	Exercises	167

13 Functions and methods — 169

- 13.1 The fundamental relation — 169
  - 13.1.1 Euler relation, Gibbs–Duhem relation — 170
- 13.2 Thermodynamic potentials — 172
  - 13.2.1 Free energy as a form of potential energy — 174
  - 13.2.2 Natural variables and thermodynamic potentials — 175
  - 13.2.3 Maxwell relations — 176
  - 13.2.4 Obtaining one potential function from another — 177
- 13.3 Basic results for closed systems — 177
  - 13.3.1 Relating internal energy to equation of state — 178
  - 13.3.2 Sackur–Tetrode equation — 182
  - 13.3.3 Complete thermodynamic information — 186
- Exercises — 186

14 Elastic bands, rods, bubbles, magnets — 188

- 14.1 Expressions for work — 188
- 14.2 Rods, wires, elastic bands — 188
- 14.3 Surface tension — 190
- 14.4 Paramagnetism — 192
  - 14.4.1 Ideal paramagnet — 195
  - 14.4.2 Cooling by adiabatic demagnetization — 197
- 14.5 Electric and magnetic work (†) — 200
  - 14.5.1 Dielectrics and polarization — 202
  - 14.5.2 Magnetic work — 207
- 14.6 Introduction to the partition function (†) — 210
- Exercises — 212

15 Modelling real gases — 216

- 15.1 van der Waals gas — 219
  - 15.1.1 Phase change — 220
  - 15.1.2 Critical parameters and the law of corresponding states — 222
- 15.2 Redlich–Kwong, Dieterici, and Peng–Robinson gas — 224
- Exercises — 226

16 Expansion and flow processes — 228

- 16.1 Expansion coefficients — 228
- 16.2 U: free expansion — 229
  - 16.2.1 Deriving the equation of state of an ideal gas — 229
- 16.3 H: throttle process: Joule–Kelvin expansion — 230
  - 16.3.1 Bernoulli equation — 231
  - 16.3.2 Cooling and liquification of gases — 232
- 16.4 General flow process — 236
  - 16.4.1 S and H: the gas turbine — 237
- Exercises — 239

17 Stability and free energy — 243

- 17.1 Isolated system: maximum entropy — 243
  - 17.1.1 Equilibrium condition with internal restrictions — 245
  - 17.1.2 The minimum energy principle — 246
  - 17.1.3 Stability — 247
- 17.2 Phase change — 249
- 17.3 Free energy and availability — 250
  - 17.3.1 Free energy and equilibrium — 253
- Exercises — 257

18 Reinventing the subject — 259

- 18.1 Some basic derivations from maximum entropy — 262
- 18.2 Carathéodory formulation of the second law (†) — 263
- 18.3 Negative temperature (†) — 265

19 Thermal radiation — 268

- 19.1 Some general observations about thermal radiation — 268
  - 19.1.1 Black body radiation: a first look — 274
- 19.2 Basic thermodynamic arguments — 275
  - 19.2.1 Equation of state and Stefan–Boltzmann law — 279
  - 19.2.2 Comparison with ideal gas — 282
  - 19.2.3 Adiabatic expansion and Wien's laws (†) — 283
- 19.3 Cosmic microwave background radiation — 288
- Exercises — 289

20 Radiative heat transfer — 291

- 20.1 The greenhouse effect — 294
- Exercises — 297

21 Chemical reactions — 299

- 21.1 Basic considerations — 299
  - 21.1.1 Reaction rate — 301
- 21.2 Chemical equilibrium and the law of mass action — 301
  - 21.2.1 Van 't Hoff equation — 305
  - 21.2.2 Chemical terminology — 306
- 21.3 The reversible electric cell (†) — 307
- Exercises — 309

22 Phase change — 311

- 22.1 General introduction — 311
  - 22.1.1 Phase diagram — 312
  - 22.1.2 Some interesting phase diagrams — 314
- 22.2 Basic properties of first-order phase transitions — 317
- 22.3 Clausius–Clapeyron equation — 320

	22.3.1	Vapour–liquid and liquid–solid coexistence lines	323
	22.3.2	Gibbs phase rule	325
	22.3.3	Behaviour of the chemical potential	326
22.4	The type-I superconducting transition (†)	326
Exercises			329

23 The third law   331

23.1	Response functions	332
23.2	Unattainability theorem	333
23.3	Phase change	334
23.4	Absolute entropy and chemical potential	335

24 Phase change, nucleation, and solutes   336

24.1	Treatment of surface effects	336
24.2	Metastable phases	338
	24.2.1 Nucleation	341
24.3	Colligative properties	347
	24.3.1 Osmotic pressure	347
	24.3.2 Influence of dissolved particles on phase transitions	350
24.4	Chapter summary	353
Exercises		353

25 Continuous phase transitions   355

25.1	Order parameter	357
25.2	Critical exponents	359
25.3	Landau mean field theory	361
	25.3.1 Application to ferromagnetism	366
25.4	Binary mixtures	370
Exercises		373

26 Self-gravitation and negative heat capacity   375

26.1	Negative heat capacity	375
	26.1.1 Jeans length	377
26.2	Black holes and Hawking radiation	378
Exercises		381

27 Fluctuations   382

27.1	Probability of a departure from the maximum entropy point	383
	27.1.1 Is there a violation of the second law?	384
27.2	Calculating the fluctuations	385
	27.2.1 More general constraints	387
	27.2.2 Some general observations	391
27.3	Internal flows	393
27.4	Fluctuation as a function of time	395

27.5	Johnson noise	398
	Exercises	401

28 Thermoelectricity and entropy flow — 403

28.1	Thermoelectric effects	403
	28.1.1 Thomson's treatment	406
28.2	Entropy gradients and Onsager's reciprocal relations	409
	28.2.1 Derivation of Onsager's reciprocal relation	411
	28.2.2 Application	416
	28.2.3 Entropy current, entropy production rate	417
	Exercises	418

Appendix A Electric and magnetic work	421
Appendix B More on natural variables and free energy	424
Appendix C Some mathematical results	428
Bibliography	431
Index	433

How to use this book

1.1 For the student

1.1 For the student	1
1.2 For the teacher	1

This is a textbook on the science of heat- and work-related processes, called thermodynamics, suitable for a university undergraduate course. The discussion begins at first year level, and extends to final year, and, in some areas, beyond. The book has a more talkative style than many, and, for those chapters that the reader uses, it is intended to be read in full. The reason for the style comes from my experience in teaching the subject. Any textbook will give to the reader an account of the correct way to proceed in a given area. However, what the student also needs is a sense of why we are even thinking this way, and why we focus our attention on one equation rather than another, and how some common misconceptions can be avoided. The book tries to offer the kind of replies to questions of this sort that I have found myself offering to small groups of students at Oxford University as I have taught the subject.

Although a professional scientist adopting this book will eventually read the whole of it, this is not what most undergraduates will either need to do or choose to do. Students wish to be efficient, and this is a good wish. However, the present subject benefits, more than most, from a thorough discussion of ideas, in order to gain a good intuition about concepts such as temperature, heat, entropy, and free energy. Therefore I would strongly urge any student to read through, in full, whatever section is relevant to their study, making their own notes of course. Also, whether or not you attempt all the exercises, at least read through them, because some introduce useful further ideas and techniques.

Table 1.1 gives some suggestions of how material can be selected to suit a given course. In the section headings, a dagger (†) is attached to a heading whenever that section can be omitted without undermining later chapters.

1.2 For the teacher

This book has been motivated by two desires. The first is to present thermodynamics as the beautiful and useful set of ideas, methods, and observations that it is. The second is to provide a resource for teachers and students.

The exercises are intended partly to give examples and practice, and partly to develop ideas in more detail. Some of the latter type should be omitted from a basic course.

Thermodynamics: A Complete Undergraduate Course. Andrew M. Steane.
© Andrew M. Steane 2017. Published 2017 by Oxford University Press.

Table 1.1 *Use of the book for a course in the physical sciences. For a basic course a selection from Chapters 21, 22 would suffice. The fuller course could also be selective in the final year.*

	Basic course	Fuller course
First year	2–9	2–9
Second year	10–20, (21, 22)	10–23
Third year or advanced		24–28

I have made one choice in the ordering of material that might be called non-standard, in that the text introduces chemical potential before introducing thermodynamic potentials such as the Helmholtz function, enthalpy, and the Gibbs function. This is because that ordering makes the logic of ideas flow more naturally, in my opinion. Another policy has been to explain the sense in which free energy is both an energy-like and an entropy-like function; this should be done in all thermal physics texts, but it is sometimes left out or left obscure.

A final policy choice has been to include mean field theory and an introduction to critical exponents, and also a preliminary treatment of fluctuations and the Onsager reciprocal relations.

In principle this whole book does not need any knowledge of statistical methods, and it could precede a course on statistical mechanics. It would make sense for students to encounter Chapter 17 and the first part of Chapter 18 before they learn the statistical method in thermal physics (but this is not crucial).

Introducing thermodynamics

2

This book is a guide to the science of energy and entropy, called thermodynamics, for students meeting the subject for the first time as part of a degree course in physics, chemistry, or engineering.

The central ideas of thermodynamics were discovered quite some time ago now (in the nineteenth century). For this reason, to young science students, it can seem as if this is part of the ancient lore of their subject, and not as 'up to date' as other areas such as relativity and quantum chemistry. I want to begin by saying that this is a false impression. If I had to choose which areas of physics contain the key ideas of the whole of modern physics, I would pick three: general relativity, quantum theory, and thermodynamics. The first of these contains the key idea of spacetime and its geometry, the second the key ideas of quantum amplitudes and symmetry principles, the third the key idea of entropy. From these key ideas much of our current understanding of the physical world follows.

The three areas on my list have in common that they deal in overarching principles, but thermodynamics differs from the other two in that it does not attempt to formulate a detailed description of the motion of individual entities such as particles. It does not deal with equations of motion. Unlike quantum physics, thermodynamics does not aim to give a detailed description of the structure of everything in the world. Rather, it offers a mathematical language for handling physical concepts such as energy and temperature in an insightful way. Thermodynamics is 'built from' a small number of general physical principles, and it aims to show what can be deduced from those principles. It deals with bulk properties such as compressibility and heat capacity, and it reveals connections between them. What is intriguing is that these connections do not depend on the details of the fundamental equations of motion, only on broad ideas such as energy conservation.

In case you think compressibility and heat capacity are dull, consider any physical system that is composed of many parts and that you do not consider to be dull. A humming bird, or a neutron star, or a DNA molecule, for example. All such systems have compressibility and heat capacity, and a number of other thermodynamic properties. Such systems cannot be understood in full without an understanding of what those properties mean and how they are connected to one another.

Thermodynamics: A Complete Undergraduate Course. Andrew M. Steane.
© Andrew M. Steane 2017. Published 2017 by Oxford University Press.

Once we formulate the so-called laws of thermodynamics and the concepts behind them, then much of thermodynamics can be deduced by a process of pure reasoning. That is the approach I will take in this book. I will present the subject as much as possible as a self-contained set of concepts, with logical and mathematical arguments leading one forward over the territory.

The fact that much of thermodynamics is a type of mathematical reasoning will help the reader understand why thermodynamics is here to stay. Ultimately thermodynamics is a theory based on physical assumptions, which are abstracted from physical observations, and which have not yet been observed to fail. Therefore at heart it is physical science not pure maths. However, the techniques of thermodynamics provide the tools needed to explore what follows from the few assumptions which need to be made. These techniques, which form the language we must acquire in order to *think thermodynamically*, will not be superseded one day when we understand the natural world better. It might be that thermodynamics cannot be applied in a useful way to important areas of science, but many of its basic arguments are beautifully and strictly correct. An example is the thermodynamic definition of temperature which we will meet in Chapter 8. In order to arrive at the definition, one has to consider some idealizations: a heat reservoir and a perfectly reversible heat engine; but having made these idealizations, and after postulating that a certain type of physical process is impossible, the central statement about temperature (Eq. 8.8) will follow inexorably. Albert Einstein, along with other notable physicists, has talked of the deep impression the theory made on him, and his conviction that it will never be overthrown. If a new fundamental theory of physics were found to be in disagreement with the principle of relativity, that would be highly surprising but does not necessarily rule out the possibility that the theory might be right. If, on the other hand, a theory were found to contradict the second law of thermodynamics, then the new theory is almost certainly fundamentally flawed.[1]

It is striking that quantum mechanics and general relativity are in complete agreement with thermodynamics. Sometimes a result which requires detailed and lengthy analysis from the starting point of Schrödinger's equation and fundamental interactions can be obtained much more simply and directly from thermodynamics. For example, by using a simple thermodynamic argument one can derive a relationship between coefficients (called Einstein A and B coefficients) describing the rate at which atoms absorb and emit light. To find the same relationship using basic laws of motion, a considerably more involved quantum mechanical argument is required, involving an integral over all possible ways an atom could emit electromagnetic waves into the vacuum. This illustrates how thermodynamics offers deep insights and useful practical methods of calculation.

Another thought-provoking example is the phenomenon of Hawking radiation. This is thermal radiation (radiant heat) predicted, via a technical argument involving general relativity and quantum mechanics, to be emitted by any astronomical black hole from a region just outside its event horizon. Here the black hole is behaving in a way that is predicted by thermodynamics for bodies which

---

[1] There have been numerous claims that existing theories, especially quantum mechanics, allow violations of the second law of thermodynamics. No such claim has ever turned out to be valid, though many have slipped through the scientific peer review process.

absorb all radiation incident upon them, in conditions of thermal equilibrium. We will meet the argument in Chapter 19. This connection was not made before the 1970s because it was assumed that a black hole could be treated as if its temperature were zero, and because black holes cannot come to stable thermal equilibrium with other things (see Chapter 26). However, it now seems clear that it is legitimate to apply thermodynamic reasoning to black holes. This could hardly be further from nineteenth century arguments about steam engines!

In both the examples just mentioned, one could say that quantum mechanics and general relativity are 'bowing to' consistency demands placed on them by thermodynamics. In return, thermodynamics gives hints that it needs a theory other than classical physics to support it. The Gibbs paradox (Chapter 9) implies strict indistinguishability of particles, a deeply non-classical concept, and thermal radiation requires a quantum theory of energy exchange to account for its distribution over frequency. The overall result is that the natural world is consistent with itself.

As previously mentioned, thermodynamics concerns what can be said about physical systems in terms of their bulk properties, without regard to the details of their microscopic construction. For example, it allows a general relationship between heat capacities and volume compressibilities (Equation 7.33) to be deduced, without any need to know what underlying structure gives rise to a heat capacity or a compressibility. However, thermodynamics does not provide the value of any heat capacity or any compressibility: those have to be found out by measurement, or by other means. The 'other means' could for example be a process of reasoning from what we know of the microscopic structure of bodies (which is in turn described by quantum mechanics). Formulating the relationship between the microscopic structure, such as the atoms and their quantum states, and the bulk properties, such as pressure and temperature, is the subject of the other great part of thermal physics, called *statistical mechanics* or *thermostatistics*. Some texts present both the thermodynamic part and the statistical part in the same book. These can be presented in an alternating sequence, or intermingled more thoroughly, since ultimately all divisions of a science into sub-fields are somewhat artificial. However, our grasp of the subject is enhanced by knowing precisely which parts rest on which foundations. I feel it is valuable therefore to regard thermodynamics as a subject in its own right, and to allow it to play by its own rules. It provides the main lines of logical structure in statistical mechanics, such as the centrality of entropy, the concept of free energy, and the definition of temperature. In this sense it underpins statistical mechanics just as surely as does quantum theory. The arguments of statistical mechanics must agree with thermodynamics in the *thermodynamic limit*, which is, roughly speaking, the limit of large systems, or systems with a large number of parts.

Although the combination of quantum theory and advanced statistical methods has been a great success story in thermal science for a hundred years, the fact remains that most systems are too complicated to be precisely modelled in this way. Thermodynamic reasoning is extremely useful because it can guide us when we do not have a complete knowledge of the microscopic details. By establishing relations which hold in the thermodynamic limit, it provides a rich set of test cases, including for example the question, 'if such and such were so, would this break the second law of thermodynamics?'

We will now turn to the subject itself. Before beginning, you might like to consider the following everyday example, which was described in one of the books of mathematical diversions by Martin Gardner. Suppose you go to a cafe to have coffee with a friend, and you are served two cups of black coffee, with a small jug of cold milk. Your friend has to leave the table briefly to make a phone call. You know your friend likes coffee white and as hot as possible. In order to help them to enjoy the drink, should you put the milk in straight away, or wait till they return?

A survey of thermodynamic ideas

3.1 Energy and entropy 7
3.2 Concepts and terminology 12
3.3 The laws of thermodynamics 29
3.4 Where we are heading 32
Exercises 33

Heat · cannot of itself · pass from one · body to a hotter · body
Heat · cannot of itself · pass from one · body to a hotter · body
Heat won't pass from the cooler to the hotter
Heat won't pass from the cooler to the hotter
You can try it if you like, but you far better not-a
You can try it if you like, but you far better not-a
'cos the cold in the cooler will get hotter as a rule-a
yes the cold in the cooler will get hotter as a rule-a

Flanders and Swann[1]

3.1 Energy and entropy

The primary concepts of thermodynamics are *energy* and *entropy*. In this book it will be assumed that the reader has some notion of the concept of energy, as it arises in classical mechanics for example, but no knowledge of entropy. There are plenty of other important thermodynamic concepts, such as temperature, heating, and thermodynamic potential, for example, and these will be introduced as we go along. However let us begin by introducing the main players through some general thoughts about phenomena in the natural world.

An important idea in classical mechanics is that of conservation of energy. For example, when two bodies collide (without interaction with any third party), we find that the total energy of the bodies after the collision is the same as it was before the collision. For simple bodies the energy could be wholly in the form of their kinetic energy, and we can write an equation such as

$$\frac{1}{2}m_1 v_1^2 + \frac{1}{2}m_2 v_2^2 = \frac{1}{2}m_1' v_1'^2 + \frac{1}{2}m_2' v_2'^2, \qquad (3.1)$$

where the unprimed symbols represent the values before the collision, the primed symbols the values after the collision, with subscripts 1 and 2 identifying the two bodies.

If for some collision, equation (3.1) were not obeyed, then we would look further into the dynamics, and we would find either that a third party was involved,

[1] 'First and Second Law' from *At the Drop of Another Hat* by Flanders and Swann, 1963. By permission of the Estates of Michael Flanders and Donald Swann. Administrator Leon Berger: http://leonberger@donaldswann.co.uk

or else that the bodies can store energy internally. Once all the avenues for energy to be passed to third parties or stored internally have been identified, we would find once again that the overall energy is conserved. The equation we could write would have more terms in it, but it would express the same principle, namely energy conservation.

In thermodynamics, this idea of conservation of energy is crucial. It is one of the linchpins of the subject, and is called the first law of thermodynamics. It may seem odd to the student that a principle that is already known in mechanics is thus 'claimed' by thermodynamics. Why give it the fancy new name? The reason is that in the natural world energy can take a form, and very often does take a form, that is hard to keep track of using the language of forces and mechanics. This 'new' form of energy is (roughly speaking) *thermal motion,* closely related to heating and cooling. For example, consider a brick falling to the ground. The dynamics as the brick falls are governed by the gravitational interaction between the brick and planet Earth. Initially the brick–Earth system has a certain amount of potential energy. As the brick falls this potential energy is flawlessly converted into kinetic energy. When the brick hits the ground, something much more complicated occurs, in which many complicated motions inside the brick and in the ground take place, carrying energy in multiple forms that are extremely hard to analyse in detail. The question arises, what is the most insightful way to discuss the energy in a situation like this? Readers are probably familiar with the idea that we say the energy is 'turned into heat,' but the question we want to ask now is, why is that a valid way to speak about it? It is not obvious from just mechanics, because the heat energy is *never* fully recovered, i.e. changed back into a simple form. This makes it difficult to prove that energy is in fact conserved in such processes, unless we learn some methods of reasoning that are provided by the science of thermodynamics. Thermodynamics offers a clear way of reasoning that shows convincingly that we should retain the notion of conservation of energy, but understand that *not all energy transformations are possible.* This will be expounded in detail in Chapters 7 and 8. Here we will discuss some simple examples in order to convey the flavour of the main ideas.

The essential idea of energy conservation is that energy can be converted from one form to another while maintaining something fixed, namely, its total amount. In the example just discussed, gravitational potential energy was converted into kinetic energy, and then kinetic energy was converted into complicated randomized oscillations of many small particles. However, these two conversions are not equivalent, because one can be completely undone, the other cannot.

If instead of hitting the floor with a resounding crash, the brick landed smoothly on a trampoline, then its kinetic energy would be converted into elastic potential energy in the springs and rubber of the trampoline, and, in the absence of friction or damping or acoustic noise and other such things, the trampoline's energy would subsequently be given back to the brick, which would rise up into the air until it regained its former height. Alternatively, one could throw a switch at the right moment, and have the trampoline transfer its energy to some other device such as an electric generator, make the generator charge a battery, have the battery run a motor, the motor drive a crane, and lift the brick that way. It seems then as if energy is freely convertible from one form to another. However, that appearance is misleading. The processes we neglected just now, namely friction and damping and things like them (electrical resistance, fluid viscosity etc.) convert energy into a form that is not freely convertible back again. This other form—the complicated random motions that we discussed above—cannot in practice be 're-claimed' and completely converted back into a simple form (such as gravitational potential energy of a brick). A partial reclaim is possible, and *quantifying this is one of the great achievements of thermodynamics*.

Next, consider the following. One often hears it stated that systems have a natural tendency to evolve towards a state of lower energy. Consider a small round ball placed in a curved china bowl (Figure 3.1a). If the ball starts somewhere away from the bottom of the bowl, then it will roll down inside the bowl, and after some time has passed it will be found motionless at the bottom of the bowl. It is sometimes said that this can be understood as an example of 'the tendency of things to move to a state of lower energy.'

For another example, consider an atom emitting a photon of light. When an atom emits light we say the atom was initially in an excited state and it evolves to a state of lower energy such as the ground state. We feel comfortable with the idea that the atom spontaneously evolves towards the lower energy state, emitting energy in the form of a small pulse of light, and this process is called 'spontaneous emission' (Figure 3.1b).

A little reflection should make us question the discussion for both of these examples: it is a little too glib. After all, didn't we just agree that energy is conserved? Far from moving towards a state of lower energy, what happens in both examples is precisely no change in total energy at all! Rather, the energy is redistributed from one form to another.

A more careful statement might be that the system under study (the ball, or the atom) moves towards a state of lower *potential* energy. This is a more useful statement but it remains insufficient. The ball rolling in the bowl involves a conversion of mechanical energy into heat; the light emission involves the conversion of electric potential energy in the atom to electromagnetic energy of the light pulse. Claiming that the atom has a natural tendency to move to a state of lower potential energy amounts to claiming that the light field has a natural tendency to move to a state of higher energy! The problem with both claims is that they fail to identify the feature which determines why the process goes one way rather than another. Why do we have spontaneous emission, but not spontaneous absorption, or re-absorption?

Thermodynamics is passionately interested in questions like this, and it offers careful and precise answers. In both of these examples, the direction of the process is determined by the most central thermodynamic concept, namely *entropy*. Without introducing a formal definition at this stage, I would like to help the student get an initial feeling for what sort of thing entropy is. Entropy is connected to

(a)

(b)

Fig. 3.1 *(a) A ball rolls in a curved bowl; (b) an atom emits a photon.*

energy, but instead of quantifying the amount of energy, it partially characterizes the form in which the energy is physically present. Roughly speaking, entropy quantifies the degree to which energy is dispersed over many things rather than concentrated in a few. The 'things' here could be types of motion, or locations in space, or inter-particle interactions. In the example of the rolling ball, initially the energy was stored (i.e. physically represented) in three physical quantities: the kinetic energy of the centre of mass of the ball, the rotational kinetic energy of the ball as a rigid body, and its gravitational potential energy. In the final conditions, the energy is to be found in the random jiggling of very many different small motions and vibrations of the matter of the ball and the bowl. The latter situation is one of higher entropy than the former, because the energy has been dispersed into many small parts.

A financial illustration might help here. Suppose I give you a single coin of value one dollar. You are one dollar the richer. But suppose instead I give you 100 one-cent coins, randomly distributed about your house. Now, in some sense you are still one dollar the richer, but you might question whether really you are any richer. It could take you many hours to find all those cents, and presumably you would not normally find it worthwhile to spend so much time earning a dollar, so in some sense you are no richer after all. (We are ignoring the possibility that you may have small children who would be delighted to find the coins for you.) Clearly, mere amount of money is not the only consideration when assessing the value of money. Similarly, in physics, mere energy, as a concept, is of limited use in trying to understand and describe what is going on in many physical phenomena. We need to think about entropy as well.

Thermodynamic entropy is not a vague notion, but a precise physical property. Its physical dimensions are energy divided by temperature (SI unit: joule per kelvin). A physical system possesses some amount of entropy just as it possesses other properties such as energy and momentum and mass. We will discuss the details in the following chapters, but here let us mention why entropy plays such an important role in physics. It is this:

> Physical phenomena do not show any tendency to prefer states of low energy (in fact they strictly conserve energy) but they do show a tendency to evolve towards states of higher entropy.

A more precise statement is

> *The entropy of an isolated system can increase or remain constant over time, but cannot decrease.*

This statement, combined with energy conservation, turns out to be sufficient to allow a vast range of physical phenomena to be understood. It is called the second law of thermodynamics.

The second law of thermodynamics is central because it represents a subtle but powerful kind of 'driving force' behind almost all natural phenomena! (The only

Fig. 3.2 *The Earth as a heat pump. The full arrows represent energy, the dashed arrows represent entropy. The surface of the Earth receives energy mainly from incident sunlight. All of this energy is subsequently emitted back into space (if this did not happen, the Earth would overheat). The role of the Sun in 'powering' life on Earth is, therefore, not so much an 'energy provider' as a 'low entropy provider'. The incident sunlight is mainly in the visible part of the electromagnetic spectrum, while the radiation emitted by the Earth is mainly infrared. The amount of entropy per joule increases with the wavelength, so this process allows the Earth to get rid of entropy without cooling down. This allows it to support processes that drive entropy out of structures on its surface, such as the processes of life.*

phenomena where it is irrelevant are those where the dynamics preserve entropy, but these are a tiny minority of all phenomena.)

A celebrated example of the 'driving force' provided by entropy is the role of the Sun in 'powering' life on Earth. Without thinking too hard about it, one might say that the Sun 'provides energy' to the Earth, and it is this energy which is needed by plants, and hence all other living things. However, it turns out that living things give off just as much energy as they get. If they are so busy getting rid of energy, then perhaps far from it being their chief requirement, energy is a problem for them, and they live in spite of it rather than because of it! Looking at the energy accounting on its own does not resolve this question one way or the other.

Now let's think about what life is; certainly a large and subtle question, not easily answered, but for our purely physical purposes we can note one striking feature: life involves the gradual organization of physical things into specific complex structures. According to thermodynamics this need not necessarily require energy, but it does require that entropy is transferred out of those organized structures into other things. However, you can only get rid of entropy by throwing away energy as well (we will explain this later). The net result is that living things should not be regarded primarily as net consumers of *energy* (they can be that for a while, but it is not the main picture). Rather, they are net exporters of *entropy*. They take in energy in a lower entropy form, and give it out in a higher entropy form.

On a larger scale, the same can be said of planet Earth as a whole. The Earth emits into space just as much energy as it absorbs from the Sun (1.2×10^{17} joules per second). However, the supplied energy (from the white-hot Sun) carries less entropy per joule than the thrown away energy (infrared radiation into space). This means that the Earth can be a net emitter of entropy, while breaking even on energy—see Figure 3.2. This is what allows it to carry out entropy-decreasing activity on its surface, and therefore support life. 'Entropy per joule delivered' will turn out to be the thermodynamic definition of inverse temperature, so the possibility of life is seen to be connected to the fact that the Earth is of intermediate temperature: colder than the Sun but hotter than the surrounding space.

We speak of 'entropy-driven processes' when a system is free to explore among many possible states, some of which have higher entropy than others. Even if the behaviour of the system is random in other respects, there will be a definite direction imposed on it by the role of entropy. This is illustrated on a small scale by the behaviour of particles in a gas, and on a grand scale in biology.

3.2 Concepts and terminology

We shall next introduce a number of concepts that will begin to form the 'language' of thermodynamics. All areas of science have some such basic language. In classical mechanics, for example, Newton's great contribution was not only to write down the laws of motion and explore their implications, but also, even before writing down laws, to establish what were the basic concepts that would allow motion to be talked about clearly. He singled out things like mass, momentum, and force. To begin thermodynamics, we have to consider several concepts that are reasonably straightforward, but it is worth defining them carefully in order to avoid making mistakes later on.

3.2.1 System

The first concept is that of a thermodynamic *system*. This is simply some physical object or collection of objects or fields that we wish to discuss. A simple enough idea, but it is important to learn the following rule: always make it clear at the start of a thermodynamic discussion, and keep it clear, what has been chosen to be the system under discussion. This can often (but not always) be done by a diagram with a box drawn around the 'system', see Figure 3.3. The surface thus separating the system from the rest of the world is called by engineers the *control surface*; it is helpful to give it a name because this helps to focus our thinking in a useful way. Everything else not included in the system is called the *surroundings*, or the *environment*.

Fig. 3.3 *A thermodynamic system (everything inside the box) and its environment (everything outside the box). Photo © danheller.com*

Table 3.1 *The three most important types of thermodynamic situation. The third column gives the main properties that cannot change for a system of the corresponding type. U refers to the total energy content of the system, called internal energy; N refers to the amount of matter in the system, which we can quantify in terms of mass or by counting particles (if each particle has a fixed mass); V refers to volume. It is understood in the examples for an isolated system that gravity is negligible and no electromagnetic fields penetrate the chamber.*

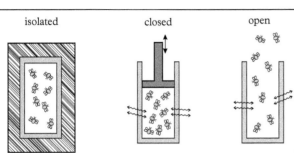

| Thermodynamic system | | Conserved |
Type	Definition and examples	quantities
Isolated	Not influenced at all by other things	U, N, V
	e.g. a gas in a rigid, insulated chamber	
	a swarm of bees in such a chamber	
Closed	Cannot exchange matter with surroundings	N
	e.g. a gas in a cylinder with movable piston	
	a swarm of bees in a thin flexible bag	
Open	Can exchange matter with surroundings	–
	e.g. a pool of water in the open air	
	a swarm of bees flying freely	

The boundary between system and environment can be of various types, constraining what sort of physical interactions can take place between the two, see Table 3.1. A system which undergoes no interaction of any type with other systems is called *isolated*. One which can interact with the surroundings so as to exchange energy with it (via heat, or work, or both), but which cannot lose or gain material is *closed*. An *open* system in thermodynamics is one which can exchange energy and also matter with the surroundings.[2]

A pool of water is an example of an open system, in ordinary circumstances, because it can evaporate. A system consisting of liquid water and water vapour, all inside a chamber whose walls are impervious to water, on the other hand, is a closed system.

Many ordinary objects, such as a brick or a rubber band, are to good approximation closed but not isolated, in ordinary circumstances.

[2] Beware: terminology in physics varies from one field to another. In quantum mechanics, the word *open* is often used to refer to systems which can exchange energy with the surroundings, whether or not they can exchange matter with the surroundings.

Closed systems are more straightforward to reason about, but the notion of an open system is important in chemical reactions and phase changes (e.g. melting, boiling).

One can choose as a thermodynamic system something quite abstractly defined, such as 'the gaseous material in region R', where R is some given volume of space with no physical boundary, into and out of which gas may freely move. The methods of thermodynamics enable us to reason correctly about the properties of even this type of open system.

To illustrate the importance of identifying a system carefully, consider the following. Suppose we would like to calculate the temperature change resulting from a slow leak of air into a rigid, initially evacuated chamber. To make a start one has to decide what to choose as the system whose energy and entropy will be considered. Is it the material of the chamber itself, or the air inside the chamber at any given moment, or the air that is in the process of moving from outside to inside? Various kinds of mistake can arise. One might reason correctly, but make a poor initial choice of system, which makes the analysis unnecessarily difficult. Or one might apply to an open system a formula that is only valid for a closed system, which would plainly be wrong. This example problem will be solved (by you!) in Chapter 7.

3.2.2 State

We discuss systems in terms of their thermodynamic 'state'.

Talking about the state of a system is common in everyday parlance. A door could be open or shut, for example: these are two different states of a door. There are also a large number of intermediate states in this example, where the door makes various angles with some other entity such as a wall. There are also other properties of a door which might be of interest. For example, wooden doors commonly expand or contract, depending on their water content and temperature. To specify the state of a door it might be necessary to give its temperature and water content, and its volume. One might go further, and start to talk about the material construction of the door, right down to the level of the molecules, or even the atomic nuclei, quarks, and electrons, so that the state of a door is changed if just one atomic nucleus changes its orientation. However, thermodynamics is a discipline which deliberately chooses *not* to discuss the microscopic construction of systems. In thermodynamics, we discuss only large-scale bulk properties, such as volume, pressure, temperature, density, mass, capacitance, magnetization, length, tension, to name a few. Although you and I know that an object such as a book is made of billions upon billions of tiny particles called atoms, thermodynamics knows nothing of that. In a thermodynamic argument, a book is just made of 'stuff', and we don't care what that stuff is like in detail. We only require that some of its bulk properties can be measured and behave in a consistent way. *Then thermodynamics will teach us how to relate these bulk properties to one another.*

It might seem that this restriction to macroscopic quantities is a limitation of the theory. Actually it is its glory. The whole point of thermodynamics is that it is possible to understand a great deal of how systems behave, and to identify constraints and laws of their behaviour, without needing to know precisely their underlying construction. This doesn't mean we can understand everything about a system through thermodynamics, but it is very useful to discover what can be done under this restriction of dealing with bulk properties alone. It is also a profound feature of the natural world that physical behaviour can be understood through this type of approach. The relationship between the underlying structure—the atoms or spins or whatever—and the bulk properties is the subject of statistical mechanics. It is equally as fascinating and important as thermodynamics, but addresses different questions.

It should be apparent to the reader that this talk of bulk properties is not perfectly well defined. For example, a familiar bulk property is volume. If you try to specify the volume of any physical object precisely, you will always run up against the difficulty that boundaries are never perfectly sharp in the real world. A book sitting on a table has a volume, but if you had to say precisely where the book finished and the table began, it might not always be straightforward: at the boundary the molecular structure overlaps. We deal with this by making an idealization: we simply suppose that there could exist an idealized book with a well-defined volume. We then discuss this idealized system. In so far as any real book were a good approximation to the ideal, then the discussion would apply to the real book in an approximate way.

This issue of idealization should not be surprising or disturbing, since it is none other than standard practice in all scientific arguments. In mechanics, for example, one deals with a body and the forces on it, but it is never the case that we account for all the forces acting in practice: most are neglected because they are very small. For a ball rolling down a plane, the main forces are the gravitational attraction to the Earth and the normal reaction from the plane. The other forces present include those due to deformation of the ball and the surface, air viscosity, van der Waals electrical forces of attraction, further gravitational forces from the Sun and other bodies, and so on. It is useful nonetheless to discuss the idealized case of a simple ball and just a few forces, in order to make progress.

We can now be more specific about what is meant by the *state* of a thermodynamic system. We first make an idealization by selecting a finite (and often small) set of macroscopic properties which will suffice to completely describe the state of affairs of our idealized system. These properties or *state variables* are sometimes called 'generalized coordinates'. They might include, for example, volume, pressure, temperature, mass, and others in the list we made previously. Then, a list of the values of some minimal set of thermodynamic properties of a system will uniquely specify the state of that system. For example, a quantity of argon gas can be idealized as a system whose state can be fully specified by three properties: mass, pressure, and volume.

3.2.2.1 State function

Consider a property such as the total volume of a system. Many different states of a given system could have the same volume (they would differ from one another in some other respect such as temperature), but it would be logically self-contradictory to suppose that a single state could have either of two or more volumes. This is expressed by saying that volume is 'a function of state', or 'a state function' for short. The relationship is

$$\text{state of system} \xrightarrow{\text{uniquely fixes}} \text{value of function of state}$$

For example, you would not say 'the door is shut' and then add 'but I don't know whether or not the door is open.' The open/shut property is a state function. However, you might say 'but I cannot tell who shut it'. The identity of the person who shut the door is not a function of the state of the door.

A function of state is any standard physical property which has a unique well-defined value for each equilibrium state of a given system. Pressure, volume, temperature, mass, and internal energy are all functions of state (not all of which have we yet thoroughly defined). Considering your own body for a moment as a thermodynamic system, most of the properties you might normally think about are functions of state, such as the position of your legs, the length of your hair, the circumference of your waist, and so on. This leads one to ask, what could we mean by something that is *not* a function of state? One example is the *average rate* at which a given state change takes place. That is a perfectly well-defined physical quantity, but it is not a function of state, because it is not determined purely by the initial and final states. Another example is a property which is shared between the system in question and its environment, such as, for example, an interaction energy between the two. Such energy can become a function of state when the system is specified in the right way, but care is needed when treating situations of this kind.

The most common example of a physical quantity which is not a function of state is a quantity that is defined in terms of a *process* rather than a state. An example in the case of your body could be the part of the water content (i.e. liquid H_2O) which was imbibed through drinking cups of coffee. The reason this is not a function of state is that one cannot tell, merely from the current state of your body, whether the H_2O got there through coffee drinks or by some other input method, such as orange juice. The total quantity of H_2O in your body *is* a function of state, but the amount *that got in via coffee* is not.

The formal definition is

> A quantity F is a function of state if and only if the change ΔF, when a system passes between any given pair of states, depends only on the initial and final states, not on the path.

The 'path' in this definition refers to the sequence of intermediate states between the starting point and the finish.

3.2.2.2 Equation of state

The properties of a system are not all independent of one another. Some are related through their very definition: such as density which is always equal to the system's mass divided by its volume. However there are also more interesting types of relationship which depend on the system in question, and tell us some important information about it. One such relationship is called the 'equation of state'. The equation of state gives the *temperature* as a function of other properties such as pressure and volume (for states called 'equilibrium states' where the temperature is well defined; this will be discussed shortly). The equation of state is specific to the system: a 1 kg bag of sugar has a different equation of state to a 1 kg bowl of water. The fact that every system has an equation of state can be deduced from the zeroth law of thermodynamics. We will discuss this in more detail in Chapter 6, but to introduce the idea let us consider a simple example.

We will be much concerned with the simplest possible type of system in which the main ideas of thermodynamics can be explored. In thermodynamics a *simple system* is defined as one whose state can be fully specified by just *two* independent properties. In a *simple compressible* system, these are pressure and volume. Consideration of such 'pV' systems is the best way to learn the subject. Such a system could be, for example, a fixed mass of gas in a chamber (see Figure 3.4), or anything at all contained in a chamber, as long as the system interacts with its surroundings only by mechanical compression and heat exchange, and the conditions give a single uniform pressure at the boundary. The equation of state applies to the *equilibrium states* of the system (see Section 3.2.4), and it is a formula relating the basic state variables or 'coordinates' to temperature. In a pV system these coordinates are the pressure p and volume V, so the equation could be something like $T = ap + b/V$, where a and b are constants. I don't know of any system having this particular equation of state. An equation of state which describes many gases quite well over a large range of parameter values is the *van der Waals* equation,

$$\left(p + \frac{N^2 a}{V^2}\right)(V - Nb) = Nk_\mathrm{B} T, \qquad (3.2)$$

where a, b are constants which depend on the gas in question, k_B is a constant which is the same for all gases, and N is the number of molecules in the gas. When the gas has low enough density we have $b \ll V/N$ and $a \ll p(V/N)^2$. In this case the a and b terms can be neglected and the equation simplifies to

$$pV = Nk_\mathrm{B} T. \qquad (3.3)$$

This is called the *equation of state of an ideal gas*.

Although a simple compressible system has a temperature as well as a pressure and volume, this does not mean it is a system having three independent properties, because the equation of state shows that the properties are not all independent. A useful way to express this fact visually is to plot a graph where the 'coordinates' p, V, T are plotted along three orthogonal axes. The equation of

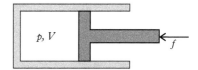

Fig. 3.4 *A an example of a simple compressible system: a fixed mass of gas in a cylinder. In order to specify what thermodynamic state the system is in (in equilibrium), it suffices to supply the values of two properties: pressure and volume.*

18   A survey of thermodynamic ideas

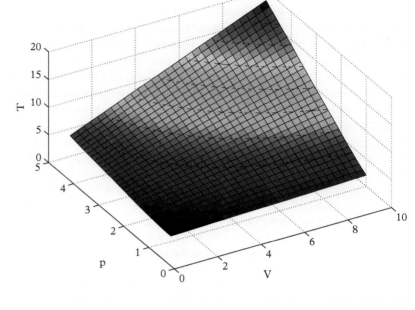

Fig. 3.5 *The equation of state defines a surface in p, V, T space. Each point on the surface is a possible equilibrium state. Points off the surface are nonsensical unless a criterion is agreed for defining p and T for non-equilibrium states; where this is possible, the points off the surface represent (a small subset of the) non-equilibrium states. The surface shown here is that given by the van der Waals equation of state, for $N^2 a = 3$, $Nb = 1/3$, $Nk_B = 8/3$.*

state then specifies a surface. Each equilibrium state of the system is represented by a point on this surface; see Figure 3.5. Points away from the surface represent non-equilibrium states which the system might be able to reach, but only in a transient way. (There are also further non-equilibrium states in which the pressure and temperature are not well-defined; these states cannot be represented on such a diagram.)

Another useful graphical method is to pick whichever two properties are most convenient (usually p and V) and consider a two-dimensional graph (see Figure 3.6). This is called an *indicator diagram*. Each point on this graph represents one state. Changes of state can be tracked by drawing an appropriate line on the graph.

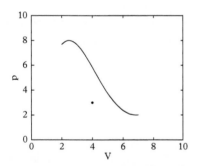

Fig. 3.6 *Indicator diagram for a simple compressible system. Each point represents one state. A line represents a set of states—for example, those through which the system passes when it undergoes a change.*

3.2.3 Extensive, intensive

Most thermodynamic quantities are either *extensive* or *intensive*.

Suppose we extend a thermodynamic system by making multiple copies of it, and placing them side by side. The properties which increase in proportion to the number of copies, such as total volume and particle number, are called extensive. The properties which do not change, such as pressure and temperature, are called intensive.

An intensive quantity has a well-defined value at each point within a system;[3] an extensive quantity is a property of the system as a whole. Volume, particle number, internal energy, and entropy are extensive; pressure, temperature, and chemical potential (introduced in Chapter 12) are intensive. These and further

---
[3] But see Chapter 27.

Table 3.2 *Intensive and extensive quantities. Those above the line are in pairs such that the product has dimensions of energy. Those below the line are in pairs such that the intensive quantity is the extensive one per unit mass or volume. C_V is the heat capacity (dQ/dT) when heat is exchanged while maintaining the volume constant; C_p is the heat capacity (dQ/dT) when heat is exchanged while maintaining the pressure constant. 'Specific' means per unit mass. The convention in thermal science is to use mostly lower case symbols for intensive quantities and upper case symbols for extensive quantities; but this is not a strict rule.*

Intensive	Extensive	
p	V	pressure, volume
T	S	temperature, entropy
μ	N	chemical potential, particle number
T	L	tension, length
σ	A	surface tension, area
$\mathbf{E}$	$\mathbf{d}$	electric field, electric dipole moment
$\mathbf{P}$	$\int \mathbf{E}\,dV$	polarization, volume integral of field
$\mathbf{B}$	$\mathbf{m}$	magnetic field, magnetic dipole moment
$\mathbf{M}$	$\int \mathbf{B}\,dV$	magnetization, volume integral of field
ρ	m	density, mass
n	N	number density, number
$\mathbf{P}$	$\mathbf{d}$	polarization, electric dipole moment
$\mathbf{M}$	$\mathbf{m}$	magnetization, electric dipole moment
c_V	C_V	specific heat capacity, heat capacity
c_p	C_p	specific heat capacity, heat capacity

examples are listed in Table 3.2. Many of these quantities come in pairs, consisting of an intensive and an extensive property whose product is an energy.

A product of an intensive and an extensive quantity is extensive; a ratio of two extensive properties is intensive.

For several types of system, the expression for work done on the system has the form of an intensive variable multiplied by a small change in an extensive variable; for example, $f\,dL$ for extension of a wire by a tension force f, and $-p\,dV$ for compression of a fluid. However, this is not a universal rule. If one uses the relative length $l = L/L_0$ of a wire compared to some fixed length L_0, for example, then the expression for work becomes $L_0 f\,dl$, which is an extensive quantity multiplied by a small change in an intensive one. Further examples arise in the treatment of electric and magnetic systems (see Chapter 7).

3.2.3.1 Non-extensive behaviour

In special circumstances properties can be neither extensive nor intensive. For example, any system of finite volume has a surface, and properties at or near the surface will differ from those in the bulk. If two such systems are placed alongside one another and joined at their border, then the volume of the system doubles but the surface area does not. A good way to understand this is to assign all surface related effects to a new notional system called the 'surface', and to regard the original system as composed of two parts: the bulk and the surface. Thermodynamic arguments can then be applied to this interacting pair of systems, and the resulting predictions offer good insight into real finite systems. An example is a water drop. If two water drops of the same size and in the same thermodynamic state are merged, then the internal pressure drops slightly owing to a change in the contribution from surface tension. This can be modelled by claiming that the pressure in the bulk of the material remains intensive, but the water bulk is in continuous interaction with its own surface, which produces forces on it (in addition to those from other systems such as the surrounding atmosphere). We will discuss this in Chapter 24.

Such a separation into 'bulk' and 'surface' works because the forces between parts of the system are short range. In the case of an electrically charged system, such as a non-neutral plasma, or a system such as a star where self-gravitation is significant, the forces are long range: the material in each spatial region interacts with the material at all other regions. This means it is no longer valid to model the system as if it were composed of many parts that could each be assigned their own pressure and temperature. Such systems are said to be 'non extensive' and thermodynamic arguments that involve an assumption of extensive behaviour do not apply to them. However, the student should regard this as a special case to be treated after we have studied the standard 'extensive' type of behaviour. Extensive behaviour captures the behaviour of ordinary matter under ordinary circumstances very well, and is assumed throughout this book, except in a few places where it is explicitly indicated otherwise.

3.2.4 Thermodynamic equilibrium

Thermodynamic equilibrium (often called *thermal equilibrium*) is the condition which a system, if left undisturbed, will approach over time.

This concept is based on the everyday observation that physical entities, if disturbed in some way, always settle down eventually to some steady state, after sufficient time has passed since the disturbance ceased (i.e. since the influence from the surroundings stopped changing). For example, consider a cup of water that is stirred. Initially the water rotates relative to the cup, but eventually this rotation dies away and the water is still. A mug of hot coffee left on a kitchen table initially changes also, by interacting with the surrounding air: the mug cools down. Eventually the change ceases, when the coffee is no longer warmer than its surroundings: an equilibrium has been reached. Another example is a plucked string of a guitar. When it is first released, the string vibrates and we hear a noise, but as time goes on the vibration diminishes until eventually the string is still (to be precise, it has tiny remaining random vibrations which, on average, neither give out nor take up energy from the surrounding air).

A system in thermodynamic equilibrium is in:

    mechanical equilibrium: no unbalanced forces
    thermal equilibrium: no temperature difference between parts in thermal contact
    chemical equilibrium: no net movement of material from one place or form to another.

The specific equilibrium state reached, for a system in equilibrium with given surroundings, will depend on the conditions set by the surroundings, called *constraints*. If the environment is at high pressure and can compress the system, then the system's equilibrium state will be one of high pressure. If the surroundings are rigid then they fix the system's volume rather than its pressure. If they are at high temperature and can exchange heat with the system, then the system's equilibrium state will be one of high temperature. For an isolated system the volume and energy are fixed.

A moment's thought will show that the states of thermodynamic equilibrium form a minute fraction of the total set of possible states available to any given system. However, they are still a fairly rich set of states and they are sufficient to allow us to develop the basic principles of thermodynamics. In fact this book is almost exclusively concerned with equilibrium states. *A large part of thermodynamics is concerned with finding what is the internal configuration of a system when it reaches equilibrium under given constraints.*

The discussion of *states* given in Section 3.2.2 mostly assumed the case of equilibrium states.

In a pV system, we can only talk about the pressure, volume, and temperature, if these can be consistently defined. In many circumstances they cannot. When a sound wave is present in a gas, for example, the wave consists of pressure variations, so there is not a uniform pressure throughout the gas. This is not an equilibrium state. This doesn't mean that one couldn't still discuss the average pressure, but this quantity will not behave in the same way (obey the same laws) as the pressure of the gas when the gas moves from one equilibrium state to another. Equilibrium thermodynamics is concerned mostly with the latter type of process. We can still use thermodynamics to understand sound waves, by applying the reasoning to small pockets of gas whose size is smaller than the wavelength of the sound.

Thermodynamic equilibrium is an idealization. In practice, the final stages of relaxation towards equilibrium typically involve an exponential damping of vibrations or currents. Roughly speaking, the departure from equilibrium varies with time t as $\exp(-t/\tau)$, where the timescale τ is called the *relaxation time*. Different processes have different relaxation times, and equilibrium is only reached after t is substantially larger than the longest relaxation time. The relaxation time for a litre of gas at room temperature and standard pressure is about ten seconds.

The relaxation time of each small region of diameter ten microns in such a gas is about 100 nanoseconds (this will be discussed in Chapter 10). The relaxation time for some properties of interstellar dust clouds is about ten million years.

When two systems are placed in thermal contact, generally changes will occur in both until a new state is reached. When there is no longer any change, each has reached a state of thermodynamic equilibrium and also the two systems are said to be *in thermal equilibrium with* one another.

3.2.5 Temperature

Temperature is defined as the property which indicates whether one body will be in thermal equilibrium with another. Two bodies which, if they were to be placed in thermal contact, would be found to be in thermal equilibrium with one another without any changes taking place, have, by definition, the same temperature. When two bodies not at the same temperature are placed in thermal contact then energy will move from one to the other. Assuming no further energy providing processes are operating, then the body which loses energy in this circumstance has, by definition, the higher temperature.

As we shall see in Chapter 6, the above goes most of the way to establishing a complete definition of temperature. The second law can be used to complete the argument by enabling a quantitative scale of temperature to be defined; this is discussed in Chapter 8.

Thermodynamics uses some concepts which are already well defined through the study of mechanics. Volume and pressure (force per unit area) are among those. However, temperature is a more subtle concept and we need to take care in defining it. The definition given above makes sense because of the zeroth law of thermodynamics, which we will discuss in Chapter 6.

3.2.6 Quasistatic

A *quasistatic process* is a sequence of equilibrium states. That is, each state in the sequence is what the equilibrium state would be if the constraints on the system had some given values. When the constraints change, so does the equilibrium state.

To realise a quasistatic process in practice, therefore, one makes small changes in one or more parameters, and then waits for the system to reach its new equilibrium; then one makes a further change, and waits again, and so on. In practice these changes are often implemented continuously but very slowly. If the process is slow enough then the sequence of states will be a sequence of equilibrium states to very good approximation. A synonym for quasistatic is *quasi-equilibrium*. The criterion for 'slow enough' is that the times involved are long compared to the thermal relaxation times of the system.

The example given previously, of a cup of coffee cooling down on the kitchen table, was a reasonable approximation to a quasistatic process. Although the hot coffee is not, initially, in thermal equilibrium with the surrounding air, it is in a state very close to what it would be in if it were placed in surroundings at a fixed temperature of approximately 90° Celsius. That is to say, its state is very similar to, and for many practical purposes identical with, a thermal equilibrium state. It is not perfectly in a thermal equilibrium state because there is convection in the coffee and the outer part is cooler than the inner part. However, to a first approximation we ignore those details. As time passes, the next state of the coffee is very close to a state in thermodynamic equilibrium at a slightly lower temperature, and so on.

Another example of a quasistatic process in a pV system would be any sufficiently slow compression or expansion of the system.

In a quasistatic process, the system moves along a path on its equilibrium surface (Figure 3.5).

A sudden process, such as a hammer striking a nail, or an explosion, is not quasistatic. Most exciting things (a firework show, bumper cars, a game of football) are not quasistatic; many dull things (paint drying, pictures fading, a game of cricket) are quasistatic. However, that is not to say quasistatic cannot be fast by everyday standards: since the relaxation time of a small cell of gas is short, much of the functioning of a car engine, for example, can be understood in terms of quasistatic processes to good approximation.

Many simple example motions studied in classical mechanics, such as a mass on a spring undergoing simple harmonic motion, can be understood as quasistatic from one point of view, and not quasistatic from another. Let L be the length of the spring, L_0 be the natural length, and k the spring constant. At each instant during the simple harmonic motion, the spring has well-defined properties such as length and potential energy. The state of the spring is characterized by L and it is almost the same as the state existing if there were no motion but simply an opposing force balancing the force $k(L_0 - L)$ exerted by the spring. In this sense, the spring proceeds through a sequence of equilibrium states, so the oscillatory process may be called quasistatic. However, during simple harmonic motion the composite system of spring and mass together is not in an internal equilibrium state, and therefore its interactions with other things are not quasistatic.

3.2.7 Reversible and irreversible

The idea of *reversibility* is important in thermodynamics. A *thermodynamically reversible* process is defined as follows:

> A reversible process is one such that the system can be restored to its initial state without any net change in the rest of the universe.

The following observation is also useful. It can be regarded as another way to define reversibility, but its application to specific examples is sometimes less straightforward:

A reversible process is one whose direction can be reversed by an infinitesimal change in the conditions.

We say that such processes have no 'hysteresis'.[4]

[4] A reversible process may instead be defined as 'one which is quasistatic and without hysteresis'. However, then one has to define the word 'hysteresis', and one thus reverts to the definition given here.

To understand the second definition, consider a piston of area A sliding in a chamber with very low friction (Figure 3.4). If the pressure in the fluid is p, and the force applied to the piston is $f = pA + \epsilon$, then for $\epsilon > 0$ the piston will move to the left. In the absence of friction, this is true even when $|\epsilon| \ll pA$. Now suppose the force is reduced by 2ϵ. This is a tiny change, but if friction is negligible, then the change will suffice to reverse the direction of the process: the piston will move to the right. This type of situation is called a reversible process. By contrast, if the piston were sticky, i.e. had friction, then a small change in applied force would merely cause the movement to stop. A larger change would be needed to overcome the friction and get the piston to move back.

Figure 3.7 shows a mechanical device that conveniently achieves the reversible compression or expansion of a gas over a large change of volume.

Another reversible example is two bodies in perfect thermal contact, with one at temperature T, the other at $T + \delta T$, for very small δT. In this situation, energy in the form of heat flows from the hotter to the colder body. A change in the circumstances by just more than δT (so as to swap which body is the hotter, which the colder) would suffice to reverse the direction of the flow of energy.

These examples show that reversible processes are smooth, slow, unrestrained, and delicate. Here 'slow' means 'slow compared to the relevant internal equilibration times.'

All reversible processes are quasistatic, but many quasistatic processes are not reversible. This relationship is shown in Figure 3.8.

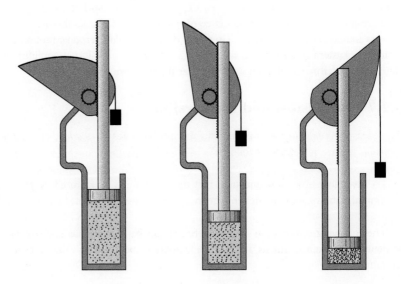

Fig. 3.7 *Reversible compression of a gas.*

The first definition, referring to the rest of the universe, draws attention to the difference between reversibility in the thermodynamic sense, and the use of the term in everyday speech. If a process is thermodynamically irreversible, it doesn't necessarily mean it can't be reversed in the everyday sense. For example, sliding a book across a table (by pushing it with your finger) is a thermodynamically *irreversible* process, because if you adjust the force by a tiny amount, the book will not start to slide back, owing to friction. However, this does not mean it is impossible to slide the book back again! It can easily be done. There is a net change in the rest of the universe, however: the table top, or the room, or the world at large, is now a little hotter than it was. To be precise, we will see later that an irreversible process involves a net increase in the total entropy of the systems involved.

Here are some more examples.

A change in a system's state is in general initiated or caused by some change in the external conditions. If, in order to reverse the transformation of the system, it is not sufficient merely to reverse the change in external conditions, then the process which took place was irreversible. Consider for example lowering the point of a pin slowly onto the surface of an inflated balloon. If the pin only pushes a small amount, then the balloon remains intact, and if the pin is raised again then the balloon returns to its initial state. This is reversible. If, on the other hand, the pin is pushed down further, until the balloon bursts, then when the pin is raised the balloon does not for that reason reassemble itself. Therefore the process was irreversible.

For a more gentle example, consider a partition between two chambers containing different inert gases at the same temperature and pressure. If the partition is removed, the gases diffuse into one another until they are thoroughly mixed. If the partition is slowly replaced, the two gases do not unmix themselves. Therefore the process was irreversible. This doesn't rule out that we could devise a way to separate out the gases and put them in the separate chambers again, but to do so would require some further process which cannot help but leave a net change in the environment.

Relaxation to equilibrium is always irreversible in the thermodynamic sense.

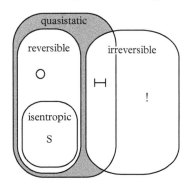

Fig. 3.8 *Venn diagram for thermodynamic terminology. The relationship of the terms* reversible, irreversible, quasistatic, isentropic *is shown, with an example process for each region: !: an explosion, H: squeezing toothpaste out of a tube, O: slowly stretching an elastic band at constant temperature, S: slowly stretching a thermally insulated elastic band. The shaded region is excluded since all processes that are not reversible are, by definition, irreversible.*

3.2.8 Adiathermal, isentropic, adiabatic, isothermal

An **isothermal** process is one taking place at constant temperature.

A **thermally isolated** system is one that cannot exchange heat with its surroundings. A process taking place in conditions of thermal isolation is called *adiathermal*. Such processes usually change the system temperature.

It is a common mistake for students to get muddled between *isothermal* and *thermal isolation*. They are *not* the same thing. In fact they are almost opposites!

Thermal isolation is typically achieved by surrounding a system with a material of very low thermal conductivity, such as air, or better still vacuum, and reflective foil (to prevent thermal radiation). One can test for thermal isolation by changing the temperature of the surroundings (without any other change) and checking to see if any change in the system follows: if not, then the system was thermally isolated. Conversely, two systems are in good thermal contact if they reach thermal equilibrium with one another rapidly.

If an ordinary gas is placed in a chamber in conditions of thermal isolation, then if the gas is compressed its temperature will rise: the process is adiathermal but not isothermal.

Conversely, if the gas is in good thermal contact with a large body at fixed temperature, then if the gas is compressed its temperature will not change, but now heat will flow out of the gas into the other body: the process is isothermal but not adiathermal.

There exist adiathermal processes in which the temperature does not change, and isothermal processes in which no heat transfer is involved, but these are the exception rather than the rule.

A reversible adiathermal process is called *isentropic*.

Another much used word in thermal science is *adiabatic*. Unfortunately there is no universal agreement on its usage. In many, especially older, treatments of thermodynamics, the word *adiabatic* was used for adiathermal. However, in modern physics, the word *adiabatic* is more often used in the stronger sense of isentropic. Even texts which announce the word *adiabatic* as meaning no heat transfer (i.e. adiathermal) often proceed to use it as a label for isentropic processes. To avoid confusion one should be aware of whether or not such an assumption has been made.

In this book *adiabatic* will be used exclusively as a synonym for *isentropic*.

The usage of the word *adiabatic* can be clarified by the following example. Let the system under discussion be a rigid thermally isolated chamber containing a firework. Suppose the firework is ignited and explodes violently. Since the chamber was thermally isolated from its environment, this process is adiathermal for the complete system of firework plus chamber. Everyone agrees on that. However, I and most physicists would *not* call such a process adiabatic (because it is not reversible), but some physicists and many older textbooks would.[5]

Since isentropic processes are by definition reversible, they can be located on the Venn diagram in Figure 3.8 as a subset of reversible processes.

Table 3.3 lists some other thermodynamic terms that apply to processes. These are all straightforward.

3.2.9 Expansion coefficients, heat capacities

For any pV system, various coefficients are defined in order to characterize the behaviour on expansion or compression. These are listed in Table 3.4, where their values for an ideal gas are also given. For example the rate of change of volume with pressure is $\partial V/\partial p$, and there are two common ways in which a system can be compressed: at constant temperature, giving $(\partial V/\partial p)_T$, or at constant entropy

---

[5] This is merely a matter of definitions of words; there is no controversy about the physics.

Table 3.3 *Types of process.*

Name	Definition	Constant
Quasistatic	Slow compared to relaxation times, hence maintaining equilibrium	–
Reversible	An infinitesimal change can reverse the direction	–
Adiathermal	In thermal isolation, i.e. no heat exchange	–
Adiabatic	See text	
Isentropic	Reversible and with no heat exchange	S
Isothermal	At constant temperature	T
Isobaric	At constant pressure	p
Isochoric	At constant volume	V

(usually called adiabatically), giving $(\partial V/\partial p)_S$. These are not the only ways (there is an infinite variety of paths on the indicator diagram) but they are the most convenient to measure and to reason about. Since for small changes the change in volume is typically proportional to the volume, it makes sense to define the coefficient by the ratio of dV to V, and the minus sign makes the coefficient positive.

Another coefficient of basic interest is the tendency of bodies to expand (or in some cases contract) when their temperature is increased. This is characterized by the 'isobaric expansivity' $(1/V)(\partial V/\partial T)_p$.

Typically it is quantities such as these which can be measured or calculated, and from them the equation of state can be derived. For example, suppose it were found that $\kappa_T = aT^3/p^2$ and $\alpha = 3aT^2/p$, where a is a constant. Then we would have

$$dV = -\frac{aVT^3}{p^2}dp + \frac{3aVT^2}{p}dT,$$

and integration (see Section 5.1.2) gives the equation of state $V = V_0 \exp(aT^3/p)$ (clearly *not* an ideal gas!).

Note also that $-1/(V\kappa_T)$ and $-1/(V\kappa_S)$ are the slopes of the isotherms and the adiabats, respectively, on the indicator diagram.

Heat capacities characterize the ability of bodies to absorb or give out heat. We have not formally defined heat yet (that will be done in Chapter 7); for the moment it suffices to say that heat is a form of *energy*, and it passes from a hotter to a colder body when the two are placed in thermal contact. For a pV system there are two standard ways to do this: either under conditions of constant pressure, or conditions of constant volume, leading to the capacities as defined in Table 3.4. A heat capacity is the amount of heat required to raise the temperature of a body by one unit.

Table 3.4 *Basic coefficients for a simple compressible system, with their values for an ideal gas. The compressibility answers the question 'if we increase the pressure, how much does the volume change'? The definition is in terms of a relative change (hence the factor $1/V$) so that the answer is independent of V for an ideal gas, and nearly so for many other systems. The signs are chosen so that κ_T and κ_S are always positive, and α is usually but not always positive. α is also called* cubic expansivity; *linear expansivity is 1/3 of this. The adiabatic compressibility refers to an adiabatic (= isentropic) change. Its value cannot be obtained from the equation of state alone. Compressibility and bulk modulus are different ways of referring to the same partial derivative; the former tends to be used for gases, the latter for solids. Heat capacities are defined as shown. For a monatomic ideal gas $C_V = (3/2)Nk_B$; for a diatomic ideal gas the value is higher and depends on temperature. The notation with a bar on small heat-related quantities is explained in Chapter 5.*

Name	Definition	Value for ideal gas
Isothermal compressibility	$\kappa_T = -\dfrac{1}{V}\left.\dfrac{\partial V}{\partial p}\right\|_T$	$\dfrac{1}{p}$
Adiabatic compressibility	$\kappa_S = -\dfrac{1}{V}\left.\dfrac{\partial V}{\partial p}\right\|_S$	$\dfrac{1}{\gamma p}$
Isobaric expansivity	$\alpha = \dfrac{1}{V}\left.\dfrac{\partial V}{\partial T}\right\|_p$	$\dfrac{1}{T}$
Heat capacity at constant volume	$C_V = \dfrac{\text{đ}Q_V}{\text{d}T}$	typically in the range 1.5–$3Nk_B$
Heat capacity at constant pressure	$C_p = \dfrac{\text{đ}Q_p}{\text{d}T}$	$C_V + Nk_B$
Isothermal bulk modulus	$B_T \equiv 1/\kappa_T$	p
Adiabatic bulk modulus	$B_S \equiv 1/\kappa_S$	γp

The phrase 'heat capacity' refers to the total capacity of some given system. This quantity is extensive. It is often interesting to consider also the heat capacity per unit volume or the heat capacity per unit mass. The latter is called the *specific heat capacity*. For example,

$$c_V = \frac{C_V}{m} = \frac{1}{m}\frac{\text{đ}Q_V}{\text{d}T}. \tag{3.4}$$

The name 'specific heat capacity' is often abbreviated to 'specific heat'. This abbreviation is not recommended when learning the subject, since it suggests a muddle of physical dimensions: a heat *capacity* (energy transfer per unit temperature change) is not a heat (energy). In SI units, specific heat capacity is measured in joules per Kelvin per kg.

The compressibilities, heat capacities, and such like are collectively known as *response functions*. They, or their partners in other types of thermodynamic system, may also be called *susceptibilities*.

3.2.10 Thermal reservoir

A *thermal reservoir* is a body of given temperature that can take in and give out heat without changing its temperature. It has infinite heat capacity.

Think of a thermal reservoir, sometimes called a 'heat bath' or just 'bath', as something like a very large quantity of water, but with perfect thermal conductivity. If another body is placed in contact with such a reservoir, then heat flow can take place between the body and the reservoir, and will continue until the two are at the same temperature. The heat energy that came from or went into the reservoir is not enough to cause any significant change to the reservoir, so the final temperature of the body is equal to the fixed temperature of the reservoir.

Of course a thermal reservoir is another idealization, but a useful one for theoretical purposes, and one which can be approximated in practice more closely than you might think. To make a thermal reservoir one does not need to use a large body: it suffices to have a 'thermostat', that is, a device that keeps the temperature of some given body constant by detecting small changes and reacting via a feedback loop to change the conditions (e.g. by pressure changes or by a heat pump) so as to restore the temperature to a fixed value.

3.3 The laws of thermodynamics

The laws of thermodynamics are ultimately distilled from experimental observations. They are organizing principles around which sense can be made of a large amount of data. Owing to the fact that the natural world is itself remarkably coherent in its physical workings, at many different levels, these principles also cohere together in a satisfying theoretical whole. So the principles also have the character of intellectually satisfying parts of a coherent intellectual model. Science is born out of the meeting-point of these two movements: the empirical data on the one hand, and the love of reason on the other. We choose to elevate to the status of a 'law' those organizing ideas which allow the essential ingredients of a wide range of physical behaviours to be captured or uncovered in a simple way.

We will now state the collection of such ideas which have come to be called the laws of thermodynamics. I think it is useful to place them here, all together, in order to get a sense of the foundations on which we wish to build. However, one does not understand the meaning of these laws (as one does not understand any law of Nature) merely by writing down brief statements. Each requires a lengthy discussion. That longer discussion is, in essence, the rest of the book.

> **ZEROTH LAW:** If two systems are separately in thermal equilibrium with a third, then they must be in thermal equilibrium with each other.

If we let the symbol '$\Leftrightarrow$' temporarily stand for 'in thermal equilibrium with' then this law states

$$\text{If } A \Leftrightarrow C \text{ and } B \Leftrightarrow C \text{ then } A \Leftrightarrow B.$$

This law will enable us to introduce the concept of *temperature* in a precise way in Chapter 6.

FIRST LAW: There is more than one way to state the first law of thermodynamics, and in the course of writing this book I switched opinion several times on which one I preferred. I settled on

Energy is conserved.

The alternative statement is

The amount of work required to change the state of a thermally isolated system depends solely on the initial and final states.

The first statement, *energy is conserved*, is the essential idea, so in the end I give it priority. In the historical development of the subject, however, the second statement was more important, because it was used to discover the first. The difficulty lies with the understanding of *heat*. In mechanics, where we apply Newton's laws to motion and forces, we are familiar with the conservation of energy, and indeed it might seem that the first law of thermodynamics is not adding anything new. However, in thermodynamics we are trying to understand heat as well as work, and Newton's laws simply do not address the issue. Ultimately, we could appeal to a microscopic understanding of heat, which shows that it is a form of energy expressed as random motions and positions. However it is valuable to note that a macroscopic thermodynamic argument also enables us to identify heat as a form of energy. This will be discussed in Chapter 7.

SECOND LAW: There is one law, but several of ways of stating it. These are all equivalent because they each imply all the others, therefore we only need to provide one. However, there is room for subjective opinion on which statement is the more basic or simple or profound. It seems to be that the Kelvin and Clausius statements are simpler in the sense of easier to grasp, and therefore a good way to learn the subject, while the Carathéodory statement is harder to grasp but is, arguably, more profound. We will follow the route of learning from Kelvin and Clausius first, and thus coming to Carathéodory. The final version (maximum entropy principle) is the subject of Chapter 18, and is the version we presented briefly in Section 3.1.

Clausius statement:

No process is possible whose sole effect is the transfer of heat from a colder to a hotter body.

Kelvin statement:

No process is possible whose sole effect is to extract heat from a single reservoir and convert it into an equivalent amount of work.

Carathéodory statement:

In the neighbourhood of any state K of a thermally isolated system, there are states K' which are inaccessible from K.

Entropy statement:

There exists an additive function of state known as the *equilibrium entropy S*, which can never decrease in a thermally isolated system.

All these statements are powerful because they are wide-ranging.

It is familiar that heat doesn't naturally tend to flow from a colder to a hotter body, but Clausius's statement goes further, and says that no matter how we try to devise a process, no matter how complicated (and people, wise or otherwise, have tried), we won't find an end result in which some heat energy came out of one body and was delivered up to a hotter one, with the rest of the apparatus and surroundings undergoing no net change. Kelvin's statement forbids a similarly wide range of processes. In the Kelvin statement the phrase 'single reservoir' means the heat is transferred at a single temperature. If more than one temperature is involved, then some work can be obtained: this is what 'heat engines' do. This will be discussed in Chapter 8.

Each of the statements not explicitly mentioning entropy can be used as the starting point for an argument to prove the existence and nature of entropy. This is the subject of Chapter 8. Equally, it is very easy to derive the other statements from the maximum entropy principle (see Chapter 18).

We separate the statement of the third law from the others because it is less important, and indeed it is not needed for much of thermodynamics. However, it is useful in chemistry and is invoked for some arguments in physics.

THIRD LAW: (Planck–Simon statement)

The contribution to the entropy of a system by each aspect of the system which is in internal thermodynamic equilibrium tends to zero as the temperature tends to absolute zero.

This law is less important than the others because it will turn out that we can get a long way by only keeping track of changes in entropy, without needing to know the absolute value or total amount of entropy. However, the third law makes a useful general observation about the entropy for one group of states (those at zero temperature). The entropy of other states can be found empirically by using thermodynamic methods, either by a combination of judicious measurements at high temperature, or by tracking entropy changes from near absolute zero temperature; this is discussed in Chapters 9 and 23. The 'absolute zero' of temperature will also need to be defined, of course. It can be defined as the temperature of a body from which no heat can be extracted. From a microscopic point of view, it is the temperature of a body whose energy is completely in non-random forms. We will find that all such bodies have the same temperature. In the Celsius scale of temperature, absolute zero is at −273.15 °C; in the more useful 'absolute scale' the absolute zero is at 0 K. Equations such as (3.2) (the equation of state of a van der Waals gas) assume that T is the absolute temperature.

The third law enables some general deductions concerning heat capacities and related quantities to be made. These are discussed in Chapter 23. It can also be used to prove the following theorem:[6]

[6] Sometimes the unattainability theorem is itself called the third law, which is wrong—see Chapter 23.

Unattainability theorem

> No process, no matter how idealized, can reduce the temperature of a system to zero in a finite number of isentropic and isothermal steps.

An example of a practical cooling method using isentropic and isothermal steps is *adiabatic demagnetization*, described in Chapter 16.

3.4 Where we are heading

The material to be discussed in the next few chapters can be summarized by the following list:

Zeroth law	→	T (temperature)
First law	→	U (energy)
Second law	→	S (entropy)

This will culminate in two crucial equations:

$$dU = TdS - pdV + \mu dN \tag{3.5}$$

and

$$\frac{Q_1}{Q_2} = \frac{T_1}{T_2}. \tag{3.6}$$

The first describes the way a small change in the energy content of a pV system can be related to changes in its entropy, volume, and particle number. The second describes the way the heats Q_1, Q_2 exchanged at two reservoirs during a reversible cyclic process are related to the temperatures of the reservoirs.

Once these equations and their meaning have been derived, we can proceed to 'play' the game of thermodynamics, applying it to all sorts of systems and processes. A very useful further idea in the game will be that of *free energy*, a quantity related to both energy and entropy. It will be introduced in Chapter 17.

EXERCISES

(3.1) State which of the terms *quasistatic, reversible, irreversible, isothermal*, apply to each of the following processes (in each case list all appropriate terms):
   (a) A gas slowly expanding in a thermally isolated vessel fitted with a frictionless piston.
   (b) A volcanic eruption.
   (c) A nail oxidizing (rusting) while immersed in a bucket of water.
   (d) Ice forming over a pool of water on a still winter evening.
   (e) A force pushing an ordinary book across an ordinary table.

(3.2) Reproduce the Venn diagram showing the relationship between the concepts reversible, irreversible, quasistatic, isentropic. Give an example of a process in each region, different from the examples given in the text.

(3.3) Give appropriate SI units for all the quantities listed in Table 3.4.

(3.4) Which has the higher bulk modulus, a gas or a solid?

(3.5) *Function of state.* Water can enter a lake by two routes: either by flowing down a river into the lake, or by falling as rain. It can leave by evaporation, or by flowing out into the outlet stream. Assume there is no difference between rain water and river water (e.g. both are pure).
   Let the physical system under discussion be the lake of water. Which of the following quantities are functions of state?:
   (a) The depth of the water.
   (b) The total amount of water in the lake.
   (c) The mass of river water in the lake.
   (d) The temperature of the water.
   (e) The mass of water that has left the lake by evaporation.
   (f) The volume of rain water in the lake.

(3.6) Suppose that the lake in the previous question empties completely during a dry spell, and then later fills up with water from the river during a period in which there is no rainfall. Does this change the answers to any parts of the previous question?

(3.7) Show that, in the limit $V \gg Nb$, the isothermal compressibility of a van der Waals gas is $(p - N^2 a/V^2)^{-1}$.

4 Some general knowledge

4.1 Density, heat capacity 34
4.2 Moles 35
4.3 Boltzmann constant, gas constant 36
4.4 Pressure and STP 37
4.5 Latent heat 37
4.6 Magnetic properties 38

In this chapter we present some general knowledge about properties of materials and definitions of fundamental constants. Although the abstract principles of thermodynamics do not require knowledge of specific material properties, it is helpful when learning to have some examples in mind. We apply the theory in the first instance to ordinary things such as a cup of water, and then when we have mastered it we can study extraordinary things such as a supernova or a living cell.

4.1 Density, heat capacity

All ordinary physical systems are made of atoms (possibly arranged into molecules), and the masses of common atoms range over about two orders of magnitude, from 1.67×10^{-27} kg (hydrogen) to 344×10^{-27} kg (lead). When atoms congregate into condensed matter, that is, solids and liquids, the spacing between the atoms does not vary greatly, but tends to be larger for the heavier atoms, with the result that the densities of common substances extend over a smaller range, and the value 1000 kg/m^3 is often a good order-of-magnitude estimate. For example, water has a density of 1000 kg/m^3, wood has densities in the range 500 to 900 kg/m^3, metals have densities in the range approximately 2000 to 20,000 kg/m^3. The number density is around 10^{28} to 10^{29} atoms per cubic metre for the solid or liquid elements, which implies the atomic spacing is in the range 0.2 nm to 5 nm. The density of solids and liquids does not vary much with temperature.

Gases, by contrast, have a density lower than solids by about a factor 1000 at ordinary temperature and pressure, and away from the boiling point the density is proportional to pressure and inversely proportional to temperature (the ideal gas law). For example, air at standard temperature and pressure has a density of 1.27 kg/m^3.

The heat capacity per unit mass (called specific heat capacity) of a given substance is typically similar in the solid and gaseous forms, and around twice as large in the liquid form (e.g. 2108 J K^{-1}kg^{-1}, 4190 J K^{-1}kg^{-1}, 2074 J K^{-1}kg^{-1} for ice, water, steam), except at very low temperature where most substances solidify and all heat capacities tend to zero—see Figure 4.1.

The specific heat capacities of some common substances at room temperature are given in Table 4.1. Water has a higher specific heat capacity than most ordinary substances. The reason for this is partly that hydrogen and oxygen are of low

Thermodynamics: A Complete Undergraduate Course. Andrew M. Steane.
© Andrew M. Steane 2017. Published 2017 by Oxford University Press.

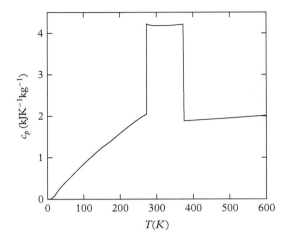

Fig. 4.1 *The specific heat capacity of H_2O as a function of temperature.*

Table 4.1 *Heat capacities of various common substances at standard temperature and pressure.*

Substance	Specific heat capacity ($JK^{-1}kg^{-1}$)	Molar heat capacity ($JK^{-1}mol^{-1}$)
Water	4186	75
Sodium	1237	28
Copper	390	25
Mercury	140	28
Lead	130	27
Graphite	699	8.4
Hydrogen gas	14,500	29
Helium gas	5250	21
Nitrogen gas	1035	29

mass, so for a given mass of material there are more molecules present in water than there are in other familiar substances such as copper. When we examine the heat capacity for a given number of molecules (called molar heat capacity, see Section 4.2), then the values for most substances are similar, with water somewhat high and graphite somewhat low.

4.2 Moles

The word 'mole' is a measure of 'amount of material'. When the word was first coined by chemists, there was not yet complete clarity about what property it measured. Nowadays the word is redundant, because we have learned that it is simply a way of referring to a certain number of particles; the number in question being approximately 6.02214×10^{23}. This number is called Avogadro's number. The use of a name here is like other names for numbers such as a 'dozen' for 12 or the old English 'score' for 20. 'A mole of eggs' means 6.02214×10^{23} eggs.

In the case of chemical substances, the word 'mole' can be ambiguous unless one is careful. 'A mole of oxygen' might mean a mole of oxygen atoms or a mole of O_2 molecules. To avoid ambiguity one should say 'a mole of oxygen atoms' or 'a mole of oxygen molecules', as the case may be.

Avogadro's number N_A is not necessarily an integer, and probably is not an integer. It is defined to be the number of carbon-12 atoms in 12 grams (0.012 kg) of unbound carbon-12 in its lowest energy electronic state. This number is currently measured to be

$$N_A \simeq 602214086000000000000000 \simeq 6 \times 10^{23}.$$

A related quantity, often of more use to physicists, is the unified atomic mass unit, defined to be one twelfth the mass of a carbon-12 atom:

$$u = 1.66053904 \times 10^{-27} \text{ kg}.$$

This is defined such that $N_A u = 0.001$ kg. That is, a mole of particles of relative atomic mass 1 will have a total mass of one gram. For this reason the mole is sometimes called the 'gram-mole'. Sometimes people use the word 'mole' for $1000 N_A$. Strictly, that number should be called either a 'kilomole' or a 'kilogram-mole'.

The idea of a mole is useful in cases where it is interesting to keep track of the *number* of constituent particles, rather than some other measure of quantity such as the mass. This makes sense when examining chemical reactions, for example, since then the numbers add up in simple ways. To make a kilogram of water you need approximately 0.112 kg of hydrogen and 0.888 kg of oxygen: not very memorable. To make a mole of water molecules you need exactly two moles of hydrogen atoms and one mole of oxygen atoms: easy!

We have already noted that heat capacities of many materials take similar values when measured per mole rather than per unit mass.

4.3 Boltzmann constant, gas constant

The ideal gas equation of state, $pV = Nk_\text{B} T$ (Eq. (3.3)) was introduced in Section 3.2.2.2. On the right-hand side, N refers to the number of particles in the gas and k_B is the *Boltzmann constant* whose value is

$$k_B \simeq 1.380649 \times 10^{-23} \text{ JK}^{-1}. \tag{4.1}$$

Since k_B is small and in ordinary circumstances N is large, the ideal gas equation of state is often written

$$pV = n_\text{m} RT, \tag{4.2}$$

where

$$n_m \equiv \frac{N}{N_A}, \qquad R \equiv N_A k_B.$$

$n_m = N/N_A$ is the number of moles of particles in the gas, and $R \simeq 8.314$ JK^{-1}mol^{-1} is called the *molar gas constant*.

4.4 Pressure and STP

The SI unit of pressure is the pascal (Pa), equal to one newton per metre-squared. Another widely used unit is the bar:

$$1 \text{ bar} = 10^5 \text{ Pa} = 100 \text{ kPa}.$$

The standard sea level pressure in Earth's atmosphere is 1 bar (often quoted as 1000 mbar), but this should not be confused with another unit called the 'atmosphere' (atm), which is equal to 1.01325 bar.

Standard temperature and pressure (STP) are either 20 °C (293.15 K) and 101.325 kPa (1 atm) (the National Institute of Standards and Technology definition), or 0 °C (273.15 K) and 100 kPa (0.986 atm) (the International Union of Pure and Applied Chemistry definition). We shall adopt the former definition, where STP corresponds to 'room temperature, 1 atm'.

The molar volume of an ideal gas, or 'standard molar volume' is

$$V_1 = \frac{RT}{p},$$

so its value depends on the definition of STP. It is

24.055 litres per mol (24.055 m^3/kmol)   at 20 °C and 101.325 kPa,
22.711 litres per mol (22.711 m^3/kmol)   at 0 °C and 100 kPa.

4.5 Latent heat

During a process such as melting or boiling, a change takes place in which much energy is transferred in the form of heat, while the temperature of the system does not change. For example, ice turns to liquid water, or liquid water turns to steam. The process is called a *phase change*, and the amount of heat energy needed to bring about the change is called a *latent heat*. The latent heat increases in proportion to the mass of the system, and therefore one normally finds quoted in tables either the specific latent heat or the molar latent heat. This quantity is a function of other variables such as pressure or temperature, but that dependence

Table 4.2 *Melting and boiling temperatures, and specific latent heats, for various substances at atmospheric pressure. Helium remains a liquid at low temperature at this pressure.*

Material	Melting point °C	L_f (kJ/kg)	Boiling point °C	L_v (kJ/kg)
Water	0	334	100	2,260
Aluminium	659	397	2327	10,500
Lead	327.3	24.5	1750	870
Hydrogen	−259	58	−253	455
Helium	N/A	N/A	−268.93	20.9
Nitrogen	−218.79	13.8	−195.81	200.1

is modest except near a condition called the *critical point*, which is discussed in Chapter 22.

The latent heat associated with the solid–liquid transition is called the *latent heat of fusion* (or *enthalpy of fusion*); the latent heat associated with the liquid–vapour transition is called the *latent heat of evaporation* or *latent heat of vaporization*. Some example values at 1 atm are given in Table 4.2.

Example 4.1

How much ice at 0 °C is required to cool 4 kg of water from 20 °C to 0 °C?

Solution.
The temperature change of the water is $\Delta T = 20$ K. The amount of heat leaving 4 kg of water as it cools by this much is $(4\,\text{kg}) \times (4186\,\text{JK}^{-1}\text{kg}^{-1}) \times (20\,\text{K}) = 334.9$ kJ. This amount of heat will melt $(334.9\,\text{kJ})/(334\,\text{kJ/kg}) = 1$ kg of ice.

4.6 Magnetic properties

We will learn the methods of thermodynamics by focussing on mechanical properties such as pressure and tension in the first instance, and then widening the discussion to magnetic, electrical, and chemical properties. Magnetic properties are especially important in low temperature physics, because when a material has magnetic properties, this is often the main way in which its energy can vary when the temperature is so low that vibrational motion of the atoms is negligible.

Some but not all atoms and molecules have a permanent magnetic dipole moment. The size of this dipole moment is of the order of one or a few *Bohr magnetons*. The Bohr magneton is defined as

$$\mu_B = \frac{e\hbar}{2m_e} \simeq 9.274 \times 10^{-24} \text{ joules per tesla}, \tag{4.3}$$

where e is the magnitude of the charge of the electron, and m_e is the mass of the electron. By multiplying the Bohr magneton by the typical number density of atoms in a solid material, one obtains an estimate of the amount of magnetic dipole moment per unit volume that may be expected when the dipoles all line up. This is

$$M = n\mu_B \simeq 4 \times 10^5 \text{ amps per metre}. \tag{4.4}$$

By multiplying this by the permeability of free space, μ_0, we obtain the order of magnitude of the magnetic field that would be produced in such a material at low temperature:

$$n\mu_B\mu_0 \simeq 0.5 \text{ tesla}. \tag{4.5}$$

This is the order of magnitude of the maximum magnetic field that is produced by the atomic and molecular magnetism in many magnetic materials.

The ratio of the Bohr magneton to the Boltzmann constant has a value of order 1 in SI units:

$$\frac{\mu_B}{k_B} \simeq 0.67 \text{ kelvin per tesla}. \tag{4.6}$$

This implies that if the magnetic potential energy $-\mu B$ is the main contribution to potential energy associated with the magnetic dipoles in some material, then the material will show significant thermal effects at a temperature of order 1 kelvin in a field of order 1 tesla. (Materials showing magnetic effects at much higher temperatures do so because the electrons also carry electric charge and so can also have a substantial electrostatic interaction; this is connected to the orientation of the magnetic dipoles through a quantum effect called exchange symmetry.)

5 Mathematical tools

5.1 Working with partial derivatives 40
5.2 Proper and improper differentials, function of state 46
5.3 Some further observations 49
Exercises 51

This chapter will introduce all the mathematical tools that will be needed for the rest of the book. The methods are concerned with functions of more than one variable, their differentiation and integration, and issues such as change of variable. It is assumed that the reader has met partial differentiation, but not yet worked with it very much.

5.1 Working with partial derivatives

To begin, let us consider some function f of two variables x and y. For example, the function could be

$$f(x,y) = 2y\log(x) + (y-1)^2. \tag{5.1}$$

This function is plotted in Figure 5.1 for x in the range 1 to 2 and y in the range 0 to 2.

Partial differentiation is the mathematical process of extracting the gradient of a function with respect to one variable, while other variables remain constant.

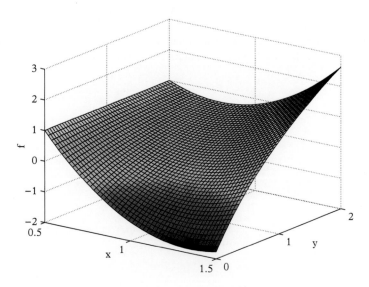

Fig. 5.1 *The function $f(x,y) = 2y\log x + (y-1)^2$, used as an example throughout this chapter.*

Thermodynamics: A Complete Undergraduate Course. Andrew M. Steane.
© Andrew M. Steane 2017. Published 2017 by Oxford University Press.

The partial derivative of f with respect to x while holding y constant is commonly indicated by the notation

$$\left(\frac{\partial f}{\partial x}\right)_y$$

so that, for example, for the given function, we have

$$\left(\frac{\partial f}{\partial x}\right)_y = \frac{2y}{x}. \tag{5.2}$$

Another notation for partial differentiation is becoming widespread, namely

$$\left.\frac{\partial f}{\partial x}\right|_y.$$

We prefer this because it is slightly less cluttered, and this is the notation adopted in this book. Occasionally we will also want a notation for 'evaluated at', which we will write using the symbol $\lfloor$, as in Eq. (5.4).

Still more succinct notations for partial derivatives are $\dfrac{\partial f_y}{\partial x}$ and $f_{y,x}$, but we will not adopt these.

Returning now to our example function, partial differentiation with respect to y at constant x yields

$$\left.\frac{\partial f}{\partial y}\right|_x = 2\log(x) + 2(y-1). \tag{5.3}$$

If we regard f as the 'height' of a hill, then $\left.\frac{\partial f}{\partial x}\right|_y$ can be interpreted as the slope (that is, the change in height divided by the horizontal distance moved) as one moves along a line of constant y. $\left.\frac{\partial f}{\partial y}\right|_x$ can be interpreted as the slope as one moves along a line of constant y. To get the total change of f as one moves from a given point (x, y) to a nearby point $(x + dx, y + dy)$, one may imagine moving first from (x, y) to $(x+dx, y)$ and then from $(x+dx, dy)$ to $(x+dx, y+dy)$ (Figure 5.2). Therefore the total change in f must be

$$df = \left.\frac{\partial f}{\partial x}\right|_y\Big\lfloor_{(x,y)} dx + \left.\frac{\partial f}{\partial y}\right|_x\Big\lfloor_{(x+dx,y)} dy, \tag{5.4}$$

where the symbol $\lfloor_a$ should be read to mean 'evaluated at a', as we have already noted, and second-order small quantities have been neglected. Now $(\partial f/\partial y)_x$ evaluated at $(x+dx, y)$ is almost the same as $(\partial f/\partial y)_x$ evaluated at (x, y). They differ by a small quantity, $(\partial^2 f/\partial x \partial y)dx$. Therefore as long as the second derivative is well behaved, we have

$$df = \left.\frac{\partial f}{\partial x}\right|_y\Big\lfloor_{(x,y)} dx + \left.\frac{\partial f}{\partial y}\right|_x\Big\lfloor_{(x,y)} dy + O(dxdy).$$

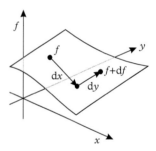

Fig. 5.2 *A small change in a function of two variables.*

Since it is understood, in equations for small quantities such as df, that later on the limit will be taken in which the small quantities tend to zero, we can omit the higher-order terms, and also take it for granted that the partial derivatives are to be evaluated at the location (x,y) where we wish to find the change in f. Therefore we may write

$$df = \left.\frac{\partial f}{\partial x}\right|_y dx + \left.\frac{\partial f}{\partial y}\right|_x dy. \tag{5.5}$$

In thermodynamics, equation (5.5) is the most widely used mathematical result concerning functions of more than one variable. It is the starting point of many arguments. We could derive the same result by making the journey from (x,y) to $(x+dx, y+dy)$ via $(x, y+dy)$ instead of $(x+dx, dy)$.

5.1.1 Reciprocal and reciprocity theorems

Equation (5.5) invites us to consider the possibility of dividing the whole equation through by dx or dy. This is legitimate as long as one is careful not to write down meaningless symbols such as df/dx. Although df/dx is well defined if f is a function of x alone, it is not well defined if f is a function of two or more variables. Unless further information is given, df/dx cannot be defined, because it refers to a change in f, but *we don't know between which two points*. For a function of a single variable, it is okay, because it is understood that the points are x and $x+dx$, but for a function of two or more variables this is insufficient. If one has in mind two points separated by dx *and having the same value of y* then one should say so; in this case the relevant ratio is $(\partial f/\partial x)_y$. However, two points separated by dx could instead have the same value of $z = x^2 + 3/y$ (say), and then for the same dx we could get a different df. Or to put it more succinctly $(\partial f/\partial x)_y$ is usually not equal to $(\partial f/\partial x)_z$.

With these cautions in mind, let's go ahead and divide equation (5.5) through by dx while holding something constant. Holding y constant yields the correct (if trivial) result

$$\left.\frac{\partial f}{\partial x}\right|_y = \left.\frac{\partial f}{\partial x}\right|_y, \tag{5.6}$$

where we have used the fact that the second term in (5.5) vanishes, since $(\partial y/\partial x)_y = 0$, as a moment's thought will show. Another way to do the reasoning is to say that (5.5) is valid for all (sufficiently small) dx, dy, and in particular it is valid when $dy = 0$ (i.e. is exactly zero) while dx is small but non-zero. Therefore, for a change at constant y the second term in (5.5) disappears, and we get (5.6).

A more interesting result is found if we divide (5.5) through by df while holding y constant. Then we have

$$1 = \left.\frac{\partial f}{\partial x}\right|_y \left.\frac{\partial x}{\partial f}\right|_y, \tag{5.7}$$

therefore

Reciprocal theorem
$$\left.\frac{\partial f}{\partial x}\right|_y = \frac{1}{\left.\frac{\partial x}{\partial f}\right|_y}. \qquad (5.8)$$

Note that it is important to notice that in this relation, both the partial derivatives have the same thing held constant. On the right-hand side the quantity $(\partial x/\partial f)_y$ appears. When using that quantity, we have in mind the fact that the formula relating f to x and y could be converted into a formula for x in terms of f and y. In the example of equation (5.1), we would have

$$x = \exp\left(\frac{f - (y-1)^2}{2y}\right), \qquad (5.9)$$

and therefore in this case $(\partial x/\partial f)_y = (2y)^{-1} \exp((f - (y-1)^2)/(2y))$. The reader can verify that this obeys the reciprocal theorem.

The result (5.7) can be extended to further variables, as long as we keep the same constraints (such as constant y) throughout, simply by linking each numerator with the previous denominator until the chain closes upon itself, e.g.

$$\left.\frac{\partial f}{\partial x}\right|_y \left.\frac{\partial x}{\partial z}\right|_y \left.\frac{\partial z}{\partial v}\right|_y \left.\frac{\partial v}{\partial f}\right|_y = 1. \qquad (5.10)$$

This is called the *chain rule*.[1]

Now return to equation (5.5) and divide through by dx while holding f constant. Then one obtains

$$0 = \left.\frac{\partial f}{\partial x}\right|_y + \left.\frac{\partial f}{\partial y}\right|_x \left.\frac{\partial y}{\partial x}\right|_f.$$

Bring the second term to the left-hand side, then divide through by $(\partial f/\partial x)_y$ and use the reciprocal relationship, to obtain

Reciprocity theorem
$$\left.\frac{\partial f}{\partial y}\right|_x \left.\frac{\partial y}{\partial x}\right|_f \left.\frac{\partial x}{\partial f}\right|_y = -1. \qquad (5.11)$$

[1] The term 'chain rule' is also commonly used to refer to other mathematical results. For example, suppose we have further variables x', y' which are related in some way to x and y. Then by dividing (5.5) by dx' while holding y' constant, we obtain

$$\left.\frac{\partial f}{\partial x'}\right|_{y'} = \left.\frac{\partial f}{\partial x}\right|_y \left.\frac{\partial x}{\partial x'}\right|_{y'} + \left.\frac{\partial f}{\partial y}\right|_x \left.\frac{\partial y}{\partial x'}\right|_{y'}.$$

This equation, and its generalization to larger numbers of variables, is also called by the name 'chain rule'.

This is called the reciprocity theorem. Note the pattern is like the chain rule in the numerators and denominators, but now for each partial derivative, a different variable is held constant (in each term it is the variable not appearing in the derivative). Note also that the result is -1 not $+1$.

This relationship can be neither shortened nor extended. That is, there is no version involving two terms or four or more terms. If there are further variables in the problem (that is, instead of having f a function of two things, we have a function of more than two, say x, y, z), then further variables can appear as 'spectator' variables which are held constant in all the terms. For example for a function $g(x, y, z)$; we could have

$$\left.\frac{\partial g}{\partial y}\right|_{x,z} \left.\frac{\partial y}{\partial x}\right|_{g,z} \left.\frac{\partial x}{\partial g}\right|_{y,z} = -1. \tag{5.12}$$

The reciprocal and reciprocity theorems appear in many thermodynamic arguments, where often what one wants to do is obtain an expression for some quantity, such as a temperature change, in terms of parameters that are accessible to laboratory measurement, such as pressure, volume, and heat capacities. For example, suppose x and y are readily controllable in a laboratory setting, but f is not. Then, one may wish to avoid terms such as $(\partial y/\partial x)_f$ in an expression. The reciprocity relationship can help to do this, by converting it into (minus) a ratio of $(\partial f/\partial x)_y$ and $(\partial f/\partial y)_x$, which is often more useful. Specific examples will be given in the rest of the book. For this reason, the most useful form in which to use the reciprocity relationship is often

$$\left.\frac{\partial x}{\partial y}\right|_f = -\left.\frac{\partial x}{\partial f}\right|_y \left.\frac{\partial f}{\partial y}\right|_x. \tag{5.13}$$

It is worth committing this formula to memory.

The reciprocity and other relations can be generalized by invoking the mathematics of *Jacobeans*, which are combinations of partial derivatives formed by taking the determinant of a matrix of partial derivatives. However, we shall not need such methods for this book.

5.1.1.1 Derivative of one extensive variable with respect to another

Suppose a function $f(x, y, z)$ has the property

$$f(\lambda x, y, z) = \lambda f(x, y, z)$$

for all values of λ. For example, this will happen when f is an extensive quantity such as volume, and x is another extensive quantity such as number of particles, and y and z are intensive properties such as pressure and temperature. Since, by assumption, the relation is true for all λ, it is true in particular for $\lambda = 1/x$, so we can deduce

$$f(1, y, z) = \frac{1}{x} f(x, y, z) \quad \Rightarrow \quad f(x, y, z) = x f(1, y, z)$$

and therefore

$$\left. \frac{\partial f}{\partial x} \right|_{y,z} = \frac{f}{x}. \tag{5.14}$$

In other words, for such functions, differentiation with respect to x while holding y and z constant is the same as dividing by x.

If we now have two such functions, f and g, then by applying this result to both of them, we find

$$\left. \frac{\partial f}{\partial g} \right|_{y,z} = \left. \frac{\partial f}{\partial x} \right|_{y,z} \left. \frac{\partial x}{\partial g} \right|_{y,z} = \frac{f}{g} \quad [f, g \text{ extensive}; y, z \text{ intensive}] \tag{5.15}$$

where the first step invokes the chain rule (note that y, z are constant throughout). This is useful for simplifying some expressions, but it should only be invoked in the right circumstances: it only holds when the correct sorts of variables are being held constant. An example is the relationship between volume and number of particles at fixed pressure and temperature:

$$\left. \frac{\partial N}{\partial V} \right|_{p,T} = \frac{N}{V}. \tag{5.16}$$

5.1.2 Integrating

Suppose we are given the partial derivatives, such as (5.2) and (5.3), and we would like to obtain the function itself by integration. In the present example, we have

$$df = \frac{2y}{x} dx + 2(\log(x) + y - 1) dy. \tag{5.17}$$

To integrate a 'two-dimensional' equation like this, one may perform two 'single-dimensional' integrals: one at constant y, the other at constant x. This is not the only possible choice, but it is convenient because in an integral at constant y the dy term is zero, and in an integral at constant x the dx term is zero. From the first one obtains $f = 2y \log(x) + g(y)$, where the constant of integration $g(y)$ is some undetermined function of y. From the second, one obtains $f = 2y \log(x) + y^2 - 2y + h(x)$, where $h(x)$ is an undetermined function of x. Hence we have the identity (i.e. true for all x, y)

$$2y \log(x) + g(y) \equiv 2y \log(x) + y^2 - 2y + h(x), \tag{5.18}$$

from which it follows that $g(y) = y^2 - 2y + c$, where c is a constant, and $h(x) = c$. Therefore the final result is $f = 2y \log(x) + y^2 - 2y + c$, in agreement with (5.1). Here, c is the constant of integration which the integral alone cannot determine. There is just one such constant here because we started from a first-order differential equation.

5.1.3 Mixed derivatives

If f is a function of x and y, then any derivative of f, such as for example equations (5.2) and (5.3), is itself a function of x and y (or it is the trivial function, zero) and can be differentiated. There are four possibilities, with examples as follows:

$$\left.\frac{\partial^2 f}{\partial x^2}\right|_y = \frac{-2y}{x^2}, \tag{5.19}$$

$$\left.\frac{\partial^2 f}{\partial y^2}\right|_x = 2, \tag{5.20}$$

$$\frac{\partial^2 f}{\partial y \partial x} = \frac{2}{x}, \tag{5.21}$$

$$\frac{\partial^2 f}{\partial x \partial y} = \frac{2}{x}. \tag{5.22}$$

The notation of the first result means y is held constant in both differentiations, and in the second x is held constant in both. The symbol $\partial^2 f/\partial y \partial x$ means we first differentiate w.r.t. x while holding y constant (Eq. (5.2)), then differentiate the result w.r.t. y while holding x constant. Note that we have

$$\frac{\partial^2 f}{\partial x \partial y} = \frac{\partial^2 f}{\partial y \partial x}. \tag{5.23}$$

This is generally true for any well behaved[2] function of x and y. To provide the proof, we use the idea that any (well behaved) function of x and y could in principle be expressed as a power series involving all powers of x, y, and combinations thereof, i.e.

$$f(x,y) = \sum_{n=0}^{\infty} \sum_{m=0}^{\infty} a_{n,m} x^n y^m. \tag{5.24}$$

Therefore, $(\partial f/\partial x)_y = \sum_n \sum_m n a_{n,m} x^{n-1} y^m$ and $(\partial f/\partial y)_x = \sum_n \sum_m m a_{n,m} x^n y^{m-1}$, from which $\partial^2 f/\partial y \partial x = \sum_n \sum_m nm a_{n,m} x^{n-1} y^{m-1} = \partial^2 f/\partial x \partial y$, Q.E.D.

[2] Here, 'well behaved' means the function does not have extreme behaviour such as an infinite value of the function or of one or more of its derivatives.

5.2 Proper and improper differentials, function of state

Consider the following small quantity:

$$x^2 \mathrm{d}x + 2xy \, \mathrm{d}y. \tag{5.25}$$

This is a perfectly well-defined small quantity, which could be evaluated for any given values of x, y and (small) $\mathrm{d}x$, $\mathrm{d}y$. One might be tempted to label it, therefore, as a small change, say $\mathrm{d}w$:

$$\mathrm{d}w \stackrel{??}{=} x^2 \mathrm{d}x + 2xy \mathrm{d}y. \tag{5.26}$$

The reason for the question-marks in this expression is that something odd is happening here. For, dividing both sides by dx while holding y constant gives $(\partial w/\partial x)_y = x^2$, and therefore

$$\left.\frac{\partial^2 w}{\partial y \partial x}\right| = 0.$$

However, dividing both sides of (5.26) by dy while holding x constant gives $(\partial w/\partial y)_x = 2xy$, and therefore

$$\left.\frac{\partial^2 w}{\partial x \partial y}\right| = 2y.$$

We now have a contradiction of equation (5.23), even though we just proved it for all well-behaved functions of x and y!

The problem here is not that there is an ill-behaved function somewhere in play. It is simply that *there does not exist any function at all* of x and y whose differential is $x^2 dx + 2xy\,dy$. Therefore when we set this small quantity equal to 'dw' in equation (5.26) we took a misleading step. We implied that we were dealing with a small change in some function $w(x, y)$; but we were not. To avoid making this mistake, while allowing a useful shorthand, we attach a bar to the differential symbol, so that we replace (5.26) by

$$đw = x^2 dx + 2xy\,dy. \qquad (5.27)$$

A quantity such as this, which is a well-defined small quantity but which is not the differential of any function, is called an *improper differential*. The terms *inexact differential* and *imperfect differential* are also used. By contrast, small quantities such as dx, dy, and df in equation (5.5), which are all differentials of some function, are called *proper differentials* or *exact differentials*.

I prefer the terminology 'proper/improper' to 'exact/inexact.'

To write the ratio of $đw$ to dx for a change in which y is constant, we shall use the notation

$$\frac{đw_y}{dx}. \qquad (5.28)$$

The subscript y recalls the partial derivative symbol $(\partial f/\partial x)_y$, but $đw_y/dx$ is not a partial derivative, strictly speaking, since there is no function $w(x, y)$ under discussion. Having said that, the practice of writing $(\partial w/\partial x)_y$ even when $đw$ is improper is fairly widespread, and is mostly harmless.

Mathematically, one can test whether one is dealing with an improper differential by obtaining the two mixed second derivatives and discovering whether they agree (as in (5.23)). If they do for all x and y, then the quantity df is a proper differential, i.e. a small change in a function of x and y. If they do not, then one is dealing with an improper differential. In thermodynamics, improper differentials continually crop up in the form of small amounts of energy passing from one place to another when work is done, or during heating and cooling.

The distinction between proper and improper differentials is closely related to another important thermodynamic concept, that of the *function of state*. This was introduced in Section 3.2.2. Recall the formal definition:

> A quantity F is a function of state if and only if the change ΔF, when a system changes state, depends only on the initial and final states, not on the path.

The definition works as follows. Suppose we have a physical quantity F, and we think we can assign it a well-defined value for each state of the system. If so, then its value when the system is in one state A could be written $F(A)$, and its value when the system is in another state B could be written $F(B)$. To find the latter value, we start with $F(A)$ and add to it the change ΔF taking place as the system moves to B. This is the only way one could consistently define $F(B)$. The system can get from A to B by many different paths. If ΔF depends on the path, then we have no way to decide what value is appropriate to use to define $F(B)$, and the assumption that F has a well-defined value for each state breaks down.

Functions of state and proper differentials are closely linked:

> A small change in a function of state is a proper differential; an infinitesimal quantity that is not the change of any function of state is an improper differential.

This shows how the physical idea (function of state) is linked to the mathematical idea (proper differential).

Let us now prove the following theorem:

Theorem: *If the integral of a function f is zero around all closed paths, then f is a function of state.*

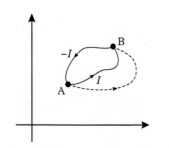

Fig. 5.3 *Construction for proof of (5.29).*

Proof: Consider some closed path which starts at A, goes via B, and returns to A (Figure 5.3). Since the integral all the way around is zero, the integral along the second segment, from B to A must equal $-I$, where I is the integral along the first segment, from A to B. Now consider another closed path which goes from A to B by some other route (dashed line in Figure 5.3), but comes back the same way as before. Since the return journey is unchanged, and the total is zero for any closed path, it follows that the integral along this new path from A to B must also equal I. The same applies to all paths between A and B, so we have proved the path-independence of the integral from A to B. Furthermore, the argument applies for any pair A and B. Therefore, by the definition of a function of state, f is a function of state. QED.

Integrals around closed loops in the state space occur in several thermodynamic arguments, and are denoted by attaching a circle to the integral sign:

$$\oint df \quad [\text{integral around a closed path}].$$

Therefore the above theorem can be conveniently expressed as

$$\oint_p df = 0 \;\; \forall p \;\; \Leftrightarrow \;\; f \text{ is a function of state}, \qquad (5.29)$$

where p indicates the path.
Function of state is often abbreviated to *state function*.

5.2.1 Integrating factor

Let us return to equation (5.27). Dividing both sides of this equation by x, one has

$$\frac{\mathchar'26\mkern-12mu d w}{x} = x dx + 2y dy. \qquad (5.30)$$

Now, the right-hand side of this equation is a proper (= perfect, exact) differential: it is the differential of the function $x^2/2 + y^2$. This means that although the value of the integral $\int \mathchar'26\mkern-12mu d w$ depends on the path taken, the value of the integral

$$\int \frac{\mathchar'26\mkern-12mu d w}{x} \qquad (5.31)$$

is independent of path. In this example, the factor x^{-1} is called an *integrating factor* for $\mathchar'26\mkern-12mu d w$. It converts an improper differential into a proper one. This idea comes up in thermodynamics, where we employ integrating factors for work from an early stage, and an important discovery is an integrating factor for heat.

5.3 Some further observations

5.3.1 Alternative derivation of reciprocal and reciprocity theorems

The derivation previously presented, in which we allowed ourselves to divide equation (5.5) through by small quantities while keeping something constant, is justified because a rigorous mathematical analysis of each of the steps (by letting dx, dy, and df be finite and subsequently taking appropriate limits) can be formulated. Here we show an alternative approach in which we avoid ever dividing one small quantity by another.

Start from (5.5), and note that since x can be considered to be a function of f and y, we have also

$$\mathrm{d}x = \left.\frac{\partial x}{\partial f}\right|_y \mathrm{d}f + \left.\frac{\partial x}{\partial y}\right|_f \mathrm{d}y. \tag{5.32}$$

Substituting (5.32) into (5.5) gives

$$\mathrm{d}f = \left.\frac{\partial f}{\partial x}\right|_y \left.\frac{\partial x}{\partial f}\right|_y \mathrm{d}f + \left(\left.\frac{\partial f}{\partial x}\right|_y \left.\frac{\partial x}{\partial y}\right|_f + \left.\frac{\partial f}{\partial y}\right|_x \right) \mathrm{d}y. \tag{5.33}$$

This equation must be valid for all (small) $\mathrm{d}f, \mathrm{d}y$, therefore the coefficients of $\mathrm{d}f$ on each side must be equal, which gives the reciprocal theorem, and the coefficients of $\mathrm{d}y$ on each side must be equal, which gives the reciprocity theorem.

5.3.2 Integration in general

In Section 5.1.2 we discussed how to integrate a proper differential. Here we will carry out a definite integration in a more cumbersome way, by choosing a specific path. This is a valuable because the method applies to improper (as well as proper) differentials.

An equation such as (5.5) can be integrated as follows:

$$\int_A^B \mathrm{d}f = \int_A^B \left.\frac{\partial f}{\partial x}\right|_y \mathrm{d}x + \int_A^B \left.\frac{\partial f}{\partial y}\right|_x \mathrm{d}y, \tag{5.34}$$

where the limits A, B on the integrals are the initial and final points $A = (x_1, y_1)$ and $B = (x_2, y_2)$. Let us carry through this integration for the example case given in (5.2) and (5.3). We have

$$\int_A^B \mathrm{d}f = \int_A^B \frac{2y}{x} \mathrm{d}x + \int_A^B 2(\log(x) + y - 1) \mathrm{d}y. \tag{5.35}$$

The two integrals can be carried out by choosing a path (and if f is a function of state then the result will not depend on the path). Let $C = (x_1, y_2)$ and let us choose the path ACB in which AC is at constant x and CB is at constant y. Then

$$\underbrace{\int_A^C \mathrm{d}f}_{\text{const. } x=x_1} = 0 + \int_{y_1}^{y_2} 2(\log(x_1) + y - 1) \mathrm{d}y,$$

$$\underbrace{\int_C^B \mathrm{d}f}_{\text{const. } y=y_2} = \int_{x_1}^{x_2} \frac{2y_2}{x} \mathrm{d}x + 0,$$

where the zeros indicate where one of the integrals is zero because either $\mathrm{d}x = 0$ or $\mathrm{d}y = 0$. Now we evaluate the terms that are not zero:

$$\int_{y_1}^{y_2} 2(\log(x_1) + y - 1)dy = 2(\log(x_1) - 1)(y_2 - y_1) + (y_2^2 - y_1^2),$$

$$\int_{x_1}^{x_2} \frac{2y_2}{x} dx = 2y_2 \log(x_2/x_1).$$

Adding these together, we find the complete integral from A to B is

$$\int_A^B df = 2(y_2 \log x_2 - y_1 \log x_1) + (y_2 - y_1)(y_2 + y_1 - 2), \qquad (5.36)$$

which agrees with $f(x_2, y_2) - f(x_1, y_1)$ using equation (5.1).

The integral of an improper differential, such as $\int đw$, can be evaluated using the same method, but now the answer will depend on the path chosen.

....

EXERCISES

(5.1) (i) Consider the following small quantity: $y^2 dx + xy dy$. Find the integral of this quantity from $(x, y) = (0, 0)$ to $(x, y) = (1, 1)$, first along the path consisting of two straight-line portions (0,0) to (0,1) to (1,1), and then along the diagonal line $x = y$. Comment.

(ii) Now consider the small quantity $y^2 dx + 2xy dy$. Find (by trial and error or any other method) a function f of which this is the total differential.

(5.2) A certain small quantity is given by

$$C_V dT + \frac{RT}{V} dV,$$

where C_V and R are constants and T and V are functions of state. (i) Show that this small quantity is an improper differential. (ii) Let $đQ = C_V dT + \frac{RT}{V} dV$. Show that $đQ/T$ is a proper differential.

(5.3) If x, y are functions of state, give an argument to prove that

$$\frac{đQ_{(p)}}{dx} = \frac{đQ_{(p)}}{dy} \frac{\partial y}{\partial x}\bigg|_{(p)}, \qquad (5.37)$$

where the subscript (p) indicates the path along which the changes are evaluated.

(5.4) A and B are both functions of two variables x and y, and $A/B = C$. Show that

$$\frac{\partial x}{\partial y}\bigg|_C = \frac{\frac{\partial \ln B}{\partial y}\big|_x - \frac{\partial \ln A}{\partial y}\big|_x}{\frac{\partial \ln A}{\partial x}\big|_y - \frac{\partial \ln B}{\partial x}\big|_y}$$

[Hint: develop the left-hand side, and don't forget that for any function f, $(d/dx)(\ln f) = (1/f) df/dx$].

6
Zeroth law, equation of state

6.1 Empirical temperature 54
6.2 Some example equations of state 59
6.3 Thermometry 66
Exercises 68

If $A \Leftrightarrow C$ and $B \Leftrightarrow C$ then $A \Leftrightarrow B$.

See Section 3.3 for the meaning of the symbols in this expression.

The zeroth law can seem self-evident, but it is valuable to give a clear statement of it nonetheless. It was not elevated to the status of a foundational idea having a formal statement until after the first and second laws, which explains why they are numbered as they are. However, some careful reasoning will show that it is, in its own way, quite useful. It offers an example of the 'getting something for nothing' explanatory power of thermodynamics. That is, we start with a premise as minimal as we can make it, such as the zeroth law, and then invest effort in the reasoning process, and find that we can deduce more than one might have guessed would be possible. In the present instance, we will deduce a meaning for temperature and the existence of the equation of state.

A good way to grasp the essence of the zeroth law is to regard system C in the above statement as a thermometer. The law says that if a given thermometer gives the same reading for A and B, then A and B must be at the same temperature. Sometimes the law is summarized informally as 'thermometers work'! But we can only read the zeroth law that way if we already have some familiarity with temperature. One aim of the present chapter is to explore carefully what is meant by temperature, building on the definition given in Section 3.2.5.

Temperature is a familiar enough concept in everyday life, so it might seem strange to say we need to find a meaning or a definition for it. However, in comparison with other physical concepts, it is not as clear an idea as some. Electrical current, for example, is essentially the passage of many tiny particles (electrons) along a wire; time is the number of oscillations of a clock (such as an atomic clock) between two events; force is a propensity to cause acceleration; and so on. It is not so easy to say what essentially temperature *is*. Therefore we are going to consider this carefully before proceeding further. In order to create a precise science we need to define a precise meaning.

The thermodynamic approach has two major steps. In the first step, we develop what might be called a practical or 'working definition' of temperature.

Thermodynamics: A Complete Undergraduate Course. Andrew M. Steane.
© Andrew M. Steane 2017. Published 2017 by Oxford University Press.

We thus obtain a measure called *empirical temperature*. This first step makes no attempt to establish what temperature *is*, it merely establishes that there is a well-defined physical property, which we can suitably call 'temperature', and we establish a way to agree on the value of the property for all the possible equilibrium states of systems. The zeroth law of thermodynamics provides this first step, to be discussed in this chapter. In essence it says that temperature is 'that property which is measured by thermometers'.

The second step is the discovery of an important law of physics, concerning the amount of heat energy going into and coming out of an idealized system called a reversible heat engine. This discovery follows from the zeroth, first, and especially the second laws of thermodynamics; it will be discussed in Chapter 8. It will enable us to define temperature in a new and more profound way, linking it to heat transfer. The quantity defined by this more profound definition is called *absolute temperature*. We can then make empirical temperature agree with absolute temperature, by appropriately calibrating whatever thermometers we were using to measure empirical temperature.

This argument might seem rather round-about, but analogous situations exist in other areas of physics. Consider the concept of length, for example. We all have an intuitive sense of what length is, but when exact science was developed it became more and more important to have a precise definition. For a long period, the best definition was 'that which is measured by rulers'. That is to say, one decided whether one body has the same length as another by aligning a third body (called a ruler) against each in turn and noting where the edges of each body met the ruler. Also a scale of length was established by associating different markings on the ruler with different length values (by some reasonable but ultimately arbitrary choice). This could be called 'empirical length'. However, the discoveries in electromagnetism, and Einstein's postulate of the independence of the speed of light of the motion of the source, made it clear that length is intimately related to temporal duration. The result is that now length is defined to be that spatial interval which would be traversed by light in vacuum in a given time. This could be called 'absolute length'. If any ruler which we had previously been using disagrees with the new definition, then so much the worse for the ruler: it has to be re-calibrated.

When we need to make the distinction between empirical and absolute temperature, we will use the symbol θ for empirical and T for absolute temperature. However, in view of the fact that eventually we are going to make sure the two agree, for much of the discussion it will not be important to make the distinction, and then we will use T without further comment.

We will describe various types of thermometer in Section 6.3 but first let us consider the general definition of empirical temperature. We will do this using a simple compressible system as an example, and then generalize afterwards. Recall that a simple compressible system could be a fixed mass of gas in a cylinder fitted with a piston (Figure 3.4). Each state is uniquely specified by pressure and volume.

6.1 Empirical temperature

Suppose we have two simple compressible systems A and R. System R will serve as a reference, while we explore A. Think of R as a thermometer, but instead of letting the thermometer change until it comes to equilibrium with the 'patient', we heat or cool the patient until he reaches equilibrium with the pre-set thermometer!

First pick a fixed state of R (i.e. one of given (p_R, V_R)). Let us call this state 1. Now bring A and R into thermal contact, and adjust the conditions of A until, with the two remaining in thermal equilibrium, R is in state 1. Make a note of the value of the state variables (p_A, V_A) of A. Now change one of the variables, say V_A, by pushing on the piston; then adjust p_A by some other method such as heating or cooling the gas, until A and R are once more in thermal equilibrium, with R again in state 1. Make a note of the new value of (p_A, V_A). By repeating this procedure, one finds a whole sequence of states of A, all of which are in thermal equilibrium with state 1 of R. This set of states can conveniently be indicated by plotting a line on an indicator diagram for A; see Figure 6.1. Such a line is called an *isotherm*. Label this line 'isotherm θ_1'.

Further isotherms of A may be found in a similar manner, by choosing different states $2, 3, \ldots$ of R. In each case one finds the states of A which are in equilibrium with the given state of R. Thus one discovers a set of isotherms θ_1, θ_2, $\theta_3, \ldots$ covering the whole indicator diagram.

The graphical procedure is exactly equivalent to assigning a value θ to each state (p_A, V_A), such that θ has the same value for all states on any given isotherm. The actual values chosen for θ are arbitrary, though it makes sense to assign them in some reasonably organized way, and one must avoid giving different isotherms the same θ. In the limit of closely spaced isotherms, one arrives at a continuous function $\theta(p_A, V_A)$, of which the isotherms are the contours.

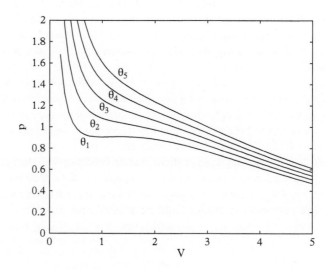

Fig. 6.1 *An indicator diagram for system A. The curve labelled θ_1 shows all those states which were found to be in equilibrium with state number 1 of system R (when A and R are in thermal contact); the curve labelled θ_2 shows all those states which were found to be in equilibrium with state number 2 of system R, etc.*

What we have discovered is that every system (we could have picked any A) has a function θ, depending on its state variables, whose value determines which reference state of R it is in equilibrium with. Now, *using the zeroth law*, we may go further and argue that *any pair of systems will be in thermal equilibrium with each other if their states have the same θ*. This follows directly, because, for any two systems A and B, θ_A identifies a state of R with which A is in equilibrium, and θ_B does the same for B; if $\theta_A = \theta_B$ then A and B are in equilibrium with the same state of R, which implies, by the zeroth law, that they must be in equilibrium with each other.

We now have a definition of temperature: it is a property whose value is given by θ. *If two systems are in states of the same θ, then they are in thermal equilibrium.* It turns out that that is all we require of an empirical definition of temperature. Note that we have not related the temperature, at this stage, to any sense in which bodies might appear to be hot or cold to our senses—that will come later.

The zeroth law also shows that the argument does not depend on which system is chosen as the reference. For, suppose we replace R with some other reference system R'. Then any state of R' in equilibrium with state 1 of R will also, according to the zeroth law, be in equilibrium with all the states of A on isotherm θ_1. Therefore R' could equally well have been used as the reference, and one would obtain the very same isotherms. As long as one picks the same values $\theta_1, \theta_2, \ldots$ the isotherms are expressible as contours of the same function $\theta(p_A, V_A)$. This new reference system R' could be of a wholly different type, such as a magnetizable body in a magnetic field, or a rubber membrane in tension, or a column of mercury, for example.

In order to compare empirical temperature measurements, it will be necessary to choose one particular system to serve as a 'standard thermometer' and then agree a specific temperature scale. The agreed scale associates temperature values (1 unit, 2 units, 3 units, etc.) with well-defined states of the standard thermometer, measured in terms of its state variables; for example pressure and volume for a pV system. Other thermometers can be calibrated against the standard, and the zeroth law guarantees that the whole procedure will be consistent. In principle it doesn't matter what system is chosen as the standard, but in practice we would do well to choose one which can be realized experimentally in an accurately repeatable way. This is discussed further in Section 6.3.

6.1.1 Equation of state

We have shown that every equilibrium state of every system has a temperature, defined by the value θ, that identifies which standard state of R it is in thermal equilibrium with. This temperature is a single-valued function of the state variables. For a pV system it can be written

$$\theta = \theta(p, V). \tag{6.1}$$

The result is an equation relating the empirical temperature to the state variables. Every system has such an equation; it is called the *equation of state*.

Simple systems have, in the first instance, two independent parameters, or two *degrees of freedom*. That is, to uniquely specify the state, it is sufficient and necessary to supply the values of two quantities. We have now introduced a third parameter, empirical temperature. However, it is not independent, since we have one equation relating the three parameters, so there are still just two degrees of freedom. As noted in Chapter 3 (Figure 3.5), we can understand this graphically by introducing a three-dimensional diagram showing p, V, θ 'space'; then the equilibrium states lie on a *surface* in this space.

Most points on the surface can be identified uniquely by specifying any two out of p, V, θ, but there can be regions where the surface runs parallel to one of the p, V, θ axes. In such regions, specifying two parameters may leave ambiguity about the value of the third. For example, at the boiling point of a fluid, if the pressure and temperature are specified, then a range of values of the volume is possible (the isotherm on the pV indicator diagram has a region which runs exactly horizontal on the diagram). For another example, consider the $pV\theta$ surface of water near its freezing point. The density of water, as a function of temperature at given pressure, goes through a maximum near 4 °C at 1 atmosphere; see Figure 6.2. Consequently, the temperature of a given mass of water at fixed pressure is not a monotonic function of volume. If a kilogram of water has a volume 1000.1 cm³ at 1 atmosphere, its temperature could be either 2 °C or 7 °C. These examples do not mean that volume or temperature has ceased to be a function of state: each point on the equilibrium surface still has a unique pressure, volume, and temperature. It is just that to pick out a unique state using just two parameters sometimes requires a judicious choice of which parameters to mention. For most

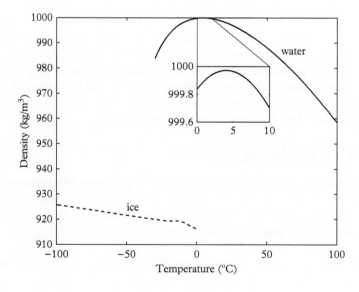

Fig. 6.2 *The density of water as a function of temperature at 1 atm pressure. The curve for liquid water extends below 0 °C because it is possible to cool pure water below the freezing point without it solidifying (a metastable condition, considered in Chapter 24).*

states of most systems, it does not matter whether one specifies p, V or p, θ or V, θ, but we have just examined some cases where more than one state can have a given p, V or p, θ.

This situation is similar to the specification of a point in a Euclidean plane using either rectangular coordinates (x, y) or polar coordinates (r, θ). One could also use (x, θ) or (y, θ) without ambiguity, but specifying (x, r) will almost always identify two points not one. For simple compressible systems the use of (p, V) to specify the state is sometimes ambiguous, but the use of (V, T) is not, and the use of (p, T) is unambiguous except in the special case of a first-order phase transition. Nevertheless, (p, V) are very useful in practice and give a good guide to understanding the general behaviour. When using (V, T) the equation of state is interpreted as an equation giving p in terms of the other variables.

We have already given an example equation of state in equation (3.2), and we shall present further examples later. For each system it is an important equation, but it is not the only important equation. It is a common misconception that the equation of state is *the* essential information about a thermodynamic system, telling us all we need to know. This is not so, because although one can derive some properties from the equation of state, such as the isothermal compressibility $V^{-1}(\partial V/\partial p)_\theta$, one cannot derive other equally important properties such as heat capacities, and one cannot use it to identify the adiabatic lines on the indicator diagram. The further information needed to gain a full description of the behaviour will be discussed in subsequent chapters.

6.1.2 Algebraic argument

We will now present an algebraic version of the above argument to define empirical temperature. The more formal expression using algebraic symbols can be useful to ensure no hidden assumptions have been made, and it provides useful practice in thermodynamic methods.

The observation that various, but not all, states of A are in equilibrium with a given state of R implies that, in conditions of thermal equilibrium between the systems, there is a formula giving the value of p_A as a function of V_A for each state (p_R, V_R) of R. That is,

$$p_A = f_{AR}(V_A, p_R, V_R). \tag{6.2}$$

The function f_{AR} describes the shape of the isotherms of A.

Similarly, another system B has isotherms described by some other function:

$$p_B = f_{BR}(V_B, p_R, V_R). \tag{6.3}$$

Each of equations (6.2) and (6.3) can be inverted to find a formula for p_R in terms of the other quantities, giving

$$p_R = f_1(p_A, V_A, V_R) \tag{6.4}$$

and

$$p_R = f_2(p_B, V_B, V_R), \tag{6.5}$$

where f_1 and f_2 are new functions. It follows that

$$f_1(p_A, V_A, V_R) = f_2(p_B, V_B, V_R). \tag{6.6}$$

That is all we know about these functions so far.

The same argument shows that, when A and B are in thermal equilibrium with each other, then p_A (for example) can be expressed in terms of the other properties:

$$p_A = f_{AB}(V_A, p_B, V_B). \tag{6.7}$$

Let the symbol $\Leftrightarrow$ mean 'is in thermal equilibrium with'. Equation (6.6) describes the situation when $A \Leftrightarrow R$ and $B \Leftrightarrow R$. Equation (6.7) describes the situation when $A \Leftrightarrow B$. However, *by the zeroth law*, the second situation must obtain when the first does. Therefore, if we solve (6.6) for p_A, getting

$$p_A = f_{ARB}(V_A, p_B, V_B, V_R), \tag{6.8}$$

we must find agreement with (6.7) for all V_R. But V_R does not even appear in (6.7). It follows that it must in fact drop out of (6.8), and since this equation followed directly from (6.6) then V_R must have cancelled out of that expression.[1] Therefore equation (6.6) must in fact have the form

$$g_A(p_A, V_A) = g_B(p_B, V_B). \tag{6.9}$$

We could introduce further systems C, D, etc., and as long as all the systems are in thermal equilibrium with each other, the same argument would show that

$$g_A(p_A, V_A) = g_B(p_B, V_B) = g_C(p_A, V_C) = g_D(p_D, V_C), \tag{6.10}$$

and so on. We have thus demonstrated that for every system (the argument generalizes readily to systems of more and different parameters), there is a function of its state parameters (possibly different for each system) whose numerical value is the same when the systems are in equilibrium. This numerical value θ is called *temperature*, and the equation

$$g(p, V) = \theta \tag{6.11}$$

is called the equation of state of the system.

---

[1] For example, f_1 might take the form $f_1(p, V, V_R) = g_A(p, V)\phi(V_R) + \eta(V_R)$ and $f_2(p, V, V_R) = g_B(p, V)\phi(V_R) + \eta(V_R)$.

6.2 Some example equations of state

We will now present some example equations of state, and in anticipation of later developments we will use the symbol T and refer to absolute temperature. This is valid because we can arrange for the empirical scale to agree with the absolute one, to be defined in Chapter 8. In the absolute scale, there is a direct correspondence between 'hotness' and temperature. Bodies which tend to give off heat have higher temperature than those that tend to receive heat.

An example equation of state was given in equation (3.2) of Chapter 3. This equation was developed from a theoretical model of a gas. It describes most gases very well when they are far from conditions of phase change (that is, from condensing into a liquid or solid), and gives a reasonable though imprecise model near to the vapour/liquid phase change. The isotherms are plotted in Figure 6.3.

6.2.1 Ideal gas

In the limit of low pressure and not too low temperature, i.e. far from condensation, it is observed in experiments that all inert gases behave alike, and exhibit **Boyle's law**:[2] at fixed temperature, the pressure (of a fixed mass of gas) is inversely proportional to the volume. This suggests an equation of state of the form $pV = f(T)$ for some function f. An empirical scale of temperature, called the *ideal gas scale*, was established by defining temperature θ such that the simple linear dependence $f = n_m R \theta$ holds, where n_m is the number of moles and R is the

[2] In order for this law to hold, some other circumstances are required, such as negligible gravitational influences.

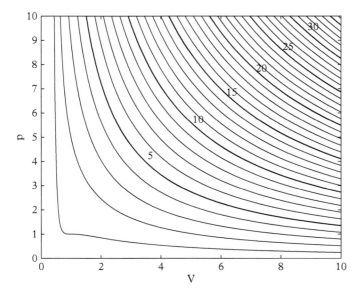

Fig. 6.3 *Isotherms of the van der Waals equation of state, equation (3.2). Every fifth isotherm is labelled by its temperature. The pressure, volume, and temperature values have been given in units of their 'critical values' (discussed in Chapter 15), which are $p_c = a/27b^2$, $V_c = 3b$, $T_c = 8a/27Nk_B b$. The main interest here is the overall form of the behaviour in the gaseous regime, at $T > 1$ in these units. The liquid–vapour phase change region is at $T < 1$ and $V \simeq 1$; it is intentionally not shown in detail here, because we shall not discuss it until the second half of the book (cf. Figure 15.5).*

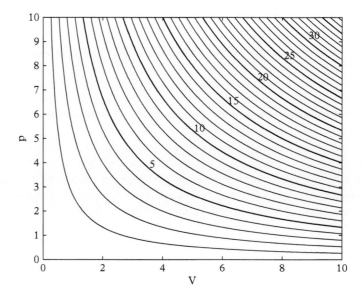

Fig. 6.4 *Isotherms of the ideal gas. Every fifth isotherm is labelled by its temperature. Each isotherm is a '1/V' function. The temperature rises as one moves from low p, V to high p, V. The case $Nk_B = 8/3$ has been chosen, to allow easy comparison with the van der Waals case (Figure 6.3).*

molar gas constant. In terms of this empirical temperature, the idea gas equation of state is

$$pV = Nk_B\theta. \tag{6.12}$$

The isotherms are plotted in Figure 6.4. This equation of state is applicable to gases at low number density (and therefore at low values of p/θ).

6.2.1.1 Definition of the ideal gas

Although equation (6.12) is an approximation for any real gas, we may *define* an idealized system called an 'ideal gas' such that its equation of state is (6.12) with $\theta = T$. Study of the ideal gas furnishes a lot of insight into thermodynamic methods, and gives results which are applicable to real gases in the relevant limit. The full definition of an ideal gas requires not just the equation of state but also a statement about its internal energy, as follows:

> *Definition of an ideal gas*
>
> (1) Boyle's law: pV is constant at fixed temperature.
> (2) Joule's law: The internal energy is independent of pressure at fixed temperature.

Sometimes it is further assumed that the relationship between internal energy U and temperature T is linear, $dU/dT = $ const. Note, however, that this is a further assumption; it does not follow from the others and it is not always made.

In particular, a gas whose heat capacities vary with temperature can still have the ideal gas equation of state.

The internal energy will be defined carefully in Chapter 7. It is, as its name suggests, a measure of the total energy content of the system. We will show later, from the laws of thermodynamics, that the two main properties (Boyle's law and Joule's law) imply that the equation of state must be $pV \propto T$ for a fixed amount of gas, where T is absolute temperature; and therefore we can write

$$pV = Nk_B T. \tag{6.13}$$

That is, the empirical temperature obtained from an ideal gas must be equal to absolute temperature.

Joule's law is sometimes written $U = U(T)$ (for a fixed quantity of gas). This mathematical statement is a shorthand meaning 'if the internal energy is expressed as a function of T and other state parameters, then it is found to depend on T alone and not on the other parameters.' It is often stated in the slightly less clear form 'the internal energy is a function of temperature alone.' The latter is less clear because the internal energy can always be written as a function of other things such as pressure and volume, and it will depend on them, but only through the combination (pV/Nk_B) which is equal to the temperature.

The indicator diagram for an ideal gas, Figure 6.4, enables one to identify readily observable aspects of the behaviour of a gas. For example, inspection of the isotherms crossed by a horizontal line on the diagram shows that the gas cools during an expansion at constant pressure. Tracing any given isotherm shows that in an expansion at constant temperature, the pressure falls (Boyle's law).

Figure 6.5 illustrates how various gases tend to the same, ideal gas behaviour in the limit $p \to 0$ at fixed T. When the van der Waals equation tends to the ideal gas equation, the constants a and b, which differ from one gas to another, have disappeared. This is both interesting in its own right, and also very useful for thermometry (see Section 6.3).

6.2.2 Thermal radiation

Another interesting simple compressible system is that consisting of the radiation contained in a cavity in thermal equilibrium. This is called *thermal radiation*. It is chiefly electromagnetic (in extreme conditions thermal radiation can contain electron–positron pairs and other particles as well as electromagnetic waves). Such radiation exerts a pressure on the walls of the chamber confining it, described by the equation of state

$$p = aT^4, \tag{6.14}$$

where a is a constant. This can be derived from basic thermodynamic principles and will be discussed in Chapter 19. It is notable that the pressure and temperature are here directly related to one another, without any dependence on volume.

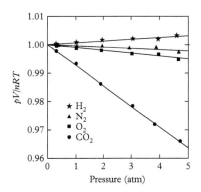

Fig. 6.5 *Gases at low pressure.*

If a chamber is expanded at constant temperature, the pressure from thermal radiation will not diminish: the isothermal compressibility is infinite.

6.2.3 Solids and wires

Many solids do not exhibit the same pressure in all directions, so a single pressure, together with another variable such as temperature, is not sufficient to specify their thermodynamic state. However some amorphous solids can be modelled to good approximation by a single isotropic pressure, and for small volume changes the equation of state is adequately approximated by

$$V = V_0(1 + \alpha T - \kappa p), \tag{6.15}$$

where V_0, α, and κ are constants.

A wire under tension can be idealized as a system described by length L and tension f. For small values of the tension, so that inelastic deformation is avoided, the equation of state is Hooke's law,

$$f = k(L - L_0), \tag{6.16}$$

where k may be temperature dependent. An adequate model for many purposes is

$$k = k_0 + \alpha T, \tag{6.17}$$

where k_0 and α are constants.

In geophysics, many important processes take place at very high pressure, and at the slow timescales involved, what we ordinarily call solids behave in many respects like fluids. A useful approximation is to assume that, at any given temperature, the bulk modulus is a linear function of pressure:

$$B_T \equiv -V \left.\frac{\partial p}{\partial V}\right|_T = a + \beta p, \tag{6.18}$$

where a and β are independent of pressure. Clearly, if $B_T(T, p)$ is regarded as a function of T and p, then $a = B_T(T, 0)$ and $\beta = (\partial B_T/\partial p)_T$. By integrating (6.18) with respect to volume, one finds

$$p = \frac{B_T(T, 0)}{\beta}\left(\left(\frac{V_0}{V}\right)^\beta - 1\right), \tag{6.19}$$

where V_0 is the volume at low pressure. This is called the *Murnaghan equation of state*. As it stands, it does not specify the dependence on temperature, but it is useful for determining how the volume and pressure are related for isothermal processes.

6.2.3.1 Ideal elastic substance

An *ideal elastic substance* is defined as one whose equation of state is

$$f = KT\left(\frac{L}{L_0} - \frac{L_0^2}{L^2}\right), \qquad (6.20)$$

where K is a constant and L_0 is a function of temperature alone. This model is suitable for materials, such as rubber, that are composed of long polymer molecules.

6.2.4 Paramagnetic material

Most materials exhibit magnetic properties, such that if a sample is placed in a magnetic field (see Figure 6.6), it then acquires a magnetic dipole moment. This is true of gases, liquids, and solids.

In the case of *ferromagnetic* behaviour, a sample possesses magnetization in the absence of any applied field: when people talk of 'magnets' in everyday life this is usually the case they have in mind. A given sample of ferromagnetic material cannot be treated as a simple thermodynamic system described by only a few parameters, but has to be treated as a collection of magnetic domains. This is because the time required for the system to reach a global equilibrium may be many years; it is much longer than the timescales we would typically like to consider. There is a large amount of hysteresis and therefore the relationship between the applied field and the average magnetization of the whole sample, at any given time, depends on the way the field was applied in the past. This means that if we use the applied field as a state parameter, then the average magnetization is not a function of state. Alternatively, if we use the average magnetization as a state parameter, then the applied field is not a function of state. The conclusion is that in order to apply equilibrium thermodynamics to ferromagnetic behaviour, one has to consider the individual domains.

In the more simple cases of *paramagnetic* and *diamagnetic* behaviour, at any given temperature there is a one-to-one relationship between the magnetization of the material and the applied magnetic field. In this case the sample reaches thermal equilibrium quickly and standard equilibrium thermodynamics can be applied. If the magnetic dipole acquired by the material is in the same direction as the applied field, the object is called *paramagnetic*. If the dipole is in the opposite direction to the applied field the material is called *diamagnetic*.

The fundamental magnetic field quantity is the field **B** which appears in the Lorentz force law for the force on a moving charged particle, $\mathbf{f} = q\mathbf{v} \wedge \mathbf{B}$. In this equation, **B** is the magnetic field whose SI unit is the tesla (or newtons per ampere-metre). When considering the effect of a magnetic field on a bulk material it is often useful to consider the amount of *magnetic dipole moment per unit volume* in the material. This quantity is called *magnetization* **M**. When speaking of the magnetization and the fields inside the material, one takes an average over

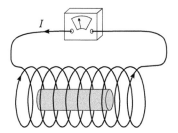

Fig. 6.6 *Apparatus for studying a paramagnetic sample.*

distance scales large compared to the atomic structure scale of the material (of order 10^{-10} m), but small compared to the dimensions of the specimen under consideration. It is then useful to introduce the quantity

$$\mathbf{H} \equiv \frac{1}{\mu_0}\mathbf{B} - \mathbf{M}, \tag{6.21}$$

where $\mathbf{B}$ is the average magnetic field inside the material (not the field outside, nor the field which would be present if the matter were removed) and $\mu_0 = 4\pi \times 10^{-7}$ H/m is a constant called the permeability of free space. Unfortunately there persists a confusion of terminology because both $\mathbf{H}$ and $\mathbf{B}$ are called 'magnetic field'. Sometimes $\mathbf{B}$ (but not $\mathbf{H}$) is called 'magnetic flux density' or 'magnetic induction', because it appears in Faraday's law $\nabla \wedge \mathbf{E} = -\partial \mathbf{B}/\partial t$. $\mathbf{H}$ appears in Ampere's law $\nabla \wedge \mathbf{H} = \mathbf{j} + \partial \mathbf{D}/\partial t$, where $\mathbf{j}$ is the current per unit area due to conduction currents. Conduction currents are those associated with motion of charge-carrying particles through the system.[3] The SI unit of $\mathbf{H}$ is amperes per metre.[4]

The magnetic behaviour of a sample of paramagnetic material can be captured by examining the way any two out of $\mathbf{B}$, $\mathbf{H}$, and $\mathbf{M}$ behave as the conditions are varied. Thus we have an example of a simple thermodynamic system. We choose $\mathbf{B}$ and $\mathbf{M}$ as the main state functions or generalized coordinates, and use (6.21) to express $\mathbf{H}$ in terms of them. The equation of state will relate the temperature to $\mathbf{B}$ and $\mathbf{M}$. We will clarify later (in Chapter 14) the relationship between $\mathbf{B}$ and the field outside the sample; for the present discussion we assume the magnetization is weak, so this distinction does not need to be made.

At low field strengths and high temperature, many materials are found to obey **Curie's law**, which is

$$\mathbf{M} = \frac{a}{T}\mathbf{H}, \tag{6.22}$$

where T is temperature and a is a constant depending on the material. a has the physical dimensions of temperature; it can be physically interpreted as the temperature which would result in $\mathbf{M} = \mathbf{H}$, but the formula only holds when $M \ll H$. For the paramagnetic materials typically used in low temperature physics experiments, a is in the range millikelvin to 0.1 kelvin.

From a thermodynamic point of view, Curie's law is an example of an equation of state: it shows how the temperature is related to state parameters (here, $\mathbf{M}$ and $\mathbf{H}$). By combining (6.21) and (6.22) we can also write it in the form

$$\mu_0 \mathbf{M} = \frac{1}{1 + (T/a)}\mathbf{B} \simeq \frac{a}{T}\mathbf{B}, \tag{6.23}$$

where the final approximation holds because in any case the law is only valid when $M \ll H$, which implies $T \gg a$. The isotherms predicted by this equation of state are shown in Figure 6.7.

---

[3] In the presence of magnetization there can also be other types of current, called magnetization current, which contribute to $\mathbf{B}$ but not to $\mathbf{H}$.

[4] The placement of μ_0 in equation (6.21) is the choice adopted in SI units; another sensible choice would be to define a field $\tilde{\mathbf{H}} \equiv \mathbf{B} - \mu_0 \mathbf{M}$. This has the advantage of giving $\mathbf{B}$ and $\tilde{\mathbf{H}}$ the same physical dimensions, which makes a lot of sense. The SI choice was made in order to get a simple relationship between $\mathbf{H}$ and current density.

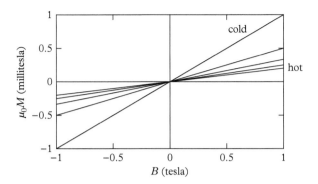

Fig. 6.7 *Isotherms of a Curie law paramagnet. Five isotherms are shown, at $T/a = \{1, 2, 3, 4, 5\} \times 10^3$. The magnetization falls as T increases. This figure should be compared with Figure 6.8.*

The value of a is given to good approximation for many paramagnetic materials by

$$a = \frac{n\mu^2 \mu_0}{3k_B}, \tag{6.24}$$

where n is the number density of atoms or molecules in the material, each of which has a permanent magnetic dipole moment of size μ.[5]

6.2.4.1 The ideal paramagnet

The essential content of Curie's law is that the magnetization is proportional to **H** and inversely proportional to temperature. This is valid over a wide range of field and temperature, but like the ideal gas, the Curie law magnet is an idealization that cannot apply in all circumstances. In practice at high field the magnetization 'saturates' and ceases to increase in proportion to **H** (see Chapter 14). Also, at low temperatures a phase transition takes place, often to a state where a non-zero magnetization exists even in the absence of any applied field (see Chapter 25).

We will present a simple theoretical model of paramagnetic behaviour in Chapter 14. The resulting model system is often called an 'ideal paramagnet' because, like the ideal gas, it ignores the possibility of interactions between the molecules, but gives a useful generic guide to the behaviour at high temperatures, away from the phase transition. The equation of state of the ideal paramagnet is

$$M = n\mu \tanh\left(\frac{\mu B}{k_B T}\right), \tag{6.25}$$

where n is the number density of magnetic particles (e.g. molecules, or sometimes atoms), each of which has a permanent magnetic dipole μ. This equation is plotted in Figure 6.8.

[5] The size of μ in this formula is related to the angular momentum quantum number j and the gyromagnetic ratio g by $\mu^2 = g^2 j(j+1)\mu_B^2$; both j and g are properties of the given atom or molecule.

6.2.5 Equations of state for other properties

Suppose we have a thermodynamic system described by some minimal set of properties that are sufficient to uniquely specify the state. Then every time we

66 Zeroth law, equation of state

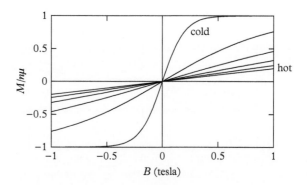

Fig. 6.8 *Isotherms of an ideal paramagnet. Six isotherms are shown, at $k_B T/\mu = \{0.2, 1, 2, 3, 4, 5\}$. (A single line as a function of B/T would suffice to show the behaviour.) The magnetization saturates at high B. The Curie law is obeyed at low B/T.*

take an interest in a further state property, there must be an equation that relates the further property to the basic properties such as p, V. Such equations may also be called 'equations of state'; this is a generalization from the equation for temperature that we have studied in this chapter. In this book we will reserve the name 'equation of state' for the equation relating temperature to the chosen basic properties, and adopt other names such as 'energy equation' or 'heat capacity equation' for the equation relating other properties such as energy and heat capacity to the basic properties.

6.3 Thermometry

We finish the chapter with a few remarks on the measurement of temperature.

Temperature measurement is part of the basic tool kit we need in many areas of science. The accurate and reliable measurement of temperature combines basic thermal physics and much ingenuity. A good thermometer must have some easily detectable property, such as length, electrical resistance, voltage, or pressure, that depends sensitively on temperature; and it should be stable and reproducible. In practice, different devices are practical in different temperature regimes, so a variety of approaches is needed.

Example thermometers in common use are the mercury-in-glass thermometer, the platinum resistance thermometer, thermocouples, thermistors, pyrometers, gas thermometers, and so on. These can be divided into two main types. *Primary thermometers* provide the calibration; *secondary thermometers* are widely used for convenience.

Primary thermometers do not need to be calibrated. They are based on a physical property whose behaviour is so well known from theory that it can be predicted with no unknown quantities. The secondary thermometers need to be calibrated against a set of known temperatures which ultimately are measured by the primary thermometers. The most important property of a secondary thermometer is that it should not drift after it has been calibrated.

The main primary thermometers are

(1) The constant-volume gas thermometer.
    A glass bulb is filled with a quantity of noble gas at low pressure and brought into contact with the system to be measured. The pressure in the gas is adjusted until the gas reaches a fixed volume. This pressure is measured and compared with the pressure p_0 obtained when the bulb is in equilibrium with a mixture of water, ice, and water vapour at the triple point. Either by repeating the measurement with various amounts of gas, or by further study of the gas, an extrapolation of both measured pressures to zero pressure can be obtained. The measured temperature is then given by

$$\frac{T}{T_0} = \lim_{p,p_0 \to 0} \frac{p}{p_0}, \qquad (6.26)$$

where $T_0 = 273.16\,\text{K}$.

(2) The acoustic gas thermometer.
    The speed of sound in a gas at low pressure is measured; this is related to the temperature by $v = (\gamma k_B T/m)^{1/2}$ (Eq. 10.32). The measured temperature is

$$\frac{T}{T_0} = \lim_{p,p_0 \to 0} \left(\frac{v}{v_0}\right)^{1/2}. \qquad (6.27)$$

(3) The total and spectral radiation thermometer.
    This detects the power in the radiation emerging from a cavity, which is given by equation (19.3).

(4) Electronic noise thermometers are based on the voltage-dependent electrical noise from a tunnel junction.

The primary thermometers are difficult to use. The calibration of secondary thermometers is mostly done by agreeing a standard set of fixed points whose temperatures are determined by primary thermometers, and then calibrating each secondary thermometer against whichever standard fixed points are in its working range. The International Temperature Scale of 1990 (ITS-90) defined 14 such points, mostly based on triple points and freezing points of the elements, covering the range $13.8033\,\text{K}$ (the triple point of hydrogen) to $1357.77\,\text{K}$ (the freezing point of copper). The range $0.65\,\text{K}$–$5\,\text{K}$ is covered by specifying the vapour pressure of helium.

At temperatures lower still, magnetic susceptibility, carbon resistance, semiconductor resistance, and direct observation of the speeds of atoms are all used, but there is not an agreed standard. At the other extreme—high temperatures, above $1000\,\text{K}$—radiation pyrometers are used. These gather thermal radiation

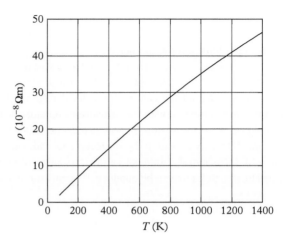

Fig. 6.9 *Resistivity of platinum as a function of temperature.*

emitted by the object under study, and compare its spectral distribution with that expected for thermal emission at a given temperature (see Chapter 19).

Gas thermometers can be used in the range $\sim 1\,\text{K}$ to $\sim 1000\,\text{K}$. The most useful secondary thermometers at 'ordinary' temperatures (the range $\sim 10\,\text{K}$ to $\sim 1000\,\text{K}$) are thermocouples and resistance thermometers. Platinum is often used for electrical resistance thermometers (Figure 6.9) because it is easy to purify, it is ductile and chemically unreactive, and it has a large temperature-coefficient of resistance.

....

EXERCISES

(6.1) Prove from the zeroth law that two different isotherms cannot intersect.

(6.2) A constant volume gas thermometer contains a gas whose equation of state is the van der Waals equation. Another contains an ideal gas. Each thermometer is calibrated at the ice and steam points (the melting and boiling point of water at standard pressure), and thereafter the pressure is used as a linear indicator of temperature. Show that with such calibration the temperature measurements provided by these two thermometers will agree at all temperatures, with no requirement to extrapolate to zero pressure.

(6.3) The *specific volume* v of a given substance is the volume per unit mass. Show that for any gaseous substance which behaves to good approximation like an ideal gas, the combination pv/T is constant for any given substance, but may vary from one substance to another.

(6.4) Plot the isotherms of a solid described by equation (6.15) on a pV diagram. Explain briefly whether you expect the slopes of these isotherms to be large or small compared to those of a gas.

(6.5) The free electrons which are responsible for conduction in metals can be regarded as an exotic kind of gas. The behaviour depends on a parameter called the Fermi temperature, given by

$$T_F = (3\pi^2 n)^{2/3} \frac{\hbar^2}{2mk_B},$$

where $n = N/V$ is the number density of free electrons, m is the electron mass, and $\hbar$ is the reduced Planck constant. When the actual temperature $T \ll T_F$, the pressure becomes a function of density alone, given by

$$p = \frac{2}{5} n k_B T_F.$$

Find the pressure of the electron gas in sodium at room temperature, given that the density of sodium is 970 kg/m^3, the atomic mass number is 23, and there is one free electron per atom.

[*Ans.* 5128 MPa.]

(6.6) The constant a in equation (6.14) has the value $a = (4/3)\sigma/c = 2.52 \times 10^{-16}$ Nm^{-2}K^{-4}, where σ is the Stefan–Boltzmann constant. Find the temperature at which thermal radiation has a pressure 1 atm.

[*Ans.* 1.4×10^5 K.]

(6.7) [Adapted from Carrington] A constant-volume gas thermometer is used to measure the temperature of a bath, using the triple point of water as a reference. The pressure in the thermometer is measured by a mercury barometer (the height of a column of mercury is proportional to the pressure), with the following results:

Reading in triple-point bath (mm Hg)	Reading in measured bath (mm Hg)
98	133.9
301	411.4
597	816.6
821	1123.4

Find the temperature of the bath, to a precision ±0.03°C.

[*Ans.* 99.98 ± 0.03 °C.]

7 First law, internal energy

7.1 Defining internal energy 70
7.2 Work by compression 74
7.3 Heat capacities 77
7.4 Solving thermodynamic problems 83
7.5 Expansion 85
Exercises 89

Energy is conserved.

In order to understand this statement of the first law we need to consider carefully the fact that energy can come in many different forms. In particular, we need to discuss the concept of heat energy. We will do this by starting with processes where there is no heat transfer, and then consider heat transfer afterwards.

7.1 Defining internal energy

When two systems exchange energy without heat flow, we say that one system 'does work on' the other. Here, *work* is defined precisely as it is in classical mechanics: it is energy, equal to the product of force $\mathbf{f}$ and the displacement parallel to the force of the point on which the force acts, $W = \mathbf{f} \cdot \Delta \mathbf{x}$. To 'do work on' a system is to push it or pull it, or its component parts, in some way, so that the agent performing the work loses energy. We will see in a moment that the system being worked on gains the energy: that is what the first law is all about.

Over the period 1843 to 1847, James Prescott Joule carried out a series of experiments in which he caused work to be performed on a thermally isolated thermodynamic system, and measured the effect. A summary of his results is given in Table 7.1. He found that, to within his experimental accuracy of about 4%, the amount of work required to cause a given change in 0.45 kg of water was the same, no matter how the work was performed.

These and other observations lead Joule and others to propose the following law of physics:

> *Principle of adiathermal work*: The amount of work required to change the state of a thermally isolated system depends solely on the initial and final states.

We shall now show that this statement is fully equivalent to the first law of thermodynamics. For this reason it, instead of the energy conservation statement given above, may be used as the fundamental statement of the first law. It is easy to prove that energy conservation implies the principle of adiathermal work. We will give a thermodynamic argument to show that the latter implies energy conservation.

Thermodynamics: A Complete Undergraduate Course. Andrew M. Steane.
© Andrew M. Steane 2017. Published 2017 by Oxford University Press.

Defining internal energy 71

Table 7.1 *Observations by Joule of the quantity of work required to raise the temperature of 1 lb of water through 1°F at constant pressure, where the work was done on the water by various methods. The amount of work energy was determined as mgh, where a mass m falling through height h provided the energy.*

Method	Quantity of work
1. Turbulent motion produced by paddle wheel	773 ft lb
2. Electric current through resistance	838 ft lb
3. Compression of gas in thermal contact with water	795 ft lb
4. Friction of metal blocks in thermal contact with water	775 ft lb

The work processes we have in mind may or may not be reversible. Figure 7.1 shows some example methods to do work on a simple compressible system such as a fluid in a chamber. In each case the process is adiathermal because the system is thermally isolated. In the case of an electrical current flowing through the system, some electrons move in at one end of the wire and out at the other, but we can use alternating current, or some other means, to arrange that the heat transported by these electrons be negligible compared to the electrical work VIt. Note that for the purpose of calculating the work done by the external agency, it is not necessary to know what goes on inside the system. In the electrical case, for example, the battery 'knows' nothing of how its electrical power is being expended: it simply pushes the charge through, in a flow of well-defined current; therefore it is doing work on, not providing heat to, the system. The thermodynamic system can be regarded as a 'black box' for the purpose of calculating the work, because the *external agency* does the work, and we only need to know what the external agency does.

Joule used various processes, and measured the system state before and after the process, by measuring the state parameters p and T (T being obtained by some empirical measure: it doesn't matter which). One finds that, for any given pair A, B of initial and final states, the work required to bring the system adiathermally from A to B does not depend on how the work was performed. For example, one could compress a fluid and then turn a crank to stir it, or one could first stir and then push, or instead decompress, stir, and pass some electrical current, etc. No matter what sequence one uses, and *therefore no matter what sequence of intermediate states the system passes through*, the work required to pass from A to B is the same.

The crucial point is that the work required does not depend on the sequence of states, i.e. the path. This path independence is the property that defines a *function of state* (see Section 5.2). Therefore *the principle of adiathermal work means that it is possible to define an energy-related function of state for any system*. The function of state U is called *internal energy* and is defined:

$U_B - U_A \equiv$ the work required to bring the system adiathermally from A to B.

(a)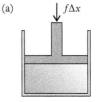

(b)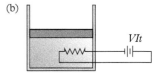

(c)

Fig. 7.1 *Performing work on a simple system. In each case the expression is given for the work done by the external agency. (a) a piston is pushed through displacement Δx by a force f: work done $= f\Delta x$. (b) A battery supplies current I for time t, thus pushing charge It through the system, across a voltage drop V: work done $= VIt$. (c) A crank is turned through angle ϕ by a torque T: work done $= T\phi$.*

The definition only needs to give the difference of internal energies of any pair of states, because nothing will depend on the absolute value of U. This is just as in classical mechanics, where the dynamics only depend on energy differences, not on how the zero of energy is chosen.

Internal energy can be thought of as the 'energy content' of a system. This is a valid concept because U is a function of state.

We can now make a satisfactory definition of *heat*. If a system is not thermally isolated, there can exist processes which bring the system from state A to state B while supplying an amount of work different to $\Delta U = U_B - U_A$. If W is the work energy supplied to the system, then the quantity of heat Q absorbed by the system is *defined* to be, for a closed system,

$$Q \equiv \Delta U - W. \tag{7.1}$$

That is, the heat energy taken in by a closed system, during a given state change, is defined as the difference between the amount of work which would have been required to cause the state change adiathermally (i.e. ΔU), and the amount of work which was supplied in fact.

We can now choose to regard internal energy U as the essential idea, and rewrite (7.1) as

$$\Delta U = Q + W. \tag{7.2}$$

This is a statement of *conservation of energy*, because it says that all the energy exchange processes we have thought of so far (namely heat and work) do not produce any net loss or gain of energy, but simply a movement of energy from one place to another. When energy is located at one place or another, it is called internal energy. When energy is moving from one place to another, the internal energy falls at the first place and grows at the second place, and the energy in transit is in the form of work or heat or both, depending on the type of process.

In differential form, the result is

$$dU = \dbar Q + \dbar W \qquad \text{[for a closed system (fixed } N\text{)]},$$

where U is a function of state and therefore has a proper (= exact) differential, but Q and W are not functions of state and therefore give improper (= inexact) differentials (Section 5.2).

For an open system the movement of particles into or out of the system can carry a further form of energy, called chemical energy, so that the total change in internal energy is

$$dU = \dbar Q + \dbar W + \dbar C. \qquad \text{[the first law]}$$

For the case of an open system it becomes harder to distinguish the three types of energy on the right-hand side of this formula, but the mathematical methods to be developed will make this clear. We will restrict attention to closed systems, where $dC = 0$, until Chapter 12.

Let us recap by summarizing the argument. The principle of adiathermal work says that when a given state change is brought about in thermal isolation, the same amount of work is always required: it doesn't depend on how the work is done, nor on what intermediate stages the system passes through, nor on anything else. This implies that a *definite amount of energy* is always associated with a *given change of state*. It follows that there is a function of state U which changes by $\Delta U = W$ when work W is performed in conditions of zero heat transfer.[1] If now the system is not thermally isolated, then any change in U not accounted for by the work done is called the heat supplied (for a closed system), or the sum of heat supplied and chemical energy (for an open system).

Note, the second part of the argument, to define heat, is not circular, because we already have a complete definition of U before heat was even considered.

[1] It might be imagined that the thermal isolation condition restricts the states we can reach, so it may not be clear whether it is correct to deduce the existence of a function of state. However, we have the option of doing irreversible things like stirring a fluid as well as compressing it, so a sufficient range of states can be reached such that we are justified in deducing the existence of a function of state.

7.1.1 Heat and work

It is common for students first learning the subject to make a basic error about heat, and talk loosely of systems as if they 'contained' heat. This is incorrect because heat is not a function of state. Bodies do contain energy, and this energy is called internal energy, but to think clearly and understand the subject properly, one must not muddle heat and internal energy.

When work is done, there is always a movement of some kind: a force acts, and the body on which the force acts is displaced. Work is inherently concerned with a *process*, and it quantifies the energy *transferred* by the process from one system to another.

Similar statements apply to heat. The movement associated with heat is a *change* in the tiny jiggling motions of the particles of the bodies concerned (or in the case of thermal radiation, of the jiggling motion of electric and magnetic fields). The particles of the hotter body fluctuate more violently, and where they meet the slower moving particles of the colder body, the former slow down on average while the latter speed up. The words 'heating' and 'cooling' refer to this energy transfer process.

The following statements may help:

> *Internal energy can be located in one physical system, heat and work cannot. Heat and work are forms of energy in transit from one system to another. Work is the orderly part of an energy transfer; heat is the randomly fluctuating part of an energy transfer.*

In small systems the distinction between heat and work can become blurred. That's okay: we have already agreed that thermodynamics is strictly accurate only in the thermodynamic limit (large systems).

A good guiding principle is to be cautious about the use of the word 'heat' as a noun—consider using it exclusively as a verb, or use the word 'heating'. Similarly, 'work' is always something to do with 'working'. Having received this caution, one can revert to the widespread use of 'heat' and 'work' as nouns; understood to refer to the amounts of energy involved in the process of transfer.

It is in general hopelessly muddled to speak of bodies 'containing' heat, because heat is not a function of state. Phrases such as 'heat storage' are therefore not allowable in careful thermodynamic discussions. One should *not* speak of a body 'losing' heat as if the heat were present in the body beforehand, but one may speak of a body 'giving off heat' (and thus losing internal energy). Equally, one should not speak of a body 'gaining' heat, but one may speak of a body absorbing heat and thus gaining internal energy. However, when discussing problems where none of the systems involved do any work or exchange particles, then all the internal energy changes come from heat exchanges, so the first law reduces to $\Delta U = Q$. In this restricted case one can speak loosely of heat as if it were like internal energy, which can be located and stored, but this sort of language is not recommended.

Another muddle that can occur is in the relationship between temperature, heat, and 'hotness'. Being hot is not necessarily associated with large quantities of heat energy. The tiny white-hot particles of metal coming from a firework 'sparkler' give off very few joules of heat, whereas many joules of heat are emitted by a pool of water as it freezes on a cold night.

We are used to examples where a body gets hotter when you supply heat to it, but this does not have to occur. Consider boiling water at constant pressure (say, 1 atmosphere), for example: one can continue to supply heat for a considerable time, while the water boils away, and the temperature of the system (water and steam) will not rise. More generally, *isothermal* processes (those taking place without a temperature change) do usually involve a flow of heat.

Processes where a body gets hotter while losing heat are also common: think of using a bicycle pump, for example. While one compresses the air in the pump, it gets hotter (that is, its temperature rises), and as soon as its temperature is above that of the surroundings it gives off heat, while continuing to get hotter because the work is being supplied faster than the heat leaves.

Similarly, decompression can result in a temperature fall while a body takes in heat.

7.2 Work by compression

We will now consider how to calculate amounts of work.

For a simple compressible system, work can only be done by compressing the system, i.e. changing its volume. If the system (e.g. a gas or liquid) is contained in a cylindrical chamber fitted with a piston, then the force required to displace the piston by a small amount will be

$$f = pA + \epsilon, \tag{7.3}$$

where A is the area of the piston, p is the pressure of the fluid, and ϵ is the contribution of friction. The work done by whoever is providing the force is

$$đW = fdx = (pA + \epsilon)dx \tag{7.4}$$

if the piston is displaced inwards by a small amount dx. This work is done *on* the fluid if the piston moves *inward*, making the volume *decrease*, so the change in volume of the fluid is

$$dV = -Adx \tag{7.5}$$

and we have

$$đW = -pdV + \epsilon dx. \tag{7.6}$$

One can extend this argument to a body of arbitrary shape by approximating the boundary by a large number of small flat surfaces (exercise for the reader). One finds that the work against the gas pressure is always $-pdV$, while the frictional term depends on the details. In the limit where the friction term is negligible, we would have a reversible process, and then

$$đW_{\text{rev}} = -pdV. \tag{7.7}$$

The total work done during some reversible change in volume is therefore

$$W_{\text{rev}} = \int -pdV. \tag{7.8}$$

This is also the area under the p verses V curve describing the change on an indicator diagram; see Figure 7.2.

In the presence of friction, the sign of ϵ is always such as to make ϵdx positive (since f must exceed pA when the piston moves in and f must be smaller than pA when the piston moves out), therefore we can write

$$đW = -pdV + |\epsilon dx|. \tag{7.9}$$

Although friction typically results in a temperature rise and heat exchange at the surfaces undergoing friction, there is no need to mention heat in this equation because the symbol $đW$ represents the energy that is supplied by the external system. It is called work because the external system did a simple, non-random thing: pushing a rod.

We now have several ways to write the first law for a closed system:

$$dU = đQ + đW \quad \text{[always]} \tag{7.10}$$

$$dU = đQ - pdV + |\epsilon dx| \quad \text{[for quasistatic process,]}$$

$$[|\epsilon dx| \text{ is energy from friction etc.]}$$

$$dU = đQ_{\text{rev}} - pdV \quad \text{[for reversible process]}$$

$$dU = TdS - pdV \quad \text{[always]} \tag{7.11}$$

76 First law, internal energy

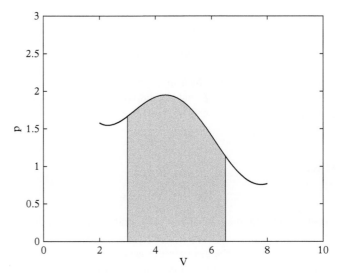

Fig. 7.2 *The work performed on a pV system when it expands or is compressed reversibly is $W = \int -p\,dV$. This quantity is given by the area under the curve describing the path of the change on an indicator diagram.*

The last version will be derived in Chapter 8 and is included here for completeness. We will not need it for any of the results to be discussed in this chapter, which is concerned with the first law. (In general, the results coming from the first law are essentially about energy, while those requiring the second law have something to do with entropy S as well as energy.)

> **Example 7.1**
>
> A lump of sodium metal of volume $10^{-4}\,\text{m}^3$ is located in a cylinder containing a mole of argon gas at temperature 295 K and pressure 10^5 Pa. Find the work done on the gas, and on the sodium metal, (i) when the pressure is increased isothermally to 10^7 Pa, and (ii) when, starting from the same initial conditions, the temperature is increased isobarically to 350 K. All processes may be considered reversible. [At STP, the isothermal bulk modulus of sodium is $B_T = 6.18 \times 10^9$ Pa and the cubic expansivity is $\alpha = 2 \times 10^{-4}\,\text{K}^{-1}$.]
>
> *Solution.*
> In all cases the work is given by equation (7.8). First consider the gas, which we treat as ideal. Then in the isothermal process,
>
> $$W = \int_{p_1}^{p_2} -p\,dV = \int_{p_1}^{p_2} -p \left.\frac{\partial V}{\partial p}\right|_T dp \qquad (7.12)$$
>
> $$= \int_{p_1}^{p_2} \frac{RT}{p}\,dp,$$

where we have used the ideal gas equation of state. Hence $W = RT\ln(p_2/p_1) = (8.31 \times 295)\ln(100)\,\text{J} = 11.3\,\text{kJ}$. In the isobaric process on the ideal gas,

$$W = \int_{T_1}^{T_2} -p\,\mathrm{d}V = -p\int_{T_1}^{T_2} \left.\frac{\partial V}{\partial T}\right|_p \mathrm{d}T = -R(T_2 - T_1),$$

which gives $W = -457\,\text{J}$. The sign indicates that the gas has expanded so has done work on other things.

Next we consider the metal. We begin as before, obtaining equation (7.12), but now the value of $(\partial V/\partial p)_T$ has to be obtained from the isothermal bulk modulus (cf. Table 3.4):

$$W = \int_{p_1}^{p_2} \frac{pV}{B_T}\mathrm{d}p.$$

Since the volume change of the solid is small, we can regard V as constant in this integral, and this pressure change will not cause much change in B_T either, so we can regard both as constant and obtain $W = (V/B_T)(p_2^2 - p_1^2)/2 = 0.8\,\text{J}$. In the isobaric process on the metal, we recall the definition of expansivity: $\alpha = (1/V)(\partial V/\partial T)_p$, so we have

$$W = \int_{T_1}^{T_2} -pV\alpha\,\mathrm{d}T = -pV\alpha(T_2 - T_1),$$

where we have treated V and α as constant. The result is $W = -0.11\,\text{J}$.

7.3 Heat capacities

The simple result

$$\mathrm{d}U = \mathrm{d}Q_{\text{rev}} - p\,\mathrm{d}V \qquad (7.13)$$

(the first law as applied to reversible processes in a pV system) has many applications.

First, consider the heat capacity at constant volume, defined as

$$C_V \equiv \frac{\mathrm{d}Q_V}{\mathrm{d}T}, \qquad (7.14)$$

where it is understood that $\mathrm{d}Q_V$ refers to a reversible process at constant volume. Dividing (7.13) by $\mathrm{d}T$ at constant V gives

78 First law, internal energy

$$\left.\frac{\partial U}{\partial T}\right|_V = \frac{\text{d}Q_V}{\text{d}T}, \tag{7.15}$$

so we have the very useful fact that the heat capacity at constant volume relates directly to internal energy changes:

$$C_V = \left.\frac{\partial U}{\partial T}\right|_V. \tag{7.16}$$

Next, by dividing (7.13) by $\text{d}T$ at constant p, one finds the slightly less simple

$$C_p \equiv \frac{\text{d}Q_p}{\text{d}T} = \left.\frac{\partial U}{\partial T}\right|_p + p \left.\frac{\partial V}{\partial T}\right|_p. \tag{7.17}$$

Most simple systems expand if the temperature is increased at constant pressure. Therefore in a constant pressure process, we should expect that more heat would be needed to achieve a given temperature change, compared to a constant volume process, because some of the energy goes to performing the work of expansion. Therefore we would expect $C_p > C_V$. We will prove later that C_p is indeed always greater than C_V, although it is not always true that $(\partial V/\partial T)_p > 0$: a counter-example is that of water near freezing point, which gets more dense when it is heated (but $C_p - C_V$ is still positive because $(\partial U/\partial T)_p$ is sufficiently larger than $(\partial U/\partial T)_V$).

The ratio of the principle heat capacities is assigned the symbol γ:

$$\gamma \equiv \frac{C_p}{C_V}. \tag{7.18}$$

This quantity is called the *adiabatic index*, because of its role in later equations (7.34) and (7.43).

For a few very simple systems the heat capacities are simply constants, unrelated to temperature and pressure etc., at least for a range of values of those parameters; but this is the exception not the rule. Usually the heat capacities depend on the conditions. For example, it can be proved from the third law that they always tend to zero at sufficiently low temperature (this means, roughly speaking, that cold things are poor at storing energy supplied as heat).

Figure 7.3 shows the principal heat capacities for a typical gas and a typical paramagnetic salt, as a function of temperature.

Difference of heat capacities of ideal gas

For an ideal gas we have $U = U(T)$ (Joule's law). Therefore for this special system, we have in U a function of a *single* variable, so

$$\left.\frac{\partial U}{\partial T}\right|_V = \frac{\text{d}U}{\text{d}T}, \quad \text{and} \quad \left.\frac{\partial U}{\partial T}\right|_p = \frac{\text{d}U}{\text{d}T}. \tag{7.19}$$

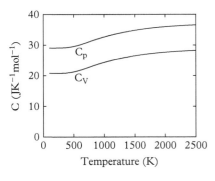
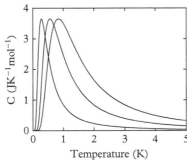

Fig. 7.3 Molar heat capacities of nitrogen gas (left) and an ideal paramagnet (right). The results for nitrogen gas illustrate equation (7.20), and show that the approximation C_V = constant is good for modest temperature changes. For the paramagnet the heat capacity at constant magnetic field is shown; that at constant magnetization is zero. The curves show C_B at $B = 0.5, 1, 1.5$ tesla, for a sample whose (spin-1/2) particles each have dipole moment equal to one Bohr magneton.

Therefore,

$$\left.\frac{\partial U}{\partial T}\right|_V = \left.\frac{\partial U}{\partial T}\right|_p.$$

Note, the reasoning in (7.19) is only valid because U is a function of T alone for the system under consideration; if it were not then the quantity dU/dT would not even be defined: it would be meaningless.

Using this result in (7.16) and (7.17), and using the equation of state to obtain $(\partial V/\partial T)_p = Nk_B/p$, we find

$$C_p - C_V = Nk_B = n_m R. \quad \text{[for ideal gas]} \tag{7.20}$$

That is, the heat capacities differ by a constant amount. As soon as we know one, we know the other, and if one has a given temperature dependence, then the other has the same temperature dependence plus an offset. Also, at given T, there will be no dependence on pressure or volume since C_V has none; which follows immediately from Joule's law because C_V can be obtained directly from $U(T)$: equation (7.16).

Example 7.2

Show that, for an ideal gas,

$$C_V = \frac{Nk_B}{\gamma - 1}. \tag{7.21}$$

Solution.
Using the definition (7.18) to replace C_p in eqn (7.20), we have $\gamma C_V - C_V = Nk_B$, and the result follows.

For a monatomic gas (one composed of single atoms not molecules), for a wide range of temperature and pressure, one finds $C_V = (3/2)Nk_B$ to good approximation, and therefore $C_p = (5/2)Nk_B$ and $\gamma = 5/3$. For a diatomic gas one finds $C_V = (5/2)Nk_B$ so $\gamma = 7/5$ at modest temperatures and $C_V = (7/2)Nk_B$ so $\gamma = 9/7$ at high temperatures (cf. Figure 7.3). These properties can be predicted by statistical mechanics, as follows. The internal energy of a monatomic gas is the total kinetic energy of its atoms; in thermal equilibrium each atom has an average kinetic energy $(3/2)k_B T$, leading to $U = (3/2)Nk_B T$ and hence $C_V = (3/2)Nk_B$. For a diatomic gas the molecules also have rotational energy, which adds a further $k_B T$ to the internal energy per molecule. This explains the observed value $C_V = (5/2)Nk_B$ at modest temperatures. At high temperatures, vibrations of each molecule can be excited, which explains the increase in heat capacity at around 800 K in Figure 7.3.

7.3.1 Energy equation

In Chapter 6 we developed the concept of temperature, and showed that every system has an equation of state, relating the temperature to the other state variables. In the present chapter we have established the existence of a further function of state, internal energy U. The question naturally arises, what is the equation relating U to the other state variables?

For a simple system there are just two degrees of freedom. Therefore, in principle, U can be expressed as a function of p and V. Such an expression is called an *energy equation of state*. We will call it simply the *energy equation* in order to distinguish it from the equation of state relating T to p and V.

Since we already have a simple relationship between $(\partial U/\partial T)_V$ and C_V (they are equal), it is natural to consider how U can be expressed as a function of T and V. Therefore let us consider $U = U(T, V)$ and write

$$dU = \left.\frac{\partial U}{\partial T}\right|_V dT + \left.\frac{\partial U}{\partial V}\right|_T dV$$

$$= C_V dT + \left.\frac{\partial U}{\partial V}\right|_T dV. \tag{7.22}$$

To make progress, we would like to express $(\partial U/\partial V)_T$ in terms of other things. We can relate the left-hand side of the equation to C_p by dividing through by dT with p constant, giving:

$$\left.\frac{\partial U}{\partial T}\right|_p = C_V + \left.\frac{\partial U}{\partial V}\right|_T \left.\frac{\partial V}{\partial T}\right|_p. \tag{7.23}$$

But the first law, equation (7.13), gives

$$\left.\frac{\partial U}{\partial T}\right|_p = C_P - p \left.\frac{\partial V}{\partial T}\right|_p, \tag{7.24}$$

so by equating these two expressions we can obtain

$$\left.\frac{\partial U}{\partial V}\right|_T = (C_P - C_V)\left.\frac{\partial T}{\partial V}\right|_p - p. \tag{7.25}$$

Finally, substitute this back into (7.22) and we have

$$\begin{aligned}dU &= C_V dT + \left((C_P - C_V)\left.\frac{\partial T}{\partial V}\right|_p - p\right) dV \\ &= C_V dT + \left(\frac{C_P - C_V}{\alpha V} - p\right) dV.\end{aligned} \tag{7.26}$$

This equation is true for all pV systems. It expresses the total derivative of U in terms of heat capacities and the cubic expansivity α. The latter can be derived from the equation of state. As long as these properties are known (i.e. we have measured or derived the expression relating them to the state variables), then we can integrate the dU equation and obtain ΔU for any change of state. After choosing some arbitrary state V, T to have $U = 0$, we thus obtain U as a function of V and T, so we have the energy equation.

To summarize, the internal energy can be obtained from the equation of state as long as both principal heat capacities are known as a function of T and V.

With the use of the second law we will later derive the remarkable fact that $(C_P - C_V)$ is not in fact independent of the equation of state: it can be derived from it! (see equation (13.26)). Therefore, U can be obtained from just one heat capacity and the equation of state. For the present we note merely that this will be a new observation which does not follow from the first and zeroth laws alone.

7.3.1.1 Energy equation of ideal gas

If we make the approximation that C_V is independent of temperature for an ideal gas, then we have constant dU/dT, and since the zero of energy can be fixed arbitrarily, we may as well pick $U = C_V T$. In this case the energy equation is

$$U = C_V T = \frac{Nk_B T}{\gamma - 1}. \quad \text{[Ideal gas having constant } \gamma\text{]} \tag{7.27}$$

(The second version uses equation (7.21).) Combining this with the equation of state gives the alternative version,

$$p = (\gamma - 1)U/V. \tag{7.28}$$

Therefore the pressure is a constant multiple of the energy density (energy per unit volume). The link between pressure and energy density is a useful insight, and it occurs in other contexts. For example, in the case of thermal radiation one has $p = (1/3)U/V$, although the equation of state (6.14) is quite different.

7.3.2 Relation of compressibilities and heat capacities

Heat capacities are concerned with 'how much heat is needed to raise the temperature' and it is natural to consider doing this either at constant pressure or constant volume (for a p, V system). The other most obvious quantity to think about is 'how hard it is to squeeze' the system, which is the quantity $\frac{\partial V}{\partial p}\big|$. This was introduced in Table 3.4, where we defined

$$\kappa_T \equiv -\frac{1}{V}\frac{\partial V}{\partial p}\bigg|_T, \qquad \kappa_S \equiv -\frac{1}{V}\frac{\partial V}{\partial p}\bigg|_S. \qquad (7.29)$$

Which of κ_T and κ_S do you expect to be larger? In an adiabatic squeeze most systems will get hotter, so it gets harder to squeeze them compared to an isothermal case, so we expect κ_S to be smaller than κ_T.

We will now prove a relation between κ_T/κ_S and C_p/C_V—a beautiful example of the 'magic' of thermodynamics. Although one should imagine that heat capacities and compressibilities would be related in some way, one might expect that the relation would depend on the system. Maybe it involves the equation of state or the energy equation. In fact the relation is simple and universal!

The diagram in Figure 7.4 helps the argument. We consider a small change of pressure and volume, and draw an isotherm (line of constant T) and an adiabat (the path for a reversible adiathermal process) through a given state A. These lines have slopes $(\partial p/\partial V)_T$, $(\partial p/\partial V)_S$ respectively (i.e. $(-V\kappa_T)^{-1}$, $(-V\kappa_S)^{-1}$).

Consider the states A, B, C, D in Figure 7.4. We will calculate the internal energy change dU in going from A to D by two routes: AD and $ABCD$.

First note that the work done in either route is the area under the corresponding curve on the pV diagram, therefore the work done in process AD is equal to the work done in AB plus the area of the triangle ABD. However, the latter is $dpdV/2$ which, being a second-order quantity, is negligible compared to pdV. Therefore, to first order the work done in the two routes is the same. Since U is a function of state, the internal energy change is the same for the two routes. From these two observations it follows that the system also loses or gains the same amount of heat in the two routes. But the route AD is adiabatic. Therefore, both routes give no net heat transfer. It follows that the heat leaving the system along AB must be equal to the heat entering the system along BD, i.e.

$$C_p(T_A - T_B) = C_V(T_D - T_B). \qquad (7.30)$$

Now, using that AC is an isotherm, we have $T_A = T_C$, so we find

$$\frac{C_p}{C_V} = \frac{T_D - T_B}{T_C - T_B}. \qquad (7.31)$$

As long as the system is not at a stationary point of temperature as a function of pressure, then to first-order approximation the temperature must be a linear

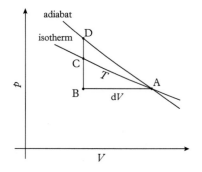

Fig. 7.4 *Relating the slopes of adiabats and isotherms to heat capacities. The argument is given in the text.*

function of pressure along the line BCD, therefore the ratio on the right-hand side of equation (7.31) is the ratio of the slopes of the two lines. In other words, we have

$$\frac{C_p}{C_V} = \frac{\left.\frac{\partial p}{\partial V}\right|_S}{\left.\frac{\partial p}{\partial V}\right|_T}. \tag{7.32}$$

In terms of the compressibilities, this is

$$\boxed{\frac{\kappa_T}{\kappa_S} = \frac{C_p}{C_V}. \tag{7.33}}$$

This result is universal—true for all pV systems—and beautifully simple. It is perhaps our first 'golden nugget' extracted after all the thinking we have invested so far. It says that, no matter how strangely behaved a simple compressible system may be, as long as it has thermal equilibrium states—and it has—then the two basic kinds of 'squashiness' are inextricably tied to the two basic kinds of 'room for heat energy'. The relationship is essentially a statement about conservation of energy. Exercise 7.7 presents a derivation with a more algebraic flavour, and we will derive it more rapidly and generally in equations (13.28)–(13.29) by making use of the concept of entropy.

It is common to introduce the symbol γ, defined in (7.18), and then equation (7.33) can be written

$$\left.\frac{\partial p}{\partial V}\right|_S = \gamma \left.\frac{\partial p}{\partial V}\right|_T. \tag{7.34}$$

7.4 Solving thermodynamic problems

Many thermodynamic problems take the form of a desire to find a final state, or an amount of energy, for some given process. The method by which such problems are solved will often involve the following steps:

(1) Identify clearly the thermodynamic system to be treated.
(2) Identify the nature of the interaction with the surroundings, and hence the type of *process*.
(3) Use the equation of state to gain information about initial and final conditions.
(4) At this stage you may well be able to calculate the heat and work inputs to the system, in terms of the state variables, although there may be some unknowns remaining in your expressions.
(5) If there remain some unknowns, use information about the heat capacity or the energy equation or both.

84   *First law, internal energy*

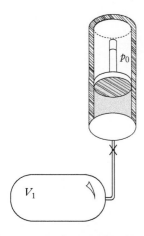

Fig. 7.5 *A rigid thermally insulated flask is connected through a narrow tube to a cylinder where the piston exerts a constant pressure p_0.*

Example 7.3

A flask holds a quantity of ideal gas at initial pressure and temperature p_1, T_1. The flask is joined by a thin capillary tube to a cylinder closed by a frictionless piston which exerts a constant pressure p_0—see Figure 7.5—the whole apparatus being thermally insulated from its surroundings. When a valve is opened, gas moves slowly between the flask and the cylinder. Find the final equilibrium temperature of the gas, assuming its adiabatic index γ to be constant.

Solution.
We follow the steps outlined above.

(1) We can choose whatever system we wish, as long as we then reason correctly about that system. Here we choose not the flask, nor the cylinder, nor both. Rather, the system is *the gas*.

(2) The gas moves between flask and cylinder, and the motion through the capillary tube may be complicated, but the important point is that the surroundings do not exchange heat with the gas, and the only place where the surroundings do mechanical work in this problem is at the piston.

(3) Let N be the number of particles and V_1, V_2 be the initial and final volumes of the gas. Then

$$p_1 V_1 = N k_B T_1, \qquad (7.35)$$
$$p_2 V_2 = N k_B T_2, \qquad (7.36)$$

where p_2, T_2 are the final conditions. T_2 is the quantity to be discovered, and at equilibrium we must have $p_2 = p_0$. Hence we have

$$\frac{T_2}{T_1} = \frac{p_0 V_2}{p_1 V_1}. \qquad (7.37)$$

The only remaining unknown is the ratio of the volumes. Note, here V_1 is the volume of the flask and V_2 is the volume of all the region filled by the gas in its final equilibrium state. Some of this is in the flask, some in the cylinder.

(4) No heat is exchanged, so $\Delta Q = 0$, but work is done by the piston. This is always at fixed pressure p_0, so the total work done on the gas is

$$\Delta W = -p_0 (V_2 - V_1).$$

Note, in this equation p_0 is the pressure exerted *by the piston*. It could be owing to the weight of the piston, or to a surrounding large atmosphere, but it is not in any simple way 'the pressure of the gas', because the gas does not necessarily have a uniform pressure or temperature during the process. However, we have assumed that the process is slow enough that the piston does not deliver energy by other means, such as by fast oscillations which create sound waves.

(5) V_2 and V_1 remain unknown, but an energy argument will now tell us what we need to know. We use the special property of ideal gases, that the internal energy depends only on the temperature, and we are told that γ is constant. This makes the energy equation very simple: $U = C_V T$ (Eq. 7.27). So, using this and the previous equation, we have

$$\Delta U = -p_0(V_2 - V_1) = C_V(T_2 - T_1).$$

Note, the presence of C_V in this equation does *not* imply that any heat transfer has taken place. Rather, it is being used simply to relate internal energy to state variables, and indeed it is more useful here to adopt the second form of equation (7.27) (which follows from (7.21)):

$$-p_0(V_2 - V_1) = \frac{Nk_B}{\gamma - 1}(T_2 - T_1) = \frac{Nk_B T_1}{\gamma - 1}\left(\frac{T_2}{T_1} - 1\right).$$

The equation of state can now be used on the right to replace $Nk_B T_1$ by $p_1 V_1$, from which we obtain

$$\frac{V_2}{V_1} = 1 - \frac{p_1}{p_0}\frac{(T_2/T_1 - 1)}{(\gamma - 1)}.$$

Substituting this into (7.37) and solving for T_2/T_1 gives

$$\frac{T_2}{T_1} = \frac{1}{\gamma}\left(1 + (\gamma - 1)\frac{p_0}{p_1}\right). \tag{7.38}$$

7.5 Expansion

7.5.1 Free expansion of ideal gas

We considered in Section 7.1 a series of experiments performed by Joule, involving doing work on an isolated system. We now consider another experiment associated with the name of Joule, and this time it is one in which no work is done nor heat exchanged!

Suppose a pV system is placed inside a closed rigid chamber whose walls are made of thermally insulating material. Then no heat can pass between the system and the chamber, and no mechanical work can be done since the latter is rigid.

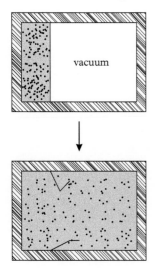

Fig. 7.6 *Free expansion of a gas into a previously evacuated space.*

It follows from energy conservation (the first law) that the energy U of the system must be constant:

$$U = \text{const.}$$

One might feel that nothing could happen to the system (for example, a sample of gas), but with the right preparation, something can. Suppose the chamber has a partition, and one part is filled with gas, while the other is evacuated. If the partition breaks, a process will take place: the gas will rush to fill the available space (see Figure 7.6). Such an expansion into vacuum is called a *free expansion* or a *Joule expansion*.

James Joule prepared such an expansion of a gas, and carefully observed the temperature of the gas before and after. Although the process involves non-equilibrium states while the free expansion is happening, the states before the partition is removed, and after the gas has settled, are both equilibrium states. Therefore they each have some well-defined temperature, pressure, and volume. Since the volume of the gas changes, and the process involves no change of internal energy, the property Joule was investigating must be the partial derivative,[2]

$$\left.\frac{\partial T}{\partial V}\right|_U . \tag{7.39}$$

Joule found that, for the gas he was using, and to the limit of his precision, there was no temperature change, so

$$\left.\frac{\partial T}{\partial V}\right|_U = 0. \tag{7.40}$$

Using reciprocity this implies $(\partial T/\partial U)_V (\partial U/\partial V)_T = 0$, from which

$$\left.\frac{\partial U}{\partial V}\right|_T = 0 \tag{7.41}$$

since $(\partial T/\partial U)_V \neq 0$ (injecting energy would change the temperature). It follows that when we write $U = U(V, T)$, a function of volume and temperature, we shall find that the volume does not appear in the function. In other words, *for a gas at low pressure, the internal energy is a function of temperature alone*. This is Joule's law (Section 6.2.1.1).

In summary, Joule's law can be deduced from the observation that the free expansion of an ideal gas does not result in a temperature change.

7.5.2 Adiabatic expansion of ideal gas

Applying equation (7.34) to an ideal gas, we find

$$\left.\frac{\partial p}{\partial V}\right|_S = -\gamma \frac{P}{V} \quad [\text{using } (\partial p/\partial V)_T = -p/V] \tag{7.42}$$

---

[2] The thermal insulation ensures there is no heat exchange, but the question of whether work is done sometimes causes confusion. One should keep in mind that 'work' always refers to work done *by an external agency* acting on the system. In the free expansion the 'system' under consideration is the contents of the *rigid* chamber. There is no movement of the point of action of any external force on this chamber, and the chamber itself does no work. Since no other system has received any work, we must conclude that the gas does none.

The only property we needed in order to deduce (7.42) was $(\partial p/\partial V)_T = -p/V$, which follows from the equation of state. Therefore the expression is valid even if the heat capacities, and therefore γ, depend on temperature.

If γ is independent of temperature (which, in view of (7.20), happens for an ideal gas if and only if C_V is independent of temperature), then (7.42) can be integrated directly. The result (Exercise 7.2) is

$$pV^\gamma = c \quad \text{[Adiabatic expansion of an ideal gas having constant } \gamma\text{]} \quad (7.43)$$

where c is a constant. Now you know why γ is called the adiabatic index. This equation shows not only that the pressure falls during an adiabatic expansion, as one should expect, but also that it falls faster than during an isothermal expansion (because $\gamma > 1$).

The approximation of constant γ is accurate for monatomic gases away from the boiling point, and remains approximately true even close to the boiling point. For diatomic gases it is a good approximation as long as the temperature change is not too great—see Figure 7.7.

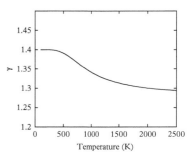

Fig. 7.7 *Adiabatic index of nitrogen gas as a function of temperature.*

Example 7.4

An ideal gas with $\gamma = 1.4$ expands adiabatically, such that the volume increases by a factor 4. By what factor does the pressure fall?

Solution.
The adiabatic expansion obeys equation (7.43), therefore if p_1, V_1 is the initial state and p_2, V_2 the final state, then $p_1 V_1^\gamma = p_2 V_2^\gamma$. It follows that $p_1/p_2 = (V_2/V_1)^\gamma = 4^{1.4} \simeq 6.96$, i.e. the pressure falls by a factor approximately 7.

7.5.3 Adiabatic atmosphere

In a gravitational field, the pressure in a fluid varies with height z according to

$$dp = -\rho g dz, \quad (7.44)$$

where ρ is the density and g the acceleration due to gravity. We can use this to find the expected variation with height of the pressure in Earth's atmosphere. Using $\rho = mn$, where m is the mass of one molecule, and treating the atmosphere as an ideal gas ($p = nk_B T$), we have

$$\frac{dp}{dz} = -\frac{mg}{k_B} \frac{p}{T}, \quad (7.45)$$

where we have left as yet unspecified the constraints that allow us to write dp/dz as a total differential. These constraints depend on what happens to parcels of air as

[3] In equilibrium we don't expect any net motion of air up or down, but we should expect this to be a dynamic equilibrium in which any given parcel of air is free to move.

they rise and fall in the atmosphere.[3] One possibility is the isothermal case, where T is independent of height. This would occur if the processes were slow enough for thermal relation, but this is a poor approximation for Earth's atmosphere. A better approximation is to assume that as any given parcel of air rises, it expands adiabatically. Consider a parcel whose volume is V at some given height. As it moves to a different height it expands adiabatically and so maintains pV^γ constant, using equation (7.43), since the temperature changes are small enough that γ is here constant to good approximation. Substituting for V using the equation of state, we have

$$p^{1-\gamma} T^\gamma = \text{constant}$$
$$\implies p = (\text{constant}) \times T^{\gamma/(\gamma-1)}$$
$$\implies \frac{dp}{dT} = \left(\frac{\gamma}{\gamma-1}\right)\frac{p}{T}.$$

Writing $(dp/dz) = (dp/dT)(dT/dz)$ and using (7.45), we obtain

$$\frac{dT}{dz} = -\left(\frac{\gamma-1}{\gamma}\right)\frac{mg}{k_B}. \tag{7.46}$$

Thus for the adiabatic atmosphere, we expect the temperature to fall linearly with height. The gradient is called the *adiabatic lapse rate*. Using (7.20), we have $C_p = R\gamma/(\gamma-1)$ for one mole, so the lapse rate can be written

$$gM_{\text{molar}}/C_p, \tag{7.47}$$

where $M_{\text{molar}} = N_A m$ is the molar mass. For dry air the adiabatic lapse rate evaluates to 9.8 K/km. The measured value in Earth's atmosphere is somewhat lower, in the region 6.4 K/km, because of latent heat effects in a moist atmosphere. When a parcel of air containing water vapour rises, it eventually reaches a height at which the water starts to condense. As the parcel rises further, its temperature falls more slowly because the heat leaving the parcel contains a significant contribution from latent heat of condensation.

7.5.4 Fast and yet adiabatic?

In the discussion of the adiabatic atmosphere, we made an assumption that needs justification. We treated processes which were faster than the timescale needed for a region of gas to exchange heat with neighbouring regions, but we nevertheless assumed the region remained in internal equilibrium with itself. It is not immediately clear that this is possible. It relies on the fact that for a modest and smooth change in the state parameters, under adiathermal conditions, the system state stays within the equilibrium state space, even for changes faster than the relaxation time. The process only needs to be slow compared to another, shorter, timescale called the *acceleration time*. This will be explored in Chapters 19 and 27.

EXERCISES

(7.1) A mole of ideal gas is taken from a state p_1, V_1 to a state p_2, V_2 along a path forming a straight line on an indicator diagram. Find an expression for the work W done on the gas. Assuming the constant-volume heat capacity C_V is independent of temperature, find also the internal energy change ΔU and the heat Q entering the system. Apply your results to find ΔU, W, and Q when the initial state is at volume 21×10^{-3} m^3 and temperature 290 K, and the final state is at 22×10^{-3} m^3, 330 K, for a gas with $\gamma = 1.4$.

(7.2) Obtain equation (7.43).

(7.3) By combining the ideal gas equation of state with (7.43), show that during an adiabatic expansion $TV^{\gamma-1}$ is constant and $T^\gamma p^{1-\gamma}$ is constant, when γ is independent of temperature.

(7.4) An ideal gas at initial pressure p_1 undergoes an adiabatic expansion from volume V_1 to volume V_2. Assuming γ is constant, find the final pressure and show that the work done is

$$W = \frac{p_1 V_1}{\gamma - 1} \left(\left(\frac{V_1}{V_2} \right)^{\gamma-1} - 1 \right).$$

(7.5) An ideal gas is taken between the same initial and final states as in Exercise 7.1, by an adiabatic expansion followed by heating at constant volume. Calculate the work done and heat absorbed.

(7.6) A container of gas is sealed by a small ball bearing which can move freely in a vertical tube. The ball is displaced vertically by a small amount and then released. Show that the period of the harmonic oscillations that result is $2\pi (V/\gamma g A)^{1/2}$, where V is the equilibrium volume and A is the cross-sectional area of the ball (and tube). [This is the basis of Rüchardt's method to measure the adiabatic index.]

(7.7) Obtain equation (7.33) as follows. First, using equation (5.37), show that

$$\frac{đQ_p}{dV} = C_p \left.\frac{\partial T}{\partial V}\right|_p, \quad \frac{đQ_V}{dp} = C_V \left.\frac{\partial T}{\partial p}\right|_V. \quad (7.48)$$

Next, argue that during a change of state by dp and dV, the total heat entering a pV system must be

$$đQ = \frac{đQ_p}{dV} dV + \frac{đQ_V}{dp} dp. \quad (7.49)$$

We must proceed carefully, because we have no guarantee that this combination is a proper differential, and in fact it is not for most changes.

However, adiabatic changes have the special property that đQ = 0. Use this to obtain an expression for $(\partial p/\partial V)_S$ in terms of the quantities in (7.48) and hence derive equation (7.33) or (7.34).

(7.8) Find expressions for κ_T and κ_S for an ideal gas of given γ.

(7.9) At one atmosphere, the specific heat capacity of water at constant pressure is given to four significant figures by the formula

$$\frac{c_p(\theta)}{c} = 0.996185 + 0.0002874\left(1 + \frac{\theta}{100}\right)^{5.26} + 0.011160\, e^{-0.0829\,\theta},$$

where $c = 4185.5\,\text{J}\,\text{K}^{-1}\,\text{kg}^{-1}$ and θ is the temperature in degrees Celsius. Find the heat energy required to raise the temperature of 50 grams of water from 0 °C to 100 °C.

(7.10) A spring of length L obeys Hooke's law $f = k(L - L_0)$, where f is the tension, L_0 the natural length, and k is the spring constant. Find the work done on the spring when its length is changed from L_1 to L_2.

(7.11) Consider the ideal elastic substance described by equation (6.20). Show that the isothermal Young's modulus E (defined as the ratio of stress to strain: $E = (\delta f/A)/(\delta L/L_0)$, where A is a cross-sectional area) is given by

$$E = \frac{KT}{A}\left(1 + 2\frac{L_0^3}{L^3}\right).$$

Calculate the work required to stretch the substance isothermally from $L = L_0$ to $L = 2L_0$.

(7.12) [From Adkins] Below 100 K the specific heat capacity of diamond varies as the cube of temperature: $c_p = aT^3$. A small diamond of mass 100 mg is cooled to 77 K by immersion in liquid nitrogen, and then dropped into a bath of liquid helium at its boiling point of 4.2 K at atmospheric pressure. In cooling the diamond, some of the helium is boiled off. The gas is collected and found to occupy a volume 2.48×10^{-5} m^3 at 0 °C and 1 atmosphere pressure. What is the value of a in the formula for the specific heat capacity of diamond? [The latent heat of vaporization of helium at 1 atm is 21 kJ/kg.]

(7.13) A machine compresses 10 mole/minute of helium, modelled as an ideal gas, from 1 to 10^6 Pa pressure. What rate of flow of cooling water, initially at 290 K, is needed if the compression is to be made isothermal at 300 K?

(7.14) A gas compressor compresses 50 mol/s of gas adiabatically from 1 bar at 15 °C to 10 bar. Treating the gas as ideal with adiabatic index $\gamma = 1.4$, find the final temperature and the power input.

(7.15) [Adapted from Endem (1938)] It is desired to heat a room at initial temperature 10 °C to 20 °C. The volume of the room is 20 m^3 and the initial pressure is 1 atm. The molar constant-volume heat capacity for air is $3.6R$.

(i) Suppose first that there is no heat loss nor movement of air in or out of the room. Determine the heat input required to warm the room.

(ii) Now suppose air moves out of the room during the heating, such that the pressure remains constant. Show that in this case the energy of the air remaining in the room at the end is the same as that of the energy of the air in the room at the beginning.

(iii) Explain where the energy supplied as heat has gone, and calculate how much heat was supplied.

(7.16) (i) Explain carefully why, when gas leaks slowly out of a chamber, the expansion of the gas remaining in the chamber may be expected to be adiabatic (that is, quasistatic and without heat exchange). [Hint: choose carefully the physical system you wish to consider.]

(ii) A gas with $\gamma = 5/3$ leaks out of a chamber. If the initial pressure is $32p_0$ and the final pressure is p_0, show that the temperature falls by a factor 4, and that $1/8$ of the particles remain in the chamber.

(7.17) In the experiment shown in Figure 7.5, a small thermometer is attached to the inside wall of the flask. The initial pressure p_1 is considerably larger than p_0. The temperature registered by this thermometer falls as the gas leaves the flask, and drops substantially below the equilibrium value given in equation (7.38), before finally rising again and settling at the equilibrium value. Explain.

(7.18) A gas cylinder is mounted vertically and sealed by a freely movable light piston. The external air pressure is negligible. A mass m_1 is placed on the piston, and initially the system is in equilibrium, such that the weight of this mass is supported by pressure forces provided by the gas. The piston is then temporarily prevented from moving, while the mass is increased to m_2, and then the piston is released. Treating the gas as ideal with $\gamma = 5/3$, find, in terms of m_1 and m_2, the ratios by which the volume and temperature of the gas change:

(i) in the case A, where the piston moves freely and the total system energy U_{tot} stays constant (where by U_{tot} we mean the sum of gravitational potential energy of the mass and internal energy of the gas)

(ii) in the case B, where the piston is released gradually, such that the change in volume is quasistatic and without heat exchange.

[Ans. $V_2/V_1 = (3m_1 + 2m_2)/5m_2$ and $(m_1/m_2)^{3/5}$; $T_2/T_1 = (3m_1 + 2m_2)/5m_1$ and $(m_2/m_1)^{2/5}$] This experiment is explored further in Exercise 7.2 of chapter 17.

(7.19) A thermally insulated chamber contains some hot gas and a lump of metal. Initially the gas and the lump are at the same temperature T_i. The volume of the chamber can be changed by moving a frictionless piston. Assuming the heat capacities of the gas and the metal lump are comparable, sketch on one diagram the pressure–volume relation for the system

(a) if the pressure is reduced to atmospheric pressure p_0 slowly enough for the temperature of the metal lump to be equal to that of the gas at all stages.

(b) if the pressure is reduced to p_0 fast enough for the metal lump not to cool at first (but the process is still quasistatic for the gas) after which the piston is further moved so as to maintain the pressure at p_0 until the metal lump and the gas attain the same temperature.

Use the first law to explain whether or not the final volume will be the same in these two processes. Explain which process finishes at the lower temperature. [Hint: consider the work done and use the fact that internal energy is a function of state for any given system such as the gas.]

(7.20) A thermally insulated and evacuated chamber is placed in a room where the pressure and temperature T_0 are maintained constant. Gas leaks slowly into the chamber through a small hole. Show that when the pressures are equalized, the temperature of the air in the chamber is γT_0, where $\gamma = C_p/C_v$ and you may assume the heat capacities are independent of temperature. [Hint: imagine placing a bag around the chamber, just large enough to enclose the chamber and all the gas that finally ends up inside the chamber, and calculate the work done by the rest of the atmosphere as this bag collapses.] Consider the case of argon gas ($\gamma = 5/3$) at 50 °C leaking into a flask made of tin. What happens to the flask?

The second law and entropy

8

Clausius statement:

> No process is possible whose sole effect is the transfer of heat from a colder to a hotter body.

Kelvin statement:

> No process is possible whose sole effect is to extract heat from a single reservoir and convert it into work.

8.1 Heat engines and the Carnot cycle	93
8.2 Carnot's theorem and absolute temperature	97
8.3 Clausius' theorem and entropy	102
8.4 The first and second laws together	105
8.5 Summary	106
Exercises	106

We begin this chapter with these two statements, not in order to suggest that they are more profound than other statements of the second law, but because they are a useful way to learn the subject.

8.1 Heat engines and the Carnot cycle

Nicolas Sadi Carnot (1796–1832) emphasized the usefulness of considering a reversible cyclic process made up of two isothermal and two adiabatic stages (see Figure 8.1). Such a cycle is called a *Carnot cycle*. In each isothermal stage, heat enters or leaves the system, and work is done. In each adiabatic stage no heat is transferred but work is done, so the internal energy changes and so does the temperature. The Carnot cycle is most easily envisaged as applied to a pV system (left-hand diagram of Figure 8.1), but since it is defined purely in terms of isothermal and adiabatic processes, a Carnot cycle exists for all types of system, whether compressible, or magnetic, or surface tension, or whatever. To illustrate this point, the right-hand diagram of Figure 8.1 shows an example of a Carnot cycle involving magnetic work on a paramagnetic sample.

The two isothermal stages define two temperatures, T_1 and T_2. Heat Q_1 enters at temperature T_1 (the heat is typically drawn from a large reservoir) and heat Q_2 leaves the system at temperature T_2. Since, by definition, after a cycle the system returns to its initial state, there is no net change in its internal energy. It follows (from the first law) that the work done by the system during one cycle is $W = Q_1 - Q_2$. The net effect, therefore, is to take some heat energy Q_1 at a certain temperature, convert some of it into work W, and deliver the rest to another reservoir at a lower temperature. This is called a *heat engine*. A general

Thermodynamics: A Complete Undergraduate Course. Andrew M. Steane.
© Andrew M. Steane 2017. Published 2017 by Oxford University Press.

Fig. 8.1 *Two examples of Carnot cycles: (a) in an ideal gas, (b) in an ideal paramagnet. Starting from A, the gas first expands isothermally while heat enters, then expands adiabatically, then it contracts isothermally while giving off heat, then it contracts adiabatically. The net work done is given by the area enclosed by the path on the indicator diagram. The paramagnet example involves changes of applied field instead of expansion and compression; it illustrates the fact that any type of system can undergo a Carnot cycle.*

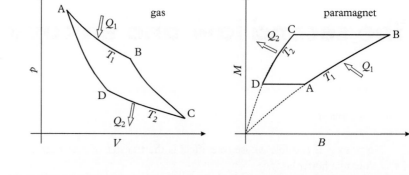

heat engine might involve a more complicated sequence of actions but the overall effect will be similar; this is indicated by a diagram such as Figure 8.2. The efficiency η of the engine is defined to be the ratio of the work extracted to the heat absorbed,

$$\eta = \frac{W}{Q_1} = 1 - \frac{Q_2}{Q_1}. \tag{8.1}$$

We might hope that Q_2 could be zero and hence $\eta = 1$, but as we shall see the second law rules this out.

It is useful to get an idea of heat engine efficiency by seeing what happens if we apply Carnot's cycle to an ideal gas. We shall use the symbol θ for temperature in this argument, because it refers to the ideal gas temperature $\theta \equiv pV/Nk_{\rm B}$. In this chapter it will be useful to keep clear the distinction between this and absolute thermodynamic temperature T, although ultimately we shall prove that they are equal.

In an isothermal expansion, for an ideal gas the internal energy does not change (Joule's law), so the heat entering the system must be given by $Q_1 = \int p\,dV$, where the path to be integrated along is one at constant temperature θ. We substitute $p = Nk_{\rm B}\theta/V$ so that the integrand is a function of V alone, and hence obtain $Q_1 = Nk_{\rm B}\theta_1 \ln(V_B/V_A)$ and $Q_2 = Nk_{\rm B}\theta_2 \ln(V_C/V_D)$. Therefore

$$\frac{Q_1}{Q_2} = \frac{\theta_1 \ln(V_B/V_A)}{\theta_2 \ln(V_C/V_D)}. \tag{8.2}$$

For convenience we take the case of constant γ, then the adiabats are described by equation (7.43), so

$$\left.\begin{array}{l}\theta_1 V_B^{\gamma-1} = \theta_2 V_C^{\gamma-1} \\ \theta_1 V_A^{\gamma-1} = \theta_2 V_D^{\gamma-1}\end{array}\right\} \Rightarrow \frac{V_B}{V_A} = \frac{V_C}{V_D},$$

and hence we find

$$\frac{Q_1}{Q_2} = \frac{\theta_1}{\theta_2}. \qquad \text{[Carnot cycle of ideal gas]} \tag{8.3}$$

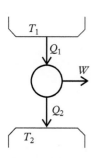

Fig. 8.2 *A generic heat engine operating between two thermal reservoirs. Heat is exchanged only at the reservoirs. If their temperatures are constant and the engine is reversible, then it must by definition be a Carnot engine.*

N.B. we have only derived this result for an ideal gas at this stage, but we will obtain the same formula in a much more general context in Section 8.2.2.

A further insight (and a neat way to derive (8.3)) can be obtained by considering a general reversible cyclic process. For any reversible change in a closed pV system, we have, from the first law,

$$\frac{đQ}{\theta} = \frac{dU}{\theta} + \frac{pdV}{\theta}, \qquad (8.4)$$

and applied to a mole of ideal gas this gives

$$\frac{đQ}{\theta} = \frac{1}{\theta}\frac{dU}{d\theta}d\theta + \frac{R}{V}dV, \qquad (8.5)$$

where we have used Joule's law in the first term, and Boyle's in the second. Note that the first term on the right is a function of temperature only, and the second is a function of volume only. It follows that both give the change in a function of state, and if we integrate around a cycle then both must evaluate to zero. It follows that

$$\oint \frac{đQ}{\theta} = 0. \qquad (8.6)$$

This applies to any reversible cycle for an ideal gas. When applied to the Carnot cycle, it gives equation (8.3).

8.1.1 Heat pumps and refrigerators

The idea behind the definition of 'efficiency' of a heat engine—equation (8.1)—is that it expresses the amount of work obtained when heat Q_1 provided by the hot reservoir is 'used up'. Equation (8.3) shows that for a Carnot cycle of an ideal gas, the efficiency is

$$\eta = 1 - \frac{\theta_2}{\theta_1}. \qquad (8.7)$$

Thus, to get the most efficient use of the supplied heat, we should make the cold reservoir as cold as possible.

A heat engine can also be run in reverse, such that work is supplied by an external agency such as a mains electricity power supply, and the engine operates to extract heat from the cold reservoir and deliver it to the hot reservoir (Figure 8.3). This type of operation is called a 'heat pump.' One example is a refrigerator, in which heat is taken from items inside the refrigerator, at a temperature near 0 °C, and delivered to the surrounding room, at a temperature in the region of 20 °C. Another example is a heat pump used to warm a house in winter, or cool it in summer, also called an air conditioning unit.

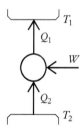

Fig. 8.3 *A generic heat pump. Work supplied by an external agency is used to drive a heat engine in reverse. The result is to extract heat from a cold reservoir and deliver heat to a hotter reservoir.*

96 The second law and entropy

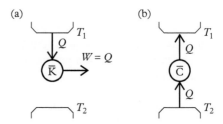

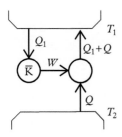

Fig. 8.4 *Two impossible heat engines. (a) A heat engine that violates the Kelvin statement. (b) A heat engine that violates the Clausius statement. (Both conserve energy so there is no violation of the first law.)*

Fig. 8.5 *A Kelvin violator combined with an allowed Carnot engine produces a net effect equivalent to a Clausius violator. This shows that the Clausius statement (forbidding the combined result) implies the Kelvin statement (the assumed $\overline{K}$ must have been impossible).*

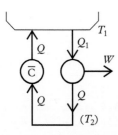

Fig. 8.6 *A Clausius violator combined with an allowed Carnot engine produces a net effect equivalent to a Kelvin violator. This shows that the Kelvin statement (forbidding the combined result) implies the Clausius statement (the assumed $\overline{C}$ must have been impossible).*

Equation (8.3), when applied to a heat pump, can be used to find how much work is required to allow a given quantity of heat Q_1 to arrive at the hot reservoir. We have

$$\frac{Q_1}{W} = \frac{Q_1}{Q_1 - Q_2} > 1.$$

In the application where the pump is used to heat something, such as to heat a house in winter, this equation shows that, as long as the cold reservoir is not too cold and the pump itself does not waste too much energy, then it is more efficient to use a pump than to convert the work directly into heat, as happens in an electric bar heater for example.

8.1.2 Two impossible things (equivalence of Kelvin and Clausius statements)

We now present a neat piece of physical reasoning to show that the Kelvin and Clausius statements each imply the other.

Kelvin's statement of the second law says that the heat engine shown in Figure 8.4a is impossible. Clausius' statement says that the heat engine shown in figure 8.4b is impossible. Let these impossible engines be called $\overline{K}$ and $\overline{C}$, but let us suppose for a moment that one or other of them were possible. Figure 8.5 shows how to combine $\overline{K}$ with an allowable heat engine, such as the Carnot cycle, so as to produce an overall effect identical to $\overline{C}$. But Clausius says such an effect is impossible, so we infer that $\overline{K}$ is impossible after all. We have thus shown that the Clausius statement implies the Kelvin statement. Next we do it the other way around. Figure 8.6 shows how to combine $\overline{C}$ with an allowable heat engine, such as the Carnot cycle, so as to produce an overall effect identical to $\overline{K}$. But Kelvin says such an effect is impossible, so we infer that $\overline{C}$ is impossible after all. Therefore the Kelvin statement implies the Clausius one.

To summarize, the Clausius and Kelvin statements are logically equivalent.

The Kelvin statement of the second law is sometimes expressed, slightly less clearly, 'No process is possible whose sole effect is the complete conversion of heat into work.' The muddle that can arise is that, in a heat engine, one has the energy $Q_1 - Q_2$ which is, one would think fair to say, heat energy, and this amount of energy is 'completely converted into work'. The statement I am recommending in this book avoids this muddle by referring to heat extracted from a *single* reservoir. What is forbidden by the second law is for a system to take in a quantity of heat at a given, fixed, temperature, and give out that amount of work, with no net change in the system. As soon as a second reservoir, at some other temperature, is introduced, we have a situation different to the one mentioned in the Kelvin statement.

8.2 Carnot's theorem and absolute temperature

8.2.1 Carnot's theorem: reversible engines are equally, and the most, efficient

Next we will propose another thing which will turn out to be impossible. The heat engine I in Figure 8.7 is supposed to be more efficient than some reversible engine R. We use the figure to show that such an engine I can be combined with R to produce an overall effect which is ruled out by the Kelvin statement. It follows that the assumption that I is more efficient than R must be invalid, and hence that no heat engine can be more efficient than a reversible heat engine operating between the same reservoirs.

Proof: We suppose the reversible engine is used as a pump, which we know to be possible. When supplied with work W it absorbs heat Q from its cold point and delivers heat $Q + W$ to the reservoir at temperature T_1. We run the engine I at such a rate that it delivers the heat Q required by the reversible engine. Let Q_1 be the heat extracted from the T_1 reservoir by I when Q is delivered at T_2. By hypothesis, the engine I is more efficient than R, therefore

$$\frac{Q_1}{Q} > \frac{Q + W}{Q}$$

(using equation (8.1)). This means we can write

$$Q_1 = Q + W + w,$$

where $w > 0$ is some positive energy. But, by the first law, the work output of I must be $Q_1 - Q = W + w$, therefore it delivers not only the work W required to run R, but also some spare work w which can be passed to other systems. The net result is a process which extracts heat from a reservoir at a single temperature T_1 and delivers an equivalent amount of work; the very thing expressly forbidden by the Kelvin statement of the second law. The reasoning is valid, therefore one of the premises must have been untrue. The only candidate is the premise that I is more efficient than R. It follows that no heat engine can be more efficient than a reversible one operating between the same reservoirs.

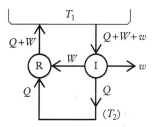

Fig. 8.7 *Proof of Carnot's theorem. R is a physically possible reversible engine, I is an engine which is supposed to be more efficient than R. We prove that the result is impossible (see text).*

Next let us make a lesser claim, merely that the efficiency of I is not known, but we know I to be itself reversible. Then the same argument, with the roles of R and I swapped, can be used to show that R cannot be more efficient than I. We have already established that I cannot be more efficient than R. It follows that the efficiencies must be equal, and we deduce that all reversible engines have the same efficiency when operating between a given pair of reservoirs.

The two results together constitute *Carnot's theorem*.

All reversible heat engines operating between given temperatures are equally efficient, and more efficient that non-reversible ones, no matter what the engines' internal construction or physical parameters may be (whether pressure, or magnetic fields, or whatever).

This is another of the remarkable universal statements arising from the basic principles of thermodynamics. Sections 8.2.2 and 8.2.3 will show that Carnot's theorem is a powerful tool in the theoretical foundations of thermal science.[1] It is also of practical importance, showing that once any given reversible engine has been analysed for some pair of temperature reservoirs, then the available maximum efficiency for those reservoirs is known: there is no need to analyse further designs of heat engine. It suffices to pick an ideal gas Carnot cycle, for example, and then one finds equation (8.7). Furthermore, one knows immediately that irreversibility in a heat engine can only make it less efficient.

8.2.2 Existence of an absolute temperature measure

Carnot's theorem allows us to define a scale of temperature in a new way, which does not appeal to any particular physical system. This scale is called the *absolute* or *thermodynamic* scale of temperature. At first meeting, the definition can seem rather abstract, so I am going to prepare you for it by discussing something more familiar, namely *mass*.

Consider Newton's second law, 'the rate of change of momentum is equal to the applied force'. This statement is limited, in that it talks about 'force' without defining what is meant. We can avoid this weakness as follows. Define 'force' to mean that which causes the momentum of a body to change (but we say nothing yet about the size of the force). Consider various physical systems which can supply a force, e.g. an elastic band (note, I don't assume Hooke's law) or a rocket motor. We argue that when a given force-providing system is in the same conditions then it must provide the same force, no matter what object it may be pushing or pulling. For example, a given elastic band of given length and temperature pulls by a given amount. (If this were not so, we could find physical situations where it would lead to impossible behaviour such as a self-accelerating closed system.)

Now imagine we have two bodies whose 'quantity of matter' is different, and we would like to have a sensible definition of 'quantity of matter'. We proceed as follows. We attach the force-providing system (elastic band) to body 1, and

---

[1] Comment: if one *assumes* Carnot's theorem, then one can derive Kelvin's and Clausius's statements of the second law (try it), so you could say Carnot's theorem is yet another way of stating the second law. However for my taste it makes better sense to regard Carnot's theorem as derived from the second law, rather than the law itself.

maintain the force-providing system in fixed conditions (e.g. by pulling on the other end of the band to keep its length constant). The body will accelerate. We measure its acceleration, a_1 (this can be measured without ambiguity because it only involves distance and time, not force). Now attach the force-providing system to the other body, 2, and repeat the experiment, making sure the force-providing system is in the same conditions in the two experiments. Thus obtain acceleration a_2. Now repeat the experiments with a number of quite different force-providing systems (e.g. rocket motor, attracting capacitor plates, surface tension, etc.). It is found that in such experiments, the ratio a_1/a_2 is independent of the force-providing system and of the speeds involved (as long as no friction or viscosity is present and the speeds do not approach the speed of light). We deduce that this ratio must be *a property of the bodies 1 and 2 alone*. This observation allows us to define inertial mass: we define inertial mass M to be such that the inertial masses of two bodies 1 and 2 are in the ratio of their accelerations when the same force is applied to each:

$$\frac{M_1}{M_2} \equiv \frac{a_2}{a_1}.$$

We can make such a definition because we have just proved that the acceleration ratio does not depend on other things, only on the bodies in question. The equation allows all masses to be defined by expressing them as a multiple of some given mass which can be taken as the unit of mass. The choice of the unit mass is arbitrary. In the SI system of units, the unit mass (1 kilogram) is defined as being that of the International Prototype Kilogram which is a platinum–iridium cylinder stored in a vault at the Bureau International des Poids et Mesures, Sèvres, France.

Note that the above argument does not require the size of any force to be known. Once it has been used to arrive at a definition of mass, we can then define momentum and then obtain force as the rate of change of momentum.

Now we present the definition of absolute temperature.

We observe, from Carnot's theorem, that for any given pair of reservoirs, the ratio Q_1/Q_2 of heats exchanged by a reversible heat engine is *universal*: it is independent of the nature of the heat engine (saving only that it must be reversible), and furthermore does not depend on anything about each reservoir except its 'degree of hotness'. That is, one could replace the upper reservoir by another (of a completely different physical makeup) in thermal equilibrium with it, or the lower reservoir by another one in equilibrium with it, and Q_1/Q_2 would not change. We conclude that we may as well use Q_1/Q_2 to *define* temperature, just as, in the mechanics argument, a_1/a_2 was used to define mass.

Absolute temperature is defined as follows. First we pick some well-defined physical situation and assign it a temperature T_0, to be called 'one unit of temperature'. Then:

> *The absolute temperature T of a body is defined to be unit temperature T_0 multiplied by the ratio Q/Q_0 of heat Q extracted at T to heat Q_0 delivered*

at T_0, if a reversible heat engine were to be momentarily operated between the body and a body at unit temperature.

Carnot's theorem guarantees that this definition will work correctly, because it guarantees that it is not necessary to mention any further properties of the systems in question: the definition is already universal as it stands. Carnot's theorem also ensures that bodies in mutual thermal equilibrium are assigned the same T. The definition has, furthermore, the desirable feature that hotter bodies (those that tend to give off heat) are assigned higher T. Thus it covers all the properties of temperature, so it is a complete definition.

The definition involves a reservoir at unit temperature. More generally, once we have the definition of T, we can derive the formula:

Ratio of heats for a reversible engine

$$\frac{Q_1}{Q_2} = \frac{T_1}{T_2}. \tag{8.8}$$

This is the ratio of heats exchanged by a reversible engine acting between reservoirs at temperatures T_1 and T_2. The result is proved by use of the process illustrated in Figure 8.8. A reversible heat engine operating between T_1 and T_2 must be equally efficient, by Carnot's theorem, as a pair of reversible engines, one which extracts heat Q_1 at T_1 and delivers Q_0 at T_0, and the other which extracts Q_0 at T_0 and delivers Q_2 at T_2. By the definition of absolute temperature,

$$\frac{Q_1}{Q_0} = \frac{T_1}{T_0} \quad \text{and} \quad \frac{Q_0}{Q_2} = \frac{T_0}{T_2}$$

and (8.8) follows immediately. It is not necessary for T_0 to be intermediate between the other temperatures for the result to be valid, since the engines can be used equally as heat pumps where necessary.

Equation (8.8) has been called by Feynman the 'heart of thermodynamics'. It is the equation which plays a comparable role in thermodynamics to that of Newton's second law in classical mechanics.

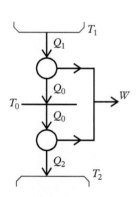

Fig. 8.8 *Proof of the heat ratio, equation (8.8) (see text).*

The efficiency of a heat engine (Eq. (8.1)) can now be expressed, for the case of a *reversible* engine, in terms of the absolute temperatures:

$$\eta = \frac{W}{Q_1} = 1 - \frac{Q_2}{Q_1} = 1 - \frac{T_2}{T_1}.$$

In the SI system of units, the physical situation used to define T_0 is an isolated container filled with water, ice, and steam (water vapour) all in equilibrium. This is called the *triple point* of water. It is found that the triple point occurs at a unique temperature (see Chapter 22), as long as the system size is large enough so that surface effects are negligible. T_0 is then equal to '1 triple point unit'. We are at liberty to define other temperature units in terms of this one. In the SI system the standard unit of temperature is the Kelvin, defined to be equal to (1/273.15) triple point units. In other words, the triple point of water occurs, by definition, at the temperature 273.15 Kelvin.

8.2.2.1 Absolute and ideal gas temperature scales

Comparing (8.8) with (8.3), we have

$$\frac{\theta_1}{\theta_2} = \frac{T_1}{T_2},$$

where θ is empirical temperature on the ideal gas scale, and T is absolute temperature. We deduce that $\theta \propto T$. As long as the appropriate proportionality constant is chosen, we have proved that the two scales of temperature agree.

8.2.3 Hot heat is more valuable than cold heat

An essential insight into thermodynamics is immediately implied by the 'heart' of the subject, Equation (8.8). 'Hot heat is more valuable than cold heat' means the following. If you supply some heat to a heat engine in order to allow it to do some work, then if a *given* quantity of heat (say, 1 Joule) is supplied at a high temperature T_1 (by isothermal transfer between a heat reservoir and part of the machine which is prepared at that temperature), then it allows the machine to do *more* work than if the *same* quantity of heat were supplied at a lower temperature. This is because the machine has to 'spit out' heat energy $Q_2 = Q_1 T_2/T_1$. Although we consider heat engines in order to make the point, of course the insight is a general one: *heat energy being delivered by a system at a high temperature is more valuable (i.e. can be used to drive a greater variety of processes) than the same amount of heat energy at a low temperature.*

Example 8.1

Two energy companies are offering heat for sale. The first company offers the price 10 euros per kilojoule of heat, and delivers the heat from a hotplate maintained at 500 K. The second company is offering heat at 8 euros per kilojoule, and delivers from a hotplate maintained at 400 K. The only practical heat sink in town is the river, temperature 280 K. Which company is offering the better deal?

Solution.
We assume we are already in possession of a reversible heat engine that can reach all of the temperatures mentioned without melting or being damaged in some other way. It may be that we have a special interest in attaining

> 500 K, but this does not necessarily imply the first company is the one we want, because we could use the cheaper energy from the second company to drive a heat pump. Therefore, the appropriate measure of value is the amount of work we can obtain per euro expenditure. The efficiency of a reversible engine operated between the hotter plate and the river is $\eta_1 = 1 - 280/500 = 0.44$, and that of a reversible engine between the colder plate and the river is $\eta_2 = 1 - 280/400 = 0.3$. Therefore using the first company we obtain 440 J of work for an expenditure of 10 euros, giving 44 J per euro. Using the second company we obtain 300 J of work for an expenditure of 8 euros, giving 37.5 J per euro. Clearly the first company offers the better deal.

We already know enough to make the claim that hot heat (i.e. heat delivered at high temperature) is more valuable than cold heat in general, but it might be objected that, in principle at least, the lower temperature reservoir of a Carnot engine might be arbitrarily close to absolute zero temperature, and therefore in principle the efficiency can approach 1 so all heat can be converted to work with approaching unit efficiency. In fact this is not true, because heat capacities tend to zero at $T = 0$, as we will explain more fully in Chapter 23. Consequently, for a Carnot engine operating near $T = 0$, both temperatures T_1 and T_2 are small, and in the limit where the lower temperature approaches zero, so does the higher.

8.3 Clausius' theorem and entropy

The next treasure which Carnot's theorem allows us to unearth is *entropy*.

It is a common experience that if you take a lump of some substance, such as cake mixture or a sponge, and knead it, it gets warmer. You are doing work on the substance, causing its internal energy to rise, which tends to increase its temperature, and it gives off heat. If after some kneading, you were to return the substance to its initial state, you would find that overall, you had done some work, and you (or the environment) had received some heat. Thus 'work is converted into heat.' This is allowed by the second law. However, what if, in the course of your manipulations, you found the material tended to extract heat from your hands, and become more springy, thus pushing back on you and doing work? This would be a very odd experience. Careful consideration of this type of physical process leads us to an important theorem.

We are going to consider a system being transported through a sequence of state changes. We want to keep track of the heat flow, so we arrange that *all* the heat Q exchanged by the system is supplied by or given to a reversible heat engine $\mathcal{C}$. $\mathcal{C}$ operates between some part of the system and a reservoir at a fixed temperature T_0; see Figure 8.9. During the sequence of state changes the system may sometimes absorb heat, and sometimes give off heat, and it may or may not be hotter than T_0. In short, it is an arbitrary sequence. $\mathcal{C}$ can either supply or take away heat, as required, since it can act as a heat pump as well as a heat engine.

Now imagine that, under these conditions, the system goes through a series of transformations and finally returns to its initial state. The transformations might be of any kind, and need not be reversible or even quasistatic. However, we can arrange that everything related to the operation of $\mathcal{C}$ is reversible. This is done by allowing $\mathcal{C}$ to interact with only one part of the system, whose temperature T is well defined during any heat exchange with $\mathcal{C}$. This part is large enough to allow heat to flow through it to the rest of the system, but small enough compared to the rest of the system that we can consider essentially any change in the system state, including cases in which the system as a whole has no well-defined temperature. Thus everything about the operation of the heat engine is reversible, though the transformations of the whole system need not be.[2]

Note that since T can change during the cycle, the system (inside the dashed box in Figure 8.9) can itself function as a heat engine, and therefore it can happen that W (the net work produced by the system over a cycle) is either positive or negative.

Let the system be called $\mathcal{A}$. Any small change in the internal energy of $\mathcal{A}$ is, by the first law, $dU = đQ + đW$. Since $đQ$ is the heat *gained* by $\mathcal{A}$, it must be the heat obtained *from* the heat engine $\mathcal{C}$, and therefore the heat extracted by the heat engine *from* the reservoir is $đQ_0 = đQ\, T_0/T$. (It is crucial to get the signs right in this argument!) We can apply the heat ratio (8.8) because we have assumed the heat flow between $\mathcal{A}$ and $\mathcal{C}$ is reversible, although whatever is happening elsewhere in $\mathcal{A}$ may or may not be reversible.

There are two heat quantities in which we might take an interest. The total heat flow into the system $\mathcal{A}$, as it goes through the sequence of transformations and returns to its initial state, is

$$\text{Net heat into system } Q = \oint đQ \qquad (8.9)$$

and the total heat taken from the reservoir is

$$\text{Net heat extracted from reservoir } Q_0 = \oint đQ_0 = T_0 \oint \frac{1}{T} đQ. \qquad (8.10)$$

Now for the clever part. It might look as though we know little about the value of either of these two integrals in general. However, in fact we can guarantee the sign of one of them: the second. How? By looking at the *combined system of $\mathcal{A}$ plus the heat engine $\mathcal{C}$*. This combined system $\mathcal{AC}$ has the special property that it only exchanges heat with its surroundings at a single temperature, namely T_0—see Figure 8.10. Therefore the Kelvin statement of the second law can be applied to it. First suppose Q_0 is negative: this would mean that overall some heat came out of $\mathcal{AC}$. By energy conservation, this energy must have come from work done on the system (since we are dealing with a closed cycle, there is no net change in

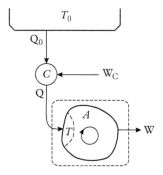

Fig. 8.9 Scenario considered in Clausius's argument. The system is indicated by the amorphous blob $\mathcal{A}$.

[2] Some presentations of Clausius' theorem use a different approach in which a mesh of adiabats and isotherms is superimposed on the state diagram. Such an approach lacks generality for two reasons. First it appears to handle only quasistatic changes, and secondly it assumes that adiabatic processes trace out a unique set of surfaces in state space, which amounts to assuming the existence of an entropy state function at the outset.

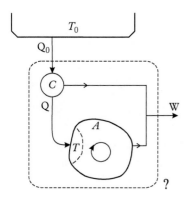

Fig. 8.10 *Clausius's theorem. By considering the system and the heat engine together as a single composite system (shown by the dashed box), we can observe that this diagram, with the total heat flow Q_0 in the direction of the arrow, shows an impossible heat engine: it violates the Kelvin statement. Therefore $Q_0 < 0$.*

the internal energy). This is perfectly possible. However, if Q_0 were positive we would have an impossible situation: it would mean that overall heat from a single reservoir went into $\mathcal{AC}$, and therefore (from energy conservation) the equivalent amount of energy came out as work, with no other net change: the very thing expressly forbidden by the Kelvin statement.

Notice the crucial role here of the 'single reservoir' part of the Kelvin statement: the integral $\oint \dbar Q$ can be of either sign, because the heat exchange at $\mathcal{A}$ can be at a variety of temperatures, allowing $\mathcal{A}$ to act as a heat engine, producing work while it goes through a cycle. This is not true of the combination of $\mathcal{A}$ and $\mathcal{C}$, because they only exchange heat with the rest of the world at a single temperature.

The conclusion is *Clausius' theorem*, first part:

$$\oint \frac{1}{T} \dbar Q \leq 0. \quad \text{[for any cycle]} \qquad (8.11)$$

(where we have used that T_0 has to be positive so we can drop it from in front of the integral).

This is always true, for any cycle of transformations of any thermodynamic system. Next, suppose that the cycle gone through by the system were actually a reversible cycle. In this case we may consider the same cycle run in reverse. The same argument would yield

$$\oint_R \frac{1}{T} \dbar Q_R \leq 0, \qquad (8.12)$$

in which $\dbar Q_R = -\dbar Q$, since we have just considered the reverse of the first process. But now (8.11) and (8.12) can only both be true if

$$\oint_R \frac{1}{T} \dbar Q_R = 0. \quad \text{[for a reversible cycle]} \qquad (8.13)$$

Now we have proved the whole of Clausius's theorem, which may be stated:

> **Clausius's theorem** The integral $\oint dQ/T \leq 0$ for any closed cycle, where equality holds if and only if the cycle is reversible.

In this theorem, the symbol T stands for the temperature at which heat was delivered to the system; it does not need to refer to the temperature of the system as a whole, which might not even have a well-defined temperature.

The result for a reversible cycle is extremely useful: we have a quantity, $\dbar Q_R/T$, which when integrated around a closed loop of equilibrium states, always gives zero, independent of the path. Using the mathematical proof given in Section 5.2 (Eq. (5.29)), it immediately follows that $\dbar Q_R/T$ is a small change in a function of state, which we will label S:

$$dS = \frac{đQ_R}{T}. \tag{8.14}$$

S is not any of the functions of state we have met in the discussion of the zeroth and first laws. It is a new one whose presence we have deduced from the first and second laws. Like internal energy, it is present in any thermodynamic system. When heat is transferred reversibly, it changes by the amount of heat transferred divided by the temperature. It therefore has dimensions of energy divided by temperature. It is the single most important quantity in thermodynamics. It is called *entropy*.

8.4 The first and second laws together

Clausius's theorem has shown us that absolute temperature is an *integrating factor* for heat (recall equation (5.31)). Putting this into the first law, $dU = đQ + đW$, we arrive at the equation which was previewed in (7.11), and which we re-state here:

Fundamental relation for a closed system

$$dU = TdS - pdV. \tag{8.15}$$

This is THE fundamental relation of the subject. All the arguments in the rest of this book take this as their starting point.

In order to derive (8.15) we considered the special case of a reversible process, where it is possible to identify $đQ_R = TdS$ and $đW_R = -pdV$. However, in the result there is no mention of either heat or work: it is purely a relationship among functions of state. If two nearby states are separated by dS in entropy and dV in volume, then their separation in internal energy must be $TdS - pdV$: the formula does not 'care' whether the system is made to move between those two states. It follows that the formula applies to the changes accompanying any process, whether reversible or irreversible.[3]

In the irreversible case, one has $đW = -pdV + |d\epsilon|$, where $d\epsilon$ is the friction term (cf. Eq. 7.9), so

$$dU = đQ - pdV + |d\epsilon| = TdS - pdV. \tag{8.16}$$

It follows that

$$đQ = TdS - |d\epsilon|. \tag{8.17}$$

This result is often expressed in the form

$$đQ \leq TdS. \tag{8.18}$$

---
[3] Here is another way to see this. It is always possible to find a reversible route between any pair of states; it is obvious that the formula applies to the results of such a reversible change; but since we are dealing with functions of state the result must apply equally, no matter what path was followed between the same two states.

I like to think of it the other way around:

$$dS \geq \frac{\text{đ}Q}{T}, \tag{8.19}$$

which can be read as 'the system's entropy increases by whatever entropy is brought in from outside by heat transfer, *plus* some positive amount spontaneously generated within the system'.

8.5 Summary

This chapter has dealt with several central concepts and results. They can be summarized as follows.

$$\{\text{heat engine, second law}\} \to \text{Carnot theorem: } \eta \leq \eta_R$$
$$\to \text{absolute temperature}:$$
$$\frac{Q_1}{Q_2} = \frac{T_1}{T_2}.$$
$$\{\text{heat ratio, Kelvin statement}\} \to \text{Clausius theorem:}$$
$$\begin{cases} \oint \frac{\text{đ}Q}{T} \leq 0 \\ \oint \frac{\text{đ}Q_R}{T} = 0 \end{cases}$$
$$\to \exists \text{ entropy!}, \, dS = \frac{\text{đ}Q_R}{T}.$$

Finally, since $\text{đ}Q_R = T dS$, the principal heat capacities can always be written in terms of entropy:

$$C_V = T \left.\frac{\partial S}{\partial T}\right|_V, \quad C_p = T \left.\frac{\partial S}{\partial T}\right|_p. \tag{8.20}$$

....

EXERCISES

(8.1) A building is maintained at a temperature T by means of an ideal heat pump which uses a river at temperature T_0 as a source of heat. The heat pump consumes power W, and the building loses heat to its surroundings at a rate $\alpha(T - T_0)$. Show that T is given by

$$T = T_0 + \frac{W}{2\alpha}\left(1 + \sqrt{1 + 4\alpha T_0/W}\right).$$

(8.2) A possible ideal-gas cycle operates as follows.

(i) from an initial state (p_1, V_1) the gas is cooled at constant pressure to (p_1, V_2).

(ii) the gas is heated at constant volume to (p_2, V_2).

(iii) the gas expands adiabatically back to (p_1, V_1).

Assuming constant heat capacities, show that the thermal efficiency is

$$1 - \gamma \frac{(V_1/V_2) - 1}{(p_2/p_1) - 1}.$$

(8.3) In an isothermal expansion of an ideal gas, an amount of work $W = \int p dV$ is done by the system on its surroundings. The internal energy U of the gas does not change, because the system draws in heat Q from a reservoir, and for an isothermal process in an ideal gas, U remains constant. Since $\Delta U = 0$ we have $Q = W$. So in this process, heat has been drawn from a single reservoir, and an equivalent amount of work has been done, 'turning' the heat totally into work. Is this a violation of the Kelvin statement of the second law?

(8.4) *Perpetual motion.* A machine that could in principle operate for ever, in spite of the presence of friction or exhaust heat, is said to be capable of 'perpetual motion', and is called a *perpetuum mobile* from the Latin. This concept no longer features much in the study of thermal physics, but it played a useful role historically in getting clarity about what processes are and are not possible. Two types of *perpetuum mobile* may be distinguished. A *perpetual motion machine of the first kind* produces more energy that it uses. A *perpetual motion of the second kind* produces exactly as much energy as it uses, and keeps running indefinitely by recycling all the output energy, whether it is produced in the form of work or heat.

(i) Explain which of the laws of thermodynamics are broken by such machines. (ii) Show how a *perpetuum mobile* of the second kind could be constructed by combining two reversible heat engines of different efficiencies (if they could be found).

9 Understanding entropy

9.1 Examples 109
9.2 But what is it? 113
9.3 Gibbs' paradox 116
9.4 Specific heat anomalies 120
9.5 Maxwell's daemon 122
9.6 The principle of detailed balance 127
9.7 Adiabatic surfaces 128
9.8 Irreversibility in the universe 131
Exercises 133

The Hammers

Noise of hammers once I heard,
Many hammers, busy hammers,
Beating, shaping, night and day,
Shaping, beating dust and clay
To a palace; saw it reared;
Saw the hammers laid away.

And I listened, and I heard
Hammers beating, night and day,
In the palace newly reared,
Beating it to dust and clay:
Other hammers, muffled hammers,
Silent hammers of decay.

Ralph Hodgson

To understand any area of science well, one must become sufficiently familiar with it such that the basic concepts and arguments become natural and intuitive. Concepts such as energy and momentum are fairly easy to grasp from our everyday experience, but entropy is much harder. The idea of entropy has gained some foothold in everyday language, but usually in an unclear and sometimes misleading form. Entropy is often said to be associated with 'disorder', but this is misleading because it is only true if one talks of 'disorder' in a rather technical sense. For example, the equilibrium state of an isolated container of gas is the one of maximum entropy, and in this state the gas is very calm and smooth, with uniform pressure and temperature; or in everyday language, very 'orderly'. This might be contrasted with gas sloshing around a container in a 'disorderly' fashion. In the latter case, if the container is left alone (i.e. isolated: no heat exchange or work), then the irregular sloshing and vibrating of the gas will gradually die away. This is an irreversible process in which entropy *increases*, until the gas is uniform and calm.

The sense in which entropy is a measure of disorder may be illustrated by a smoke ring. The ring is a form of order (or, better, *structure*) in the position and motion of the smoke particles of which it is constituted. When the ring dissipates into the surrounding air this structure is lost and entropy increases.

It is generally unhelpful, at least when learning the subject, to speak of entropy when everyday macroscopic objects such as socks or playing cards get separated or shuffled. A pack of cards can be restored to order by reversible processes without any heat transfer. In this case the information about the ordering of the cards is acquired by the sorting machine, and one can argue that entropy is then moved from the cards to the machine. However, this type of argument involves some subtlety and is not a good way in to the subject. In short, the rearrangement of macroscopic objects in space is not a good model to invoke when first learning about entropy.

Clearly, from the definition (8.14), entropy is, or can be, something to do with heat transfer. In reversible heat transfer, the two bodies are at the same temperature (apart from a tiny difference to make the direction of quasistatic heat flow well defined), so one can immediately see that any entropy lost by one body is gained by the other. This makes it legitimate to think of entropy as 'flowing' from one body to another. In reversible processes, entropy is a conserved quantity, and this conservation is achieved locally. That is, it is not true that an entropy decrease in one place is balanced by an increase in some other unconnected place. Rather, if the amount of entropy is falling somewhere, then this is because entropy is flowing away from that place, and towards some other place where the amount of entropy is increasing. This is discussed more fully in Chapter 28.

In irreversible processes, entropy is created: it is not a conserved quantity. This follows from Clausius's theorem (8.11) and the fact that entropy is a function of state, satisfying $\oint dS = 0$. Equation (8.11) says the total entropy entering the system from outside is less than zero for an irreversible closed cycle. To balance the books it follows that entropy must have been spontaneously generated within the system. Therefore the entropy change dS in any process involving heat transfer is equal to the entropy arriving or leaving through the borders, plus the entropy created within the system. It follows that for a thermally isolated system, where no heat flows in or out, *the entropy of the system can only increase or stay constant*, not decrease:

$$dS \geq 0 \quad \text{[For a thermally isolated system.]} \quad (9.1)$$

9.1 Examples

To gain some familiarity, let's calculate a few examples of entropy change. This is done by identifying a reversible process which could take a system between given initial and final states, and integrating equation (8.14).

The latent heat of melting of ice is 333 kJ/kg. This means that it requires 333 kilojoules of heat to melt a one kilogram block of ice. Consider such a block held in a plastic bag whose temperature is maintained very slightly above $0\,°C$ while the ice melts. Since all the heat enters at 273 K, and the heat exchange is reversible, the entropy of the contents of the bag increases by $333000/273 = 1220\,\text{JK}^{-1}$.

Something else, such as the air in the room, has lost this amount of entropy during the heat exchange, and may have gained entropy in other ways.

A brick is tossed into a lake. By how much does the entropy of the brick and the lake change? Suppose the brick and the lake are at the same temperature of 290 K, and the brick (of mass 2 kg) falls from a height of 2 m. It therefore hits the lake with kinetic energy $mgh = 39$ J. It is quickly slowed down by the water, to a speed of about 10 cm/s, so that its final kinetic energy is negligible. The energy goes into the lake, initially as water flow and wave motion, but is soon dissipated entirely into random motion. Since this all takes place at 290 K, the entropy of the lake has increased by at least $\Delta Q/T = 0.13$ J/K. The entropy of the brick has not changed at all, since no heat flows in or out of it, and no other significant internal change takes place.

Now suppose a hot potato is tossed into a lake. We shall assume the potato is initially at a temperature of 350 K, and the kinetic energy of the potato is negligible compared to the heat it exchanges with the lake. The heat exchange process is irreversible, because it takes place across a finite temperature difference (of 350 − 290 = 60 K in the first instance). Therefore we have to apply the formula $dS = đQ_R/T$ with care. We argue that the potato loses just as much entropy as it would lose if it were cooled from 350 K to 290 K reversibly. This is

$$\Delta S = \int_{350}^{290} \frac{đQ}{T} = \int_{350}^{290} \frac{1}{T}\frac{đQ}{dT}dT = C\ln\frac{290}{350}, \qquad (9.2)$$

where in the final step we have assumed that the heat capacity of the potato, C, is independent of temperature. This is a good approximation for the temperatures involved here. Taking $C = 800$ J/K (a reasonable estimate for a 200 g potato) the entropy change is a decrease by 158 J/K. Meanwhile, the lake acquires the $800 \times 60 = 48,000$ J of heat at 290 K, so its entropy increases by $48,000/290 = 166$ J/K. This more than makes up for the loss of entropy from the potato; the net effect is an increase in the entropy of the universe by 8 J/K.

A power station needs to get rid of excess entropy, while minimizing its loss of energy. How can it do this? Answer: emit the entropy in the form of heat at as low a temperature as possible. This is why engineers go to the trouble of providing power stations with large ugly cooling towers. The entropy flow in a perfect heat engine is illustrated in Figure 9.1.

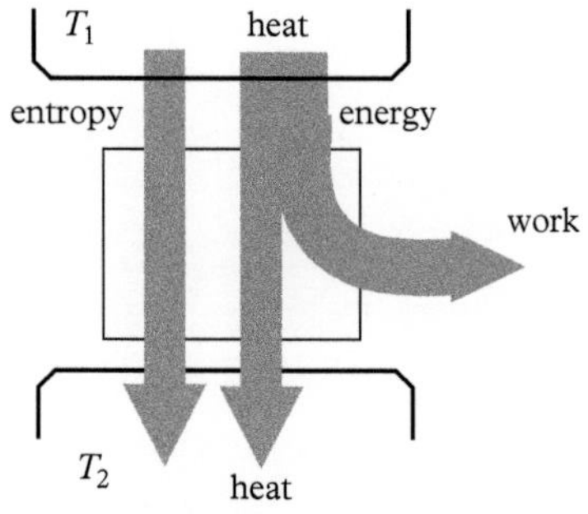

All the above examples involve the simple but important idea presented in Figure 9.2. When two systems can exchange heat, then some energy ΔQ moves out of one and into the other. One system therefore loses entropy, the other gains entropy. The total entropy of the pair increases whenever the gainer acquires more entropy than the loser loses. The system acquiring ΔQ thereby receives entropy $\Delta S_1 = \Delta Q/T_1$ and the system emitting ΔQ thereby loses entropy $\Delta S_2 = \Delta Q/T_2$. The total entropy of the pair of systems therefore increases whenever

$$\frac{\Delta Q}{T_1} > \frac{\Delta Q}{T_2}.$$

Fig. 9.1 *The flow of energy and entropy in a perfect heat engine. This diagram summarizes the energy and entropy exchanges going on in a cycle of a Carnot engine. Heat energy extracted from the upper reservoir is shared out between the work output of the engine, and the heat ejected into the lower reservoir. All the entropy flowing from the upper reservoir is passed to the lower. Although the diagram uses separate arrows to indicate energy and entropy flows, there is just one process at each reservoir, in which entropy and energy are given out or taken in together.*

In other words, the system with lower temperature acquires the energy. In equilibrium, the inequality becomes an equality: this occurs when the temperatures are equal.

9.1.1 Entropy content

So far we have discussed entropy changes. One can also ask the question, 'what is the total entropy content?' This is a perfectly well-posed question. Once we have specified our thermodynamic system, then it has many well-defined properties such as volume, mass, internal energy, and so on, and also entropy. Thermodynamics on its own cannot tell you the volume of your system: you must measure it, or derive it from other measurements. Similarly, thermodynamics on its own cannot tell you how much entropy a given system has. But it can tell you how to deduce the total entropy content from measurements you might make.

We use a closed simple compressible system to convey the essential idea. First prepare the system at fixed volume and suppose for a moment that it could also be prepared at absolute zero temperature. Then provide known amounts of energy to the system in the form of heat, being sure that the heat flow is reversible at every stage. After this has been going on for a while the system temperature will be non-zero, and the total entropy that has entered the system is

$$S(T_f) - S(0) = \int_0^{T_f} \frac{\text{d}Q_{\text{rev}}}{T}. \tag{9.3}$$

Since the whole process is reversible, the system entropy has not changed for any other reason, so this is the entropy of the system in its final state. This follows because we can invoke the third law in order to claim that $S(0) = 0$, i.e. the system entropy starts from zero when the temperature is at absolute zero.

Of course, the method just proposed is impractical, especially in the requirement to start from absolute zero temperature. In practice one would instead perform a sequence of experiments, accompanied by theoretical studies, to establish the heat capacity of the system as a function of temperature as accurately as possible. If one has reason to believe that no further effects will intervene below the temperatures accessible to experiments, then one can extrapolate the known C_V down to absolute zero, or trust a theoretical model, and then use

$$S(V, T_f) = \int_0^{T_f} \frac{C_V(V, T)}{T} \text{d}T + \sum_i \frac{L_i}{T_i}, \tag{9.4}$$

where it is understood that the volume is held constant in the integral, and the second term is a sum over all stages encountered when there is a finite heat supplied in (9.3) without a change in temperature. This happens during a *phase change*, such as melting or evaporation, and the heat supplied is called a *latent heat*. Note that in practice heat capacities tend to zero at least as fast as T at low temperatures, so the integral in (9.4) is finite. However, until one has discovered what the

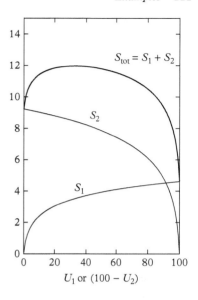

Fig. 9.2 *Entropy changes when a pair of otherwise isolated systems exchange heat. In the example shown, the total energy available to be shared between the systems is 100 units. The maximum entropy occurs when the gradients of the S_1 and S_2 lines are equal and opposite.*

Table 9.1 *Molar entropy of some gases at 1 atm, 298 K, in units $J K^{-1} mol^{-1}$.*

H	115	H_2	130.5
He	126	N_2	192
Ne	146	CO	197
Ar	155	F_2	203
Kr	164	O_2	205
Xe	170	Cl_2	223
CH_4	186	C_2H_6	229
$H_2O(g)$	187	n-C_3H_8	270
CO_2	213	n-C_4H_{10}	310

Table 9.2 *Entropies of some solid elements at 298 K.*

C (diamond)	2.5 $JK^{-1}mol^{-1}$
C (graphite)	5.7
Fe	27.1
Pb	51.0
Na	64.9
S (rhombic)	32.0
Si	18.9
W	33.5

heat capacity does at low temperature, one cannot perform the integral, because the low temperature part may contribute a substantial amount to the total.

There are also ways to obtain the total entropy more directly, without recourse to low temperature measurements and integration. This is discussed in Chapter 23.

The total entropy content of a system is sometimes called *absolute entropy* in order to distinguish it from the entropy differences that we are mostly concerned with in thermal physics. Once the absolute entropy of a few example systems has been established, one can find further results by studying a large range of physical phenomena, especially chemical reactions, which involve entropy changes as heat is absorbed or emitted. This enables one to fix the entropy difference between one system and another very accurately, and thus construct tables of absolute entropy by bringing all the information together. Tables 9.1 and 9.2 give some examples.

9.1.2 Entropy production and entropy flow

Now consider heat flow through a piece of material of finite thermal conductivity (Figure 9.3). For example, suppose two metal blocks at different temperatures are separated by a piece of plastic. We assume the heat flow through the plastic is slow enough, and the thermal conductivity of metal high enough, that each block has a well-defined temperature. The temperature gradient then appears wholly across the plastic. When a small amount of heat energy $đQ$ leaves the hotter block, that block's entropy falls by $dS_h = đQ/T_h$. When this *same* amount of heat energy arrives at the colder block, the latter's entropy increases by $dS_c = đQ/T_c$. This increase is by more than dS_h, since $T_c < T_h$, so we have again a net increase of entropy in the universe. This example illustrates the fact that heat always flows in such a direction as to increase entropy overall.

If the temperatures of the two ends of the piece of plastic are constant and the heat flow has reached a steady state, then the rate of entropy production inside the plastic is

$$\frac{dS}{dt} = \frac{J}{T_c} - \frac{J}{T_h}, \tag{9.5}$$

where J is the heat current, i.e. the heat energy flowing through the plastic in unit time. Let $T = T_c$ and $\Delta T = T_h - T_c$, then if $\Delta T \ll T_h$, we have

$$\frac{dS}{dt} = \frac{J \Delta T}{T^2}.$$

The quantity J/T is an entropy current. It is the entropy carried in to the plastic by the heat flow at the hot end, and also the entropy leaving the plastic in the heat flow out of the cold end. Therefore, when heat flows through a heat conducting material in steady state, we find that entropy flows through the material at a rate $I_S = J/T$ and entropy is generated within the material at a rate

$$\frac{dS}{dt} = I_S \frac{\Delta T}{T}. \tag{9.6}$$

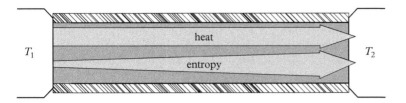

Fig. 9.3 *The flow of heat and entropy in a rod made of material of finite thermal conductivity. We consider the case where there is no heat loss out of the sides of the rod, and no work is done, so that all the heat entering one end leaves from the other end (conservation of energy). This heat corresponds to a larger amount of entropy delivered to the low temperature reservoir than was extracted from the high temperature reservoir. This is indicated by the width of the entropy arrow. It follows that entropy is being continuously generated within the rod. In contrast to Figure 9.1, we can't reverse the directions of the arrows because that would give a net entropy reducing process.*

When an electric current I flows through a resistance R, the electrical power $\Delta V I$ is dissipated as heat, where ΔV is the electric potential difference between the ends of the resistor. Therefore if the resistor is in steady state at temperature T, then entropy is being generated within it and passed to the surroundings at the rate

$$\frac{dS}{dt} = \frac{\Delta V I}{T} = I \frac{\Delta V}{T}.$$

This result has a similar structure to equation (9.6).

So far we have assumed that the electrical resistor had no temperature gradient along its length, so there is no heat flow along the resistor, only out of its sides. An interesting case is that of a wire that conducts both a heat current and an electrical current. Then the arguments given previously must be combined, and we find that the total rate of entropy production inside the wire is given by

$$\frac{dS}{dt} = I_S \frac{\Delta T}{T} + I \frac{\Delta V}{T}. \tag{9.7}$$

This result will be developed further in Chapter 28 and applied to a thermocouple; it is introduced here as an example of the validity (and fruitfulness) of taking the point of view of 'entropy flow' when reasoning about processes involving heat. To preview the later discussion a little, one may guess that some entropy generating processes inside a wire make contributions to both terms in (9.7), so that the two terms are not independent. We thus have a situation of *coupled flows*. One result is that the electric current through a wire depends not only on the electric potential difference, but also on the temperature difference between the ends of the wire—the *thermoelectric effect*.

9.2 But what is it?

We have already commented that the idea of entropy as 'disorder' can be misleading. A better insight is expressed by the following ideas:

(1) Entropy is *the absence of structure*.
(2) Entropy is *freedom to explore*.
(3) Entropy increase is the *dispersal of energy*.
(4) Entropy is that which separates the future from the past.

One of the great contributions of Ludwig Eduard Boltzmann (1844–1906) was to show how to quantify the first statement, and thus provide the foundation of statistical mechanics. Boltzmann quantified the notion of structure and its absence by connecting it to the notion of *freedom to explore*, in a sense we will now explain.

Boltzmann's idea was to ask the question, 'what if we tried to specify the microscopic details of the system, such as the momentum and position of all its particles?' You could imagine, for example, writing down a long list of positions and momenta (they would not need to be specified with perfect precision, and indeed by the Heisenberg uncertainty principle that would be impossible anyway; it suffices if the precision is sufficient that no further structure emerges at smaller scales). The state specified by such a list is called a 'microstate'. Boltzmann realized that a given set of macroscopic properties, such as pressure, volume, and temperature, could be produced by any one of a large number W of different microstates. For example, if one molecule speeds up a little and another slows down, perhaps as a result of a collision, then nothing happens to the pressure or volume of the gas. So the two microstates (before and after the collision) both correspond to the same thermodynamic state, specified, for example, by a given pressure and volume. It follows that when the macroscopic properties are fixed, the system still has freedom to explore a large number of different internal configurations or microstates. It can (and does) 'wander' among the microstates, because no constraint is preventing it from doing so. In fact, at ordinary temperatures, it turns out that the number W of microstates corresponding to each macroscopic situation is enormous. Boltzmann proposed that entropy can be understood as a measure of the size of this group of microstates:

$$S = k_B \ln W. \tag{9.8}$$

This equation and the distinction between microstates and macrostates is quite subtle. It is the subject of statistical mechanics, which merits a book in its own right, so we shall not explore it fully here. However, we shall comment on how this equation captures the idea of 'absence of structure.'

The most structureless thing one can easily imagine is a gas in thermal equilibrium. One could adjust positions and momenta of molecules in a gas in many ways without having any impact on the large-scale properties. It follows that the gas has freedom to wander among a very large number of microstates within the constraints of a given total energy and volume. Therefore W is large and the entropy is high. Something with considerably more structure is a crystalline solid. Now swapping molecules among the lattice positions would have no effect, but moving a molecule to a non-lattice position would require added energy. As a result, such internal movements are not available to the system. It is possible for some molecules to move to non-lattice positions, but only by acquiring energy from other molecules, and consequently reducing their freedom to move. The overall result is that the total number of possible allocations of position and momentum to the particles in a solid is greatly reduced compared to the gaseous case. This is a reduction in W and hence, according to the Boltzmann hypothesis, a smaller entropy.

According to this idea, a given collection of molecules simply explores all those microstates that are energetically available, and the reason the system is gaseous rather than solid at high temperature is simply that most of the microstates correspond to a gaseous configuration. At low temperatures the energy per particle is not enough to allow many of the particles to escape the attraction of their neighbours, so now the gaseous arrangement is not energetically available and the molecules have no choice but to stay close to one another. A crystalline structure results because it allows the molecules to find positions of small potential energy, and thus gain more kinetic (vibrational) energy. The increased momentum contribution to W more than makes up for the reduced position contribution.

To take another example, a living cell 'cares about' (i.e. its macroscopic behaviour depends on) the placement of molecules within it much more than a dead cell containing the same material. Therefore there are fewer microstates per macrostate for the living cell, and it has the lower entropy.

Of course there is more to Boltzmann's hypothesis than these qualitative examples. It is a quantitative statement. A more precise example of its use is offered in the following.

9.2.1 Entropy increase in a free expansion

We introduced the free expansion or Joule expansion in Section 7.5.1. Using the fundamental relation (8.15) it is easy to find the entropy change in such an expansion. From (8.15) we find, for any process in a closed pV system,

$$\Delta S = \int \frac{1}{T} dU + \int \frac{p}{T} dV. \tag{9.9}$$

In a free expansion U is constant, so this simplifies to

$$\Delta S = \int \frac{p}{T} dV, \tag{9.10}$$

and for an ideal gas the equation of state allows the integral to be performed:

$$\Delta S = \int_{V_1}^{V_2} \frac{Nk_B}{V} dV = Nk_B \ln(V_2/V_1). \quad \text{[ideal gas]} \tag{9.11}$$

This is a useful insight into the properties of an ideal gas. Now let's notice the connection with equation (9.8). If the volume doubles, $\Delta S = k_B(N \ln 2)$. This can be understood as arising from the fact that each of the particles in the gas now has twice as much room in which to move, and therefore the number of available positions has doubled, while the number of available momentum values has not changed because the temperature is constant in a free expansion of an ideal gas. Consequently the number of microstates W available to the whole collection of particles grows by a factor 2^N, and therefore Boltzmann's hypothesis (9.8) predicts the result correctly.

Note, we derived (9.11) from (8.15) by purely thermodynamic reasoning, given the definition of the ideal gas. The discussion in terms of microstates (merely sketched here) is a discussion of a system specified in terms of its microscopic structure, namely a collection of non-interacting freely moving particles. The agreement between the two predictions shows that a system obeying Joule's and Boyle's laws, which are simple statements about macroscopic parameters, might possibly consist of such a collection of particles. This is not the only microscopic model that can lead to ideal gas behaviour, however—another possible model involves a large number of waves in a chamber, and no particles at all!

9.3 Gibbs' paradox

Josiah Willard Gibbs (1839–1903) drew attention to the following important paradox.

Suppose an isolated chamber is prepared with a partition down the middle, and an equal quantity of inert gas is introduced into each side of the chamber—see Figure 9.4.

When the partition is removed, what happens? And, if anything does happen, does the entropy increase?

First suppose the gases introduced on the two sides are different, say helium and neon. Then when the partition is removed the two gases mix, and thus an irreversible process takes place: diffusion of two different gases into one another. The entropy of the system increases. To calculate the entropy increase, we argue that at low enough density, each set of molecules behaves independently of the other, so it is just like two free expansions. If the total number of particles is N, then, by equation (9.11), the increase in entropy of the helium is $(N/2)k_B \ln 2$, and the increase in entropy of the neon is $(N/2)k_B \ln 2$. The total increase in entropy of the system is

$$\Delta S = \frac{N}{2}k_B \ln 2 + \frac{N}{2}k_B \ln 2 = Nk_B \ln 2. \tag{9.12}$$

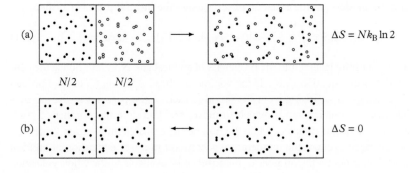

Fig. 9.4 *Gibbs' paradox. When two distinct gases mix by diffusion, an entropy increase takes place (a). However, a single type of gas mixing with itself does not constitute an entropy increase (b). Only non-classical microscopic laws of motion can correctly account for this.*

Now suppose the same type of gas is introduced on both sides of the partition. Now when the partition is removed, does anything take place or not? Certainly the two sets of molecules intermingle just as before, but this sort of intermingling is going on all the time in any gas. It is just part of the essential nature of a gas; it does not represent an increase in entropy. If it did then we would have to infer that the entropy of any gas is permanently increasing, even though there is no change of thermodynamic state, which is a contradiction. Therefore the only satisfactory answer is

$$\Delta S = 0. \tag{9.13}$$

There is no paradox in the fact that equations (9.12) and (9.13) are different. They describe different experiments and both are correct. The paradox lies in the difficulty in understanding why the intermingling of a gas with itself (the second experiment) does not represent an increase in entropy.

One could imagine, for example, attaching labels to all the gas molecules, say 'L' for those starting on the left, 'R' for those starting on the right. Then it appears as if essentially the same process is happening at a microscopic level in the two experiments, namely a diffusion of 'L' molecules throughout the chamber, and a diffusion of 'R' molecules throughout the chamber. Why, then, are the entropy changes so different? Also, one could imagine two sets of molecules or atoms that are almost alike, say both neon, but those on the left having an internal charge distribution with a slightly elliptical shape, or a small vibration, while those on the right are spherical and non-oscillating. In this way we could imagine making a continuous change from the case of different gases to the case of identical gases: how would the entropy then behave? How different is 'different enough' for formula (9.12)?

These questions opened up for the first time one of the signs that classical physics is profoundly incomplete. The classical picture, of particles that could in principle be singled out and labelled, cannot provide a microscopic model for the entropy of a gas. The paradox is resolved by the concept of *genuine indistinguishability*, a basic part of quantum theory having no classical counterpart. In quantum physics two entities having the same internal structure and internal state, such as two neon atoms in their ground state, are not just similar but completely indistinguishable. This fact on its own does not resolve the paradox, because the idea of two things that are exactly alike is well defined in classical physics also. However in classical physics one can in principle keep track of which entity is which by plotting their trajectories. In quantum physics this book-keeping is not possible, owing to the wave-like nature of all things: when waves overlap and then separate again, it is not possible to keep track of which wave went where. Because of this, the notion of particle labelling ('L' and 'R') that seemed to be so natural and innocent is in fact deeply unphysical: if there is no physical property corresponding to the label, then the label is meaningless, and is in fact profoundly misleading.

The conclusion is striking: if we are to reason correctly, we are not allowed to picture atoms in a gas as little balls whose position may be tracked. Such a picture leads to an incorrect conclusion about entropy increase. We must conclude instead that *all states of a system that differ only by a permutation of identical particles must be considered one and the same state*. This resolution is worked out fully, and in fact becomes self-evident, when one develops quantum field theory. The Gibbs paradox thus provides a link from entropy and thermodynamics to wave mechanics and the exchange principle.

(In quantum field theory one does not in the first instance need to talk about particles at all. Rather, what we call 'matter' is a pattern of excitation of a collection of wave-like fields. The microscopic state of a system is specified by stating the degree of excitation of all the available fields. This amounts to saying how many particles are present in each state of motion, without ever having to label the particles or single out one from another in any way. This is similar to the way we treat energy: the thousand millijoules of energy that are present in a joule of energy do not need to be singled out and labelled. We should similarly resist the temptation to label particles such as atoms in a gas.)

The case of the neon atoms with an elliptical shape or a vibration can also be handled by quantum theory. The answer to the question 'how different is different enough?' is 'when the quantum states are orthogonal'. In an intermediate case, where the quantum states have some finite overlap, then (assuming some interference effects are washed out and the states are stable),[1] the gas in each chamber would behave as a mixture of two types 'ordinary neon' and 'excited neon' which can be distinguished from one another.

9.3.1 Entropy of mixing

Equation (9.12) describes the total entropy increase when equal quantities of different gases mix. It is readily generalized to the case where xN particles of one gas mix with $(1-x)N$ particles of another. We wish to treat the case where no temperature or total pressure changes are involved, so we suppose the first group of particles to be initially enclosed in a chamber of volume $V_1 = xV$ and the second group in a chamber of volume $V_2 = (1-x)V$, where V is the final volume of both. Then, using equation (9.11), we have the total entropy change,

$$\Delta S = \Delta S_1 + \Delta S_2 = xNk_B \ln\left(\frac{1}{x}\right) + (1-x)Nk_B \ln\left(\frac{1}{1-x}\right)$$
$$= -Nk_B \left(x \ln x + (1-x)\ln(1-x)\right). \tag{9.14}$$

---

[1] Any distribution of charge that differs from the ground state distribution will involve excited states that can decay. However, there exist metastable states, for example different hyperfine components of the ground electronic state, which survive long enough for this thought-experiment to be realistic.

This function is plotted in Figure 9.5. It is an important function because it enters into any discussion of entropy where just two different types of entity are involved. The proportions x and $1-x$ can usefully be regarded as the probabilities P_1, P_2 that a particle chosen at random from the final mixture would be of each type, so that equation (9.14) can be written

$$\Delta S = -Nk_B \sum_i P_i \ln P_i. \tag{9.15}$$

This is a positive quantity because $0 \le P_i \le 1$.

Equation (9.14) is commonly called the 'entropy of mixing', but beware! This is a misleading name. Mixing of gases is a perfectly reversible process when it is done without allowing either gas to expand, as we now discuss. The reader should contemplate Exercise 9.13 after reading Section 9.3.2.

9.3.2 Reversible mixing

Let system (1) consist of a single chamber of volume V, filled with a mixture of inert gases of two types, A and B, one mole of each, at temperature T (see Figure 9.6a). Let system (2) consist of a pair of chambers, each of volume V, with one mole of A in the first chamber and one mole of B in the second (Figure 9.6b), both at temperature T. Which system, (1) or (2), has the more entropy?

To answer this question, we propose a sequence of two steps that modifies system (2) until it becomes equivalent to system (1), and calculates the entropy changes. First suppose the pair of chambers in system (2) are placed side by side and the wall between them removed. In the subsequent irreversible mixing by diffusion, the system entropy grows by $\Delta S = 2Nk_B \ln 2$, where N is Avogadro's number. Now subject the system to a reversible isothermal compression from volume $2V$ to V. The temperature does not change, and therefore (for an ideal gas) neither does the internal energy. Therefore the heat leaving the system is equal to the work done, which is $\int -p\,dV = 2Nk_B T \ln 2$. The entropy leaving the system is therefore $2Nk_B \ln 2$, which is exactly equal to what was gained during the irreversible mixing, therefore the net effect is no change of entropy. We conclude, therefore, that in their initial states systems (1) and (2) had the same entropy.

This conclusion might seem surprising at first, because system (2) appears to have more structure, with the gases separated out, while in system (1) they are mixed. However, the fact that the gases are of different type is enough to indicate which is which, whether or not they share the same region of space, so there is no reduction of entropy associated with separating them into different chambers.

The fact that there is no net entropy change of system (2) in the experiment just described suggests that it should be possible to move reversibly between the

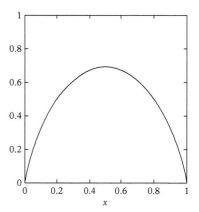

Fig. 9.5 *The function* $-x \ln x - (1-x) \ln(1-x)$.

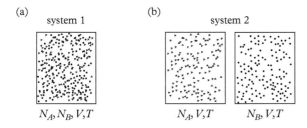

Fig. 9.6 *Which has the higher entropy, a single chamber containing a mole of A and a mole of B mixed together in a volume V, or two chambers each containing one type of gas, of volume V each, all at the same temperature?*

Fig. 9.7 *Reversible mixing of gases (first version). The cylinders all have the same volume. The mixed gas can be separated by moving the pistons as shown. The upper pipe has a membrane permeable to A but not B; the lower pipe has a membrane permeable to B but not A. The process is reversible and without heat transfer.*

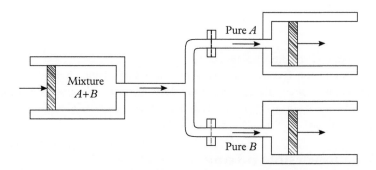

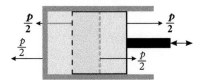

Fig. 9.8 *Reversible mixing of gases (second version). The black dashed membrane is permeable to gas A (left and central regions) and is attached to the black moveable piston. The grey membrane is permeable to gas B (right and central regions) and is fixed in the centre of the large chamber. Each gas moves completely freely through the membrane that is permeable to it, and exerts pressure on the membrane that is not. All the forces are balanced and do not change as the piston is moved. The forces shown in bold font are owing to B, the forces in normal font are owing to A.*

initial and the final state. Two methods to achieve this are shown in Figures 9.7 and 9.8. We will describe the second method.

We construct an open-ended chamber of volume $2V$ with a central partition made from a membrane through which gas B can pass freely, without any drop in pressure, but which is impermeable to gas A. The construction of such a membrane is discussed in the box. Put gas A into the left-hand region of this chamber, contained behind the membrane. We close the chamber using as 'piston' a second chamber of volume V, one end of which consists of a membrane through which gas A can pass freely, but which is impermeable to gas B. In the initial state, this 'piston' contains all of gas B and none of gas A. The right-hand chamber or 'piston' can now be moved reversibly without doing any work. There is no net force on it because only gas B exerts any force on its end walls, and these forces remain equal and opposite at all times. There is no viscous damping because gas A passes freely through the membrane. Thus by moving the piston left and right the gases can be mixed and unmixed adiabatically.

9.4 Specific heat anomalies

The general form of the heat capacity as a function of temperature, for any system, is a gradually increasing function at most temperatures, with, superimposed on this, occasional relatively narrow peaks, such as that shown in Figure 9.9, or that shown for a magnetic system in Figure 7.3b.

Such peaks are called *specific heat anomalies*. The area under the heat capacity curve as a function of temperature is the total heat that enters the system when the temperature rises through the anomaly (or that leaves the system if it cools). If we first subtract the contribution of the slowly changing background, then we have the heat associated with the physical process causing the anomaly. The integral of C/T is the change in entropy (cf. Eq. 9.4).

The rapid change in entropy at such a peak indicates either that a parameter such as volume is changing rapidly, or else that some sort of reordering or reorganization of the structure of the system is going on. The fact that the heat capacity goes through a maximum as a function of temperature, and thereafter decreases

Specific heat anomalies 121

> **Semi-permeable membranes** A semi-permeable membrane is, in general, a membrane through which some but not all fluid substances can diffuse. For the argument in this section we invoke a special type of semi-permeable membrane where the passage of one gas through the membrane is completely free, while the other is completely blocked. By 'completely free' we mean there is no drop in chemical potential; to be discussed later. For a mixture of ideal gases at the same temperature, it means there is no drop in pressure for the gas that passes through.
>
> If the molecules of gas A are larger than those of gas B, one can imagine in principle that such a membrane could be constructed from a fine mesh with holes big enough for B but not A to pass.
>
> More generally, one can invoke the internal structure which distinguishes A molecules from B molecules. For example, a sheet of intense light can present an energy barrier to particles passing through it, because they acquire an induced electric dipole moment which interacts with the light field. If the light is tuned to a frequency in between resonant frequencies of A and B, then one type of molecule will be repelled by the sheet (and therefore bounce off), the other attracted (and therefore pass straight through, running down one side of the potential valley and up the other).
>
> In short, there is nothing thermodynamically wrong with the idea of a perfect semi-permeable membrane.

for a while as the temperature increases, is owing to the finite amount of energy associated with this reordering (a falling heat capacity is a sign that the system is struggling to accommodate more energy).

The peak associated with the magnetic degrees of freedom and having the characteristic shape shown in Figure 7.3b is called a *Schottky anomaly*. This occurs at a temperature that depends on the size of the applied magnetic field, and it is typically a low temperature, of order 1 K. A peak having the shape shown in Figure 9.9 is called a *lambda point*, on account of the similarity between the shape of the curve and the Greek letter λ. The temperature at a lambda point can vary greatly from one system to another: 742 K in beta brass; 2.17 K in liquid helium at 1 atmosphere. But the entropy changes in all these cases are similar.

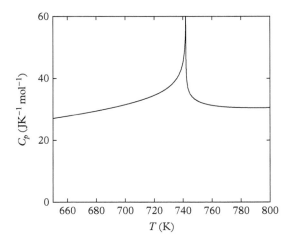

Fig. 9.9 *The molar heat capacity of beta brass as a function of temperature.* [Based on data from Salamon and Lederman, Phys. Rev. B **10**, 4492 (1974).]

The molar entropy change in the Schottky anomaly for the simplest case is $N_A k_B \ln 2$ (cf. Eq. 14.34). The entropy change associated with the lambda point in beta brass is also close to $N_A k_B \ln 2$ (one must carefully separate out the background and integrate over a wide temperature range to ascertain this). In both cases this is owing to a two-fold increase in the number of states energetically available to each particle in the system. There are also magnetic systems such as chromium potassium alum in which the molar entropy increases by $1.39R$. This is $N_A k_B \ln 4$, which can be understood from the fact that the chromium ions have angular momentum $\mathcal{J} = 3/2$, and hence $2\mathcal{J} + 1 = 4$ states associated with orientation of their magnetic dipole moment.

The Schottky anomaly is associated with the alignment of magnetic dipoles within the magnetic material. In the presence of an applied magnetic field there is no phase transition in this case, but a steady increase of alignment as the temperature falls. If each dipole is associated with a spin 1/2 particle, then it has two distinct quantum states and at high temperature is free to occupy either. At low temperature, by contrast, it occupies the lowest energy state alone. This accounts for the change in S. At zero applied field there is, in most magnetic materials, a phase transition which also gives a peak in the heat capacity, and it is found that the net entropy change is independent of the means by which it occurs.

The lambda point is associated with a *continuous phase transition*, a notion which we shall discuss more fully in Chapter 22. Beta brass is a 50/50 copper–zinc alloy. The crystal structure is body-centred cubic, and this consists of two interlaced simple cubic structures, analogous to the two-dimensional centred square lattice shown in Figure 9.10. The chemical bonds are such as to favour unlike neighbours, so that the state of least energy is one where all the Cu atoms, say, occupy sites on the A lattice, and all the Zn atoms occupy sites on the B lattice. This is the state at low temperature. At high temperature, by contrast, either type of atom is equally likely to be found at either type of site. This is a higher entropy situation, and hence the transition from one situation to the other must involve an increase in entropy and therefore an associated contribution to the heat capacity. Transitions of this kind are called *order–disorder transitions*.

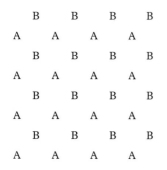

Fig. 9.10 *The centred square lattice, an easy-to-visualize two-dimensional analogue of the body-centred cubic structure.*

9.5 Maxwell's daemon

James Clerk Maxwell (1831–1879) made great contributions to thermal physics as well as to electromagnetism. He proposed the following thought-experiment. Imagine a gas chamber with a partition down the middle, and a small trap-door in the partition. Suppose a little creature sits next to the trap-door, observing the particles of the gas (Figure 9.11). Whenever a fast particle approaches from the left, or a slow one from the right, he opens the door. When a slow particle approaches from the left, or a fast one from the right, he keeps the door closed. In this way he causes the temperature of the gas on the left to fall, and that on the right to rise, without doing any work (the door can be assumed frictionless, and if it has a spring then the work required to open it can be reclaimed when closing it). This is in violation of the Clausius statement of the second law of thermodynamics.

Maxwell originally called the creature simply a 'finite being'. The phrase 'Maxwell's daemon' was coined by William Thomson (Lord Kelvin); the word 'daemon' means 'one that mediates a process'. However as the word 'daemon' fell out of use in the English language, the word 'demon' was more and more applied to the finite being considered by Maxwell. It seemed appropriate to regard Maxwell's daemon as a mischievous creature, busy trying to overturn one of the foundations of physics, or else to confuse us. The essential point is that the behaviour ascribed to Maxwell's daemon appears at first sight to be physically possible, but it leads to a breakdown of the second law. Therefore either thermodynamics is wrong, or the behaviour is impossible, or something has been overlooked.

It took approximately one hundred years before the physics of Maxwell's daemon was clearly understood. On the way some false avenues were explored, such as the idea that it takes finite work to acquire information. This is not true. Leo Szilard introduced a useful way to reduce the problem to its essential ingredients, and when the physics of information was better understood by Rolf Landauer and others, the resolution was finally achieved.

9.5.1 Szilard engine

Szilard considered a minimal version of Maxwell's thought-experiment, with just a single gas particle in a box (Figure 9.12). The 'daemon' should be thought of as a physical mechanism which can determine on which side of a chamber the molecule lies, and in response to this, insert a piston on the empty other. Since the piston only meets vacuum, this requires no work. The gas particle then does work against the piston during an isothermal expansion, in which heat is allowed to enter the chamber. $k_B T \ln 2$ joules of heat enter, and this amount of work is extracted (the internal energy of the one-particle 'gas' depends only on T and does not change). After this the process repeats. The net result is a pure conversion of heat into work... or is it?

The paradox is resolved by paying attention to something that has been overlooked. This is the physical state of the 'daemon', i.e. the mechanism that receives information and inserts the correct piston. The information received must have a physical form. It does not matter what form it takes (a mechanical switch, a current, a magnetic domain), but it certainly changes the physical state of the daemon. Therefore at the end of the process the daemon is not in the same physical state as before, therefore we do not have a true thermodynamic cycle, so there is no violation of the second law. To complete the cycle the daemon must restore its internal state, or in other words reset its 'memory'. This requires it to eject into its surroundings the information it has acquired. This ejection or erasure of information has also to take a physical form, and depending on the surroundings, it may require energy. Rolf Landauer argued that to erase the information (i.e.

Fig. 9.11 *Maxwell's daemon. Maxwell considered a finite being that operated a trap-door so as to allow faster molecules to pass in one direction, slower ones in the other, and so build up a temperature difference without doing work. The illustration depicts another type of process, in which the door is opened when a molecule approaches from the left, but not when one approaches from the right, thus building up a density difference without doing work. In either case the finite being or 'Maxwell daemon' is performing a sorting operation that results in an entropy decrease of the gas. The resolution of the paradox involves careful consideration of the physical state of the daemon itself, depicted here by the snapshot of its memory, represented as a series of black or white squares.*

Fig. 9.12 *The Szilard engine: a simple (and physically allowed) heat engine. A chamber possessing two pistons contains a single molecule. The chamber is initially isolated. (1) A detector (not shown) determines on what side of the chamber the molecule lies. (2) A piston is depressed on the other side of the chamber—the molecule is not there so no work is required to do this. (3) The molecule subsequently hits all sides of the chamber, including the piston. The piston is allowed to move, so the molecule does work against it and loses energy. This energy is resupplied by allowing thermal contact between the chamber and a single heat reservoir. After this the chamber and molecule have completed a cycle—they have been returned to their initial thermodynamic state—and some work has been obtained. The detector meanwhile has not completed a cycle until the record it holds, indicating the outcome of the measurement in step (1), has been erased. Such erasure involves heat output (see text).*

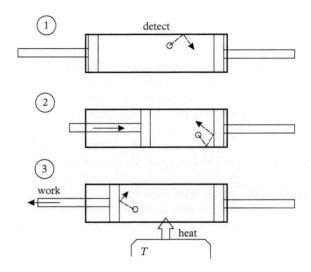

move from either of two possible initial states to a single final state) by reversible dynamics, the information must be transferred elsewhere, but in such a way that it gets dispersed. It means that some other system (the surroundings) must evolve, during the erasure process, into a thermodynamic state that has the following features. First, the interaction must be such that the daemon's state loses the information, and this requires a significant (i.e. macroscopically detectable) change in the surroundings. Secondly, the final state of the surroundings must be the same thermodynamic state, irrespective of which measurement outcome is obtained, otherwise the information has not been erased but simply moved, which implies that the cycle of Szilard's engine is still not complete.

Let W be the number of microstates available to the surroundings in its thermodynamic state before the erasure process. Then the first condition is met if there are W microstates at the end of the process that are consistent with one measurement outcome, and a further W that are consistent with the other measurement outcome. The second condition is met if both these sets of microstates are subsumed into the set of states available to the surroundings during its random exploration of available microstates. In other words the number of microstates doubles. Note that only when this doubling happens has the information been fully transferred, and if the environment behaves as a thermal reservoir then it does not retain a record of which outcome has occurred; it simply gains an enlargement of its set of available microstates by a factor two. It follows that the entropy of the surroundings increases by $k_B \ln 2$. This exactly matches the loss of entropy from the thermal reservoir during the isothermal expansion of Szilard's engine, therefore there is no overall reduction in entropy of the universe when the cycle is truly complete.

Since we have assumed the process of erasure to be reversible, the $k_B \ln 2$ entropy increase in the surroundings requires that the daemon export $k_B T \ln 2$ of heat energy to the surroundings. It can either obtain this directly from the isothermal expansion stage, in which case it acquires no work energy, or else it must convert the work back into heat. In either case there is zero net work acquired per cycle.

This study makes clear that an essential feature of Maxwell's daemon is the ability to store and make use of information. Maxwell's daemon is sometimes misunderstood to mean anything that selectively allows the passage of particles, such as various kinds of filter. A filter works by throwing away information immediately; a Maxwell daemon has the option of postponing the erasure stage.

9.5.2 The Feynman–Smoluchowski ratchet

Another thought-experiment designed to explore the second law of thermodynamics was discussed by Feynman in his *Lectures on Physics* and has become known as the Feynman ratchet. This terminology may perhaps be unfair to Marian Smoluchowski, who proposed and investigated the idea in 1912. It is a good teaching aid, and it also illustrates some important processes in biology.

Feynman's (or Smoluchowski's) ratchet is an ordinary ratchet and pawl, that is, a toothed wheel with a stopper that can be lifted to allow the wheel to rotate, and which is restored by a spring or gravity (Figure 9.13). The 'sawtooth' angles of the teeth on the wheel are such that gentle nudges in one direction will cause the teeth to lift the pawl so that the wheel can rotate; while gentle nudges in the other direction simply embed the pawl more firmly against the steep edge of a tooth, preventing rotation.

All we need to do is attach such a ratchet to a set of vanes in a gas in thermal equilibrium. Random motion of the molecules of the gas against the vanes will cause random forces, sometimes in one direction, sometimes in another. Every now and then the force will be sufficient to rotate the wheel by one tooth, after which the pawl engages so the wheel does not rotate back. We can even imagine allowing the wheel to raise a slight load: occasionally the random jiggles will be sufficient to overcome the weight of the load. Thus the ratchet slowly does work,

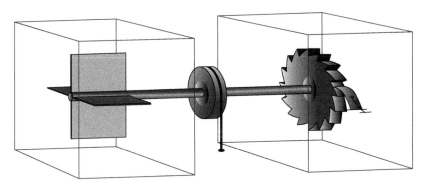

Fig. 9.13 *The Feynman–Smoluchowski ratchet.*

while attached to only a single heat reservoir. There goes the second law again: now the Kelvin statement is violated... or is it?

The resolution once again requires a careful look at the physical mechanism. When the pawl falls down to re-engage against the wheel, what happens? Either the process is reversible, in which case it bounces off again, or else there is damping and it comes to a halt. In the first case the pawl is in a continual state of oscillation, so the wheel can rotate in either direction and there is no ratchet effect. Therefore we must have the second case. However, damping is an energy conserving process, so it must convert the kinetic energy of the pawl into heat. Furthermore, unless the ratchet and pawl are themselves at absolute zero temperature, it is also subject to thermal fluctuations. Therefore, the pawl can randomly rise and release the wheel, allowing it to rotate back.

A simple analysis proceeds as follows. Let ϵ be the energy required to lift the pawl so that the toothed wheel can rotate. Let E be the energy delivered from the gas to the vanes when they rotate just enough to advance the wheel by one tooth. Note, this same energy will be given back to the gas if the vanes rotate in the other direction, because the pawl mechanism and the load exert the very torque which the gas has to overcome to rotate the vanes. So we have the following processes, for a rotation by one tooth:

Forward rotation	Energy taken from gas	E
	Energy given to ratchet	ϵ
	Work available to raise a load	$E - \epsilon$
Backward rotation	Energy to lift the pawl	ϵ
	Work released as the load is lowered	$E - \epsilon$
	Energy passing from vanes to gas	E

The second set of energies is the same as the first, but with opposite sign. To find out which, if any, process tends to happen the most, we need to calculate the rates of the two types of rotation. Here we simply quote a general fact about thermal systems, which we will explore more fully in Chapters 14 and 27. This is, the probability that a given part of a system in thermal equilibrium gets a given energy ϵ_i is proportional to the *Boltzmann factor* $e^{-\epsilon_i/k_B T}$. The important point for the present calculation is that this factor depends on the ratio of energy to temperature, and any other multiplying factors are the same for the two ends of the shaft (i.e. at the vanes and at the ratchet), because in both cases the same physical variable is affected, namely the angle of rotation of the shaft. Note, we have not given the proof of these details; we are simply outlining what is in fact found.

We will apply the Boltzmann factor to both the vanes and the pawl. First, note that by increasing the load, we can always arrange that the weight of the load 'wins' and overall the load is lowered and work is done on the gas. Conversely, if there is some value of the load where the process is reversible, then for a lighter load the vanes will 'win' and raise the load. Intermediate between these two cases there is a value of the load where the whole situation is reversible. In this case, the forward rate will agree with the backward rate, so we must have

$$\frac{E}{k_B T_1} = \frac{\epsilon}{k_B T_2}, \qquad (9.16)$$

where T_1 is the temperature of the gas, and T_2 is the temperature of the ratchet and pawl. Now we can finish the argument by invoking a very simple piece of thermodynamic reasoning. All we need to do is interpret the energies E and ϵ appropriately. The whole system should be regarded as a type of heat engine. E is the heat Q_1 passing to the vanes; ϵ is the heat Q_2 passing out to the ratchet and pawl. Equation (9.16) then tells us that the ratio of these heats is

$$\frac{E}{\epsilon} = \frac{Q_1}{Q_2} = \frac{T_1}{T_2}.$$

But this is none other than the standard result for a reversible heat engine. Therefore the ratchet and pawl is simply an unusual type of heat engine. It can do work if there is a temperature difference between the gas and the ratchet, and then it will eject the standard amount of heat and thus conserve entropy. If there is no such temperature difference, then the pawl is as likely to rise randomly and release the wheel as it is to be lifted by the sloping surface of a tooth as the wheel rotates owing to a fluctuation in the gas. Then no work is done on average; the wheel just rotates in either direction, randomly jiggling about.

Some biological processes take advantage of temperature differences in a way quite reminiscent of the Feynman–Smoluchowski ratchet. The process fluctuates too and fro, but on average makes progress while dissipating small amounts of energy. The ratchet can also be used to explore effects away from thermal equilibrium.

9.6 The principle of detailed balance

An important property of thermal equilibrium is illustrated by Figure 9.14. If three bodies A, B, C are in thermal equilibrium with one another, then the question arises, whether there could be a continuous flow of energy from body A to B, and from B to C, and from C to A, such that each body has overall a steady state, but there is this flow around, forming a loop. In thermal equilibrium, this is not possible, because if it were, then one could place a device in the loop, between A and B for example, and use the flow to drive some process so as to obtain work. The energy would be extracted from the ensemble of bodies, so the first law is not violated, but since the ensemble is at a single temperature, the extraction process would violate the Kelvin statement of the second law.

It is the orderly nature of the flow of energy here which allows it to be used to drive a work related process. For example, if the energy flow is in the form of

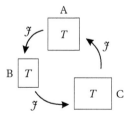

Fig. 9.14 *Three bodies in mutual equilibrium. The energy flow around a loop, as indicated by the arrows, is forbidden by the second law.*

electromagnetic radiation, then it will exert a pressure which can be used to turn a paddle wheel. More generally, any energy flow carries momentum with it.

From this observation one can deduce the absence of vortices in a fluid in thermal equilibrium, and by generalizing to more abstract definitions of the three bodies, one deduces the absence of more abstract types of 'vortex', such as one-way circular sequences of chemical reactions, or loops of energy flow among ensembles of atoms in one internal state or another. The general observation is thus useful and important. It may be stated thus:

> **The principle of detailed balance**. In an ensemble of systems in thermal equilibrium, each internal process that could in principle cause a change of state of one or more of the systems proceeds at the same rate as the reverse process.

In a thermodynamic discussion such as at the present, this principle can only be stated in the thermodynamic limit. It has many uses in that limit. It is found that the same principle also applies to rates of all processes, including microscopic, such as molecular collisions, and it thus becomes a very useful principle in the consideration of all sorts of microscopic processes as well.

9.7 Adiabatic surfaces

We have discussed several physical examples, and carried out quantitative calculations of entropy. We now present a further basic mathematical idea.

The main mathematical consequence of our arguments from the second law is that there exists a function of state, entropy S, that stays constant during reversible adiathermal processes (i.e. adiabatic processes). It follows that, in a state space of n dimensions, the set of states at any given value of S fall in a region of $n-1$ dimensions called a *hypersurface*. This is usually abbreviated to 'surface' and we talk of *adiabatic surfaces*.

For example, a simple closed system has a state space of two dimensions (p and V) and the states of a given entropy fall on a *line*. An open simple system has a state space of three dimensions (p, V, N) and the states of a given entropy now fall on a *surface*. If the system is also paramagnetic then it has another independent generalized coordinate (the magnetic field B); now the state space is four dimensional and the adiabatic 'surfaces' are three dimensional. And so on.

The significance of these surfaces is only revealed once the state space has more than two dimensions. This is because we can infer the existence of adiabatic lines in the simple case without appealing to the second law. For a simple closed system, an adiabatic line is a line followed by a system which undergoes a reversible process without exchanging heat. From the first law it follows that, for a such a process,

$$\mathrm{d}U = -p\mathrm{d}V.$$

But for any process,

$$dU = \left.\frac{\partial U}{\partial p}\right|_V dp + \left.\frac{\partial U}{\partial V}\right|_p dV,$$

so we have, in the case where dp and dV are along an adiabatic line,

$$\left.\frac{\partial U}{\partial p}\right|_V dp + \left(\left.\frac{\partial U}{\partial V}\right|_p + p\right) dV = 0,$$

so the gradient of the line is

$$\frac{dp}{dV} = -\left(\left.\frac{\partial U}{\partial V}\right|_p + p\right) \left.\frac{\partial U}{\partial p}\right|_V^{-1}. \qquad (9.17)$$

This is a differential equation for the line which has a unique solution for any given starting state. So we have deduced the existence of these lines running through the state space without appealing to the second law.

If we now proceed to a three-dimensional state space, the above argument breaks down. Let's see why. The first law now takes the form

$$dU = đQ - pdV - qdR, \qquad (9.18)$$

where q, R are some further variables that are needed to specify the state, such as $q = m$, the magnetic dipole moment, and $R = B$, the magnetic field, or $q = \mu$, the chemical potential, and $R = N$, the number of particles. Let's examine the situation in terms of p, V, R. For an adiabatic process we now have

$$dU + pdV + qdR = 0,$$

and by considering $U = U(p, V, R)$, we find that for such a process,

$$\frac{\partial U}{\partial p} dp + \left(\frac{\partial U}{\partial V} + p\right) dV + \left(\frac{\partial U}{\partial R} + q\right) dR = 0, \qquad (9.19)$$

where it is understood that the partial derivatives take as constant in each case the appropriate pair of variables from the set $\{p, V, R\}$. The problem is that there is no guarantee that (9.19) describes a surface. To see this, it suffices to give an example.

Consider the ordinary three-dimensional space described by Cartesian coordinates x, y, z. Write down the equation of some simple surface, such as $x^2 + 4y^2 - 3z^2 = $ const. Differentiating, we find that this surface is described by

$$xdx + 4ydy - 3zdz = 0.$$

So if we start at some point (x_0, y_0, z_0) and make small steps dx, dy, dz such that they satisfy this condition, then we shall find ourselves to be tracing out a surface. But now consider another condition, such as

$$dx - z\,dy + dz = 0. \tag{9.20}$$

The problem is that this condition does not restrict us to a surface. In fact, starting out from any (x_0, y_0, z_0) you can move to anywhere in the space while always satisfying (9.20). By writing $dx = z\,dy - dz$ you can easily prove that dx here is not a proper differential. There is no function $x(y, z)$ whose differential it is.

More generally, the condition that a form

$$X\,dx + Y\,dy + Z\,dz = 0$$

describes motion on a surface is that the functions X, Y, Z satisfy

$$X\left(\frac{\partial Y}{\partial z} - \frac{\partial Z}{\partial y}\right) + Y\left(\frac{\partial Z}{\partial x} - \frac{\partial X}{\partial z}\right) + Z\left(\frac{\partial X}{\partial y} - \frac{\partial Y}{\partial x}\right) = 0. \tag{9.21}$$

For us, the essential point is that the first law is not sufficient on its own to guarantee that reversible adiathermal motion lies on a surface in state space. But the first and second laws together do make that guarantee. By furnishing a function of state S that is constant for reversible adiathermal motion, they show that such motion lies on the surface $S = $ const. Or, if you prefer, they tell us that (9.19) satisfies (9.21) (after translating the notation from x, y, z to p, V, R and defining the functions X, Y, Z so as to match (9.19)).

In fact we can arrive at the entropy via adiabatic surfaces in a rather neat argument that gets at the essence of Clausius' theorem by focusing on one particular cyclic process. It is a process consisting of two adiabatic stages and one heat-exchange.

Consider, for the sake of making the argument, a three-dimensional state space. We will map this space using energy and those other parameters which change when reversible work is done, namely V and R in equation (9.18). We start at A in Figure 9.15. We will show that the Kelvin statement of the second law implies that it is not possible to reach both states B and C from A by a reversible adiathermal process, when C differs from B only in energy.

To prove this, suppose the converse, with a view to showing a contradiction. Then we can start at A, move to B without exchanging heat, then move to C by bringing in some heat, then move back to A without exchanging heat, since, by assumption, the path from A to C is reversible and adiathermal. The movement from B to C can be achieved by input of heat alone because C differs from B only in energy: no change of V or R is required. The net result is no net change of state, so, by the first law, the energy brought in as heat must have been sent out as work during the adiathermal stages. But the heat could have been provided by a heat engine working in a cyclic manner and operating between the system and a *single* reservoir, just as in Figure 8.10, without any other change in the rest of the universe. So we have a violation of the Kelvin statement.

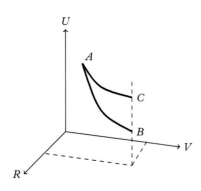

Fig. 9.15 *Argument to prove the existence of adiabatic surfaces (see text).*

The implication is that, starting out from A on our reversible adiathermal explorations, we shall only encounter one state at each V, R. In other words, we are on a surface. From other states above and below A there extend other surfaces. These surfaces never cross (Exercise 9.14), and none of them ever thickens out into a higher-dimensional entity, nor folds back on itself. It also follows that a vertical line at any V, R intersects each surface only once. By labelling each surface with a single number, S, we have a function of state which satisfies the equation $S(U, V, R) = $ const. on each adiabatic surface. This function is the entropy.

The adiabatic surfaces encourage us to think of entropy simply as a property of the system. It can be calculated from the other properties—there will be an entropy equation, just as there is an energy equation. It is just there. Get used to it. But there are plenty of other functions of state of no great importance, such as Up^3 and $p^{1/2}/V$ and an infinite number of others. What is so special about S? The physics comes in when we add one further thought about these adiabatic surfaces. For an isolated system, each adiabatic surface has a further ingredient: on one side of it lies the future, on the other lies the past!

9.8 Irreversibility in the universe

The claim that entropy does not decrease in an isolated system is one of the most basic statements about *time* that science has to offer. It is called the *thermodynamic arrow of time* because it gives a sense of direction to time. It asserts that things don't merely change, they change in a certain direction—the entropy increasing direction—and it is tempting to extend that statement to the universe as a whole. All statements about the universe as a whole involve an extrapolation and consequently a need for caution to check whether concepts are still applicable; but this extrapolation seems to make sense, and in any case we can always content ourselves with examining part of the universe, such as a billion or so galaxies. In that part, we claim, total entropy increases, as long as not much heat is being emitted to other parts of the universe.

However, at present our most general description of microscopic phenomena, namely quantum theory as it is expressed in the standard model of particle physics, is based on reversible equations of motion. According to Schrödinger's equation, the motions of the fundamental particles and fields are always reversible in principle, although it would be extremely difficult to reverse them all in practice. This implies that thermodynamic irreversibility, and the associated idea of entropy increase, are not built into the basic equations, but 'merely' express the fact that we can't keep track of the details of the microscopic motions. By looking into this more fully, one finds that in order for the reversible equations of particle motion to be consistent with the observed increase of entropy, one must make strong assumptions about the initial state of whatever system is under consideration. In the case of the universe as a whole, or any large part of it, one is forced to the conclusion that the initial state is extraordinary, with an astonishingly low entropy.

The previous paragraph captures much of what many physicists think about entropy increase, but there are several further issues that should also be born in mind. First of all, the other component of basic physics, namely general relativity, suggests that there can be strictly irreversible phenomena, namely the growth of the event horizons of black holes (it is an open question whether quantum field theory modifies this conclusion). Next, attempts to arrive at a unified theory that incorporates both general relativity and quantum mechanics sometimes suggest that such a merger must produce a description in which irreversible dynamics are built in (this is another open question). Finally, whether or not this is related somehow to gravity, there are in any case questions about reversibility that are raised by the subtle and still unresolved issue called the *quantum measurement problem*.

The measurement problem in quantum mechanics is the problem that the physical implications of the mathematical results of quantum theory are not universally agreed. The question is how to connect the equations of motion—a set of mathematical statements about abstractions such as wavefunctions, differential equations, and Hilbert space—to the physical behaviour, which involves physical things moving about and exchanging physical energy etc. One point of view is that the mathematical quantum state is in one-to-one correspondence with the physical state of the universe at all times, and both evolve as prescribed by Schrödinger's equation. This may be called a 'many worlds' view because it implies that physical reality consists of a vast collection of macroscopically differing situations, all present at once in a vast quantum superposition. Another point of view is that the evolution of the physical universe is not mapped by the mathematical apparatus of quantum theory quite so directly. Rather, quantum theory describes what future states are possible, and their probabilities, and the cosmos evolves by taking random steps from one such possible state to another. In order that this second type of interpretation can make sense, it is necessary to have a criterion to rule out some configurations, and irreversibility can be used to furnish such a criterion. One is thus led to conjecture that some types of physical behaviour are intrinsically irreversible. On this model, it is not the case that a reversal of all velocities would simply reverse all motions, because it is in the nature of fundamental physical processes to be probabilistic, not deterministic. If this is the case, then entropy is part of the fundamental dynamics of the physical world. The elaboration of this idea is an area of ongoing research.

In any case, some processes, such as an avalanche, or death, are so large and complicated that it is as certain as anything else we know that they will never be reversed by natural causes. This is true whether or not the equations of microscopic motion are reversible. Therefore we live in an irreversible universe.

At the largest scale, the cosmos as a whole is on a trajectory from low entropy in the past towards high entropy in the future. The very long term prognosis is a 'heat death' in which eventually no stars shine nor do galaxies exist, everywhere is randomly filled with dispersed matter and radiation in random motion. However, one should not make such sweeping statements without also commenting that they involve an extrapolation of our knowledge of the cosmos very far beyond the timescale on which it has ever been tested.

Ozymandias

I met a traveller from an antique land
Who said: "Two vast and trunkless legs of stone
Stand in the desert ... Near them, on the sand,
Half sunk, a shattered visage lies, whose frown,
And wrinkled lip, and sneer of cold command,
Tell that its sculptor well those passions read
Which yet survive, stamped on these lifeless things,
The hand that mocked them, and the heart that fed:
And on the pedestal these words appear:
'My name is Ozymandias, king of kings:
Look on my works, ye Mighty, and despair!'
Nothing beside remains. Round the decay
Of that colossal wreck, boundless and bare
The lone and level sands stretch far away."

Percy Bysshe Shelley

....

EXERCISES

Consult Tables 4.1 and 4.2 for the data you need in the following.

(9.1) Give a careful statement of the relationship between entropy and heat. Is entropy extensive or intensive? What dimensions does it have? Under what circumstances can it make sense to think of entropy 'flowing' from one place to another?

(9.2) Why are the cooling towers of a power station so large?

(9.3) A mug of tea has been left to cool from 90 °C to 18 °C. If there is 0.2 kg of tea in the mug, and the tea has specific heat capacity 4200 J K^{-1} kg^{-1}, show that the entropy of the tea has decreased by 185.7 J K^{-1}. Comment on the sign of this result.

(9.4) Consider inequality (8.19). Is it possible for dS and $đQ$ to have opposite signs? If it is, then give an example.

(9.5) 1 kg of water is warmed from 20 °C to 100 °C (a) by placing it in contact with a reservoir at 100 °C, (b) by placing it first in contact with a reservoir at 50 °C until it reaches that temperature, and then in contact with the reservoir at 100 °C, and (c) by operating a reversible heat engine between it and the reservoir at 100 °C. In each case, what are the entropy changes of (i) the water, (ii) the reservoirs, and (iii) the universe? (Assume the heat capacity of water is independent of temperature.)

(9.6) Calculate the change in entropy of 1 kg of water when it is heated from 10 to 100 °C and completely vaporized, all at 1 atm pressure. Check your answer by consulting the chart in Figure 16.7.
Does the change in entropy imply any irreversibility in the process?

(9.7) Two identical bodies of constant heat capacity C_p at temperatures T_1 and T_2 respectively are used as reservoirs for a heat engine. If the bodies remain at constant pressure, show that the amount of work obtainable is

$$W = C_p(T_1 + T_2 - 2T_f),$$

where T_f is the final temperature attained by both bodies. Show that if the most efficient engine is used, then $T_f^2 = T_1 T_2$.

(9.8) A heat engine operates between a tank containing 1000 m³ of water and a river at a constant temperature of 10 °C. If the temperature of the tank is initially 100 °C, what is the maximum amount of work which the heat engine can perform? Answer the problem algebraically before you substitute in the numbers. Show that your result can be expressed in the form $W = \Delta U - T_0 \Delta S$, and interpret the symbols physically.

(9.9) Calculate the changes in entropy of the universe as a result of the following processes:

(a) A copper block of mass 400 g and heat capacity 150 J K⁻¹ at 100 °C is placed in a lake at 10 °C.

(b) The same block, now at 10 °C, is dropped from a height 100 m into the lake.

(c) Two similar blocks at 100 °C and 10 °C are joined together (hint: save time by first realizing what the final temperature must be, given that all the heat lost by one block is received by the other, and then reuse previous calculations).

(d) A capacitor of capacitance 1 μF is connected to a battery of e.m.f. 100 V at 0 °C. (N.B. think carefully about what happens when a capacitor is charged from a battery.)

(e) The same capacitor after being charged to 100 V is discharged through a resistor at 0 °C.

(f) One mole of gas at 0 °C is expanded reversibly and isothermally to twice its initial volume.

(g) One mole of gas at 0 °C is expanded reversibly and adiabatically to twice its initial volume.

(9.10) A system consists of two different volumes of ideal gas, either side of a fixed thermally insulating partition. On one side there are 2 moles of gas at temperature $T_{1i} = 500$ K. On the other side there are 4 moles of gas at temperature $T_{2i} = 200$ K. The molar capacity at constant volume is $C_{V,m} = (3/2)R$. (i) If the partition is made thermally conducting, find the final temperature throughout the system when equilibrium is attained, and find the net change in entropy of the complete system. (ii) Now return to the initial conditions, and suppose the partition is made thermally conducting again, but only for a short time, so that some heat transfer takes place, but not enough to reach equilibrium, after which the thermal insulation is reinstated. Show that the final temperatures on the two sides are related by

$$T_{1f} + 2T_{2f} = 900 \text{ kelvin}$$

and find an expression for the net change in entropy ΔS as a function of T_{1f}. Plot ΔS on a graph as a function of T_{1f}, and find where it reaches a maximum.

(9.11) In a free expansion (also called Joule expansion), U does not change, and no work is done. However, the entropy must increase because the process is irreversible. Is this a counter-example to the relation $dU = TdS - pdV$? Explain.

(9.12) The number of ways of choosing N_1 items from a collection of $N > N_1$ items without respect to ordering is

$$W = \frac{N!}{N_1! N_2!} \qquad (9.22)$$

where $N_2 = N - N_1$. Calculate $\ln W$, making use of Stirling's approximation $\ln n! \simeq n \ln n - n$, which is very accurate for large n. Express your result in terms of N and $x \equiv N_1/N$, and compare with equation (9.14).

(9.13) In view of Section 9.3.2, in what sense, if any, is it the case that the mixing of gases is itself a cause of entropy increase? What is the right physical interpretation of equation (9.14)? Discuss the following two perspectives: (i) Mixing has strictly nothing to do with it. Each gas expands just as it would if the other were not there (not in the detailed dynamics, but in the sense of the overall change in its thermodynamic state). The process amounts to a pair of free expansions that happen to be sharing the same space. The entropy change is strictly, fully, and precisely given by that consideration. It is unequivocally unrelated to the mixing itself. (ii) Considered as a whole, we started with a chamber containing $N_A + N_B$ particles arranged in a rather special way, with those of type A all in one half of the chamber, and those of type B all in the other. By contrast, at the end any given particle could be in either place, but nothing has happened overall to the volume or temperature of the complete system. Therefore the entropy increase is strictly owing to the increase in the number of ways the particles could be spatially configured inside the system, consistent with the macroscopic properties. In short, it is wholly owing to the mixing.

(9.14) Extend the argument of Figure 9.15 to prove that adiabatic surfaces never cross.

10 Heat flow and thermal relaxation

10.1 Thermal conduction; diffusion equation   136
10.2 Relaxation time   145
10.3 Speed of sound   146
Exercises   148

This chapter is concerned with heat conduction and relaxation times. We turn to these subjects at this stage because they can be treated without the need for further concepts such as free energy, which will be introduced in subsequent chapters. Therefore this chapter is a break from the development of the main themes of the book. The study of heat flow requires almost no thermodynamic reasoning: it is mostly just a study of certain differential equations and their solutions. The only essentially thermodynamic part is that we have some understanding of temperature, that heat is a form of energy, and that energy is conserved.

10.1 Thermal conduction; diffusion equation

Heat can pass from one place to another primarily by three mechanisms: conduction, convection, and radiation. In conduction heat moves from one part of a material medium to another without any net displacement of the material of the medium. Convection can take place in fluids, it refers to the displacement of the matter of the fluid, which carries internal energy along with it. For example, material which absorbs heat in a hot region may subsequently move to a cold region and there give off heat. The net effect is as if heat were transported, though strictly speaking it is internal energy which is transported in convection. We will not discuss convection further in this book, except to say that a very crude model, appropriate in non-turbulent conditions, is given by *Newton's law of cooling*, presented in Exercise 10.3 of this chapter.

Radiation refers to the energy transport by waves, especially electromagnetic waves. When electromagnetic waves pass from one body to another the resulting energy transport can contain both zero entropy ('ordered') and non-zero entropy ('disordered') parts, hence both work and heat. This will be discussed in Chapter 19.

In the following we discuss purely conduction.

The heat flux $\mathcal{J}$ is defined as the amount of energy crossing a given plane, per unit time per unit area of the plane. For modest temperature gradients, $\mathcal{J}$ is proportional to the (negative of the) temperature gradient, and the constant of proportionality is called the *thermal conductivity* κ:

$$\mathbf{J} = -\kappa \nabla T. \tag{10.1}$$

The box on page 138 gives some information about the thermal conductivity of a monatomic ideal gas.

When the material conducting the heat is not itself generating heat, and no work or chemical process is involved, then the conservation of energy implies that $\mathbf{J}$ satisfies a continuity equation:

$$\nabla \cdot \mathbf{J} = -\frac{\partial \rho}{\partial t},$$

where ρ is the internal energy density. Hence

$$\nabla^2 T = \frac{1}{\kappa} \frac{\partial \rho}{\partial t}. \tag{10.2}$$

In this equation, $\delta \rho$ is an internal energy change (per unit volume) associated with the arrival of heat, therefore it is given by

$$\delta \rho = c \delta T,$$

where c is the heat capacity per unit volume,[1] and δT is the temperature change of the material. Substituting this into (10.2), we obtain

---

The thermal diffusion equation
$$\nabla^2 T = \frac{c}{\kappa} \frac{\partial T}{\partial t}. \tag{10.3}$$

This equation is an example of the *diffusion equation* which also arises in other branches of physics, such as diffusion of particles in gases and propagation of electromagnetic fields in conductors, so techniques for its solution can be applied to all, and similar phenomena result.

In steady state, the diffusion equation reduces to the Laplace equation.

When a material both conducts heat and has heat generated inside it (for example, think of a current-carrying electrical resistor) then we need to add a term H to the continuity equation, equal to the rate per unit volume at which heat is being generated:

$$\nabla \cdot \mathbf{J} = -\frac{\partial \rho}{\partial t} + H.$$

The thermal diffusion equation then becomes

$$c \frac{\partial T}{\partial t} - \kappa \nabla^2 T = H. \tag{10.4}$$

The term H makes the equation inhomogeneous and is said to be a *source term*.

In steady state this equation becomes the Poisson equation.

---

[1] Clearly $c = C/V$, where C is the heat capacity of some small volume V of material, but which heat capacity? C_V, C_p, or another? The answer depends on the circumstances of the heat transport. In practice we often have in mind the case of constant pressure. For solids and liquids the distinction is of small significance since the two capacities are almost the same.

Values of κ, c, D.
A theoretical treatment of a monatomic ideal gas using *kinetic theory* yields the formulae

$$c = \frac{3}{2} n k_B, \tag{10.5}$$

$$\kappa \simeq c \lambda \bar{v} \propto T^{1/2}, \tag{10.6}$$

$$D \equiv \kappa/c \simeq \lambda \bar{v} \propto T^{1/2}, \tag{10.7}$$

where n is the number density, λ is the mean free path, and $\bar{v}$ the mean speed of atoms in the gas. The formula for c has high accuracy; the other formulae have around 20% accuracy.

10.1.1 Steady state

A steady state situation of heat flow occurs when there is a temperature difference maintained by some external agency between two places, and we are interested in the distribution of temperature and the heat flux. The agency maintaining the temperature difference obviously has to provide the heat flux at the hot point, and remove it from the cold point.

The simplest case is a one-dimensional geometry such as heat flow down a uniform bar, thermally insulated except at the ends. For $H = 0$ and κ independent of T the solution is a uniform temperature gradient.

The following example gives another often encountered case.

Example 10.1

A cylindrical wire is surrounded by a jacket made of material of thermal conductivity κ. The radius of the wire is a, the outer radius of the jacket is b, and the outer surface of the jacket is maintained at temperature T_o. If the wire generates heat at the rate P per unit length, then what is the temperature at the surface of the wire?

Solution.
To answer a problem like this, rather than solving the diffusion equation, it can be simpler to consider the thermal conduction directly in terms of conservation of energy. Consider the heat flowing across a cylindrical surface of length L and radius $r > a$ (see Figure 10.1). The curved surface area is $2\pi r L$ so the rate of heat flow across this surface is $2\pi r L \mathcal{J}$. By conservation of energy, in steady state this must equal the power generated in length L of the wire:

$$2\pi r L \mathcal{J} = PL. \tag{10.8}$$

Now use (10.1) and we have a simple first-order differential equation for T:

$$2\pi r \kappa \nabla T = -P\hat{\mathbf{r}}.$$

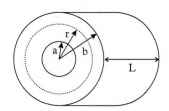

Fig. 10.1 *A cylindrical wire.*

The cylindrical symmetry implies that T only varies in the radial direction, and the radial component of ∇T is just dT/dr, hence

$$\frac{dT}{dr} = -\frac{P}{2\pi\kappa r}.$$

The solution is

$$T(r) = \text{const.} - \frac{P}{2\pi\kappa}\ln r.$$

The given boundary condition was $T(b) = T_o$, so we find

$$T(r) = T_o - \frac{P}{2\pi\kappa}\ln\frac{r}{b}.$$

It is not difficult to alter the reasoning to treat the interior of a wire which both generates heat and conducts it outwards. The heat flow out through a cylindrical surface is still $2\pi rL\mathcal{J}$, as in equation (10.8); now we equate this to the heat generated within the cylindrical region bounded by that surface:

$$2\pi rL\left(-\kappa\frac{dT}{dr}\right) = \pi r^2 LH. \tag{10.9}$$

This equation is even simpler to solve than the previous one (exercise for the reader!).

Steady-state spherical problems fall to similar methods. When $H = 0$ the general solution in the case of spherical symmetry is of the form

$$T = A + \frac{B}{r},$$

where A and B are constants. This is a solution of Laplace's equation (except at $r = 0$), and indeed it should be familiar since the same solution appears in electrostatics for the electric potential of a point-like charge.

10.1.2 Time-dependent

For time-dependent problems, it is best to consider the diffusion equation itself and general methods of solution. Assuming the material properties κ and c are independent of temperature, the equation is linear and we can use the superposition principle.

In the case of a linear differential equation, it is often useful to enquire whether wave-like solutions are possible. With this in mind, we propose the trial solution

$$T(x, t) = Ae^{i(kx-\omega t)}, \tag{10.10}$$

where A, k, and ω are constants (the physical temperature will be the real part of T). Substituting this into (10.3), we obtain

$$k^2 = i\omega \frac{c}{\kappa}. \tag{10.11}$$

This is an example of a *dispersion relation* (i.e. an equation relating wave vector to angular frequency for wave motion). There are two types of solution, depending on whether ω is real or imaginary. First let's consider real ω. To reduce clutter we introduce the *thermal diffusivity* $D \equiv \kappa/c$, then we have

$$k = \pm\sqrt{\frac{i\omega}{D}} = \pm(1+i)\sqrt{\frac{\omega}{2D}}$$

(using $\sqrt{i} = \pm(1+i)/\sqrt{2}$). Our trial solution is therefore valid and appears in one of two forms:

$$Ae^{-\alpha x}e^{i(\alpha x - \omega t)} \quad \text{or} \quad Ae^{\alpha x}e^{i(-\alpha x - \omega t)},$$

where $\alpha = \sqrt{\omega/2D}$. The first solution is a wave propagating towards positive x and decaying as it goes, the second solution is a wave propagating towards negative x and decaying as it goes. The characteristic decay length (called *skin depth*) is $1/\alpha$. Since this is equal to the inverse of the wave vector, the spatial oscillation of the wave is mostly suppressed: after one wavelength the amplitude has decayed by a factor $\exp(-2\pi) \simeq 0.00187$ (see Figure 10.2).

The other type of solution, which is expressed by (10.10) when ω is imaginary, is more conveniently written

$$T(x,t) = Ae^{\alpha x - \gamma t}, \tag{10.12}$$

where the decay rate γ is real. Then one obtains $a^2 = -\gamma/D$ so a is pure imaginary. Therefore, we have

$$T(x,t) = Ae^{ikx}e^{-\gamma t}, \tag{10.13}$$

with $\gamma = k^2 D$. These solutions oscillate in space and decay in time, whereas the first set of solutions oscillate in both space and time and decay in space.

In three dimensions, there are plane wave solutions behaving similarly to those exhibited above, and also cylindrical and spherical wave solutions. The solution obtaining in any given case depends on the boundary conditions. We shall consider two standard cases which illustrate the main themes.

10.1.2.1 Conduction through a flat wall

First suppose we have a thick flat wall, the outside of which is subject to periodic temperature fluctuation:

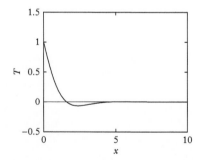

Fig. 10.2 The function $T = e^{-\alpha x} \cos(\alpha x - \omega t)$ at $\alpha = 1$, $t = 0$. This is a solution of the diffusion equation. It applies, for example, to the case where the temperature at one end of a bar is oscillating.

$$T(0, t) = T_0 + A\cos(\omega t).$$

Since this is the real part of $T_0 + A\exp(i\omega t)$, we may immediately observe that a solution to the diffusion equation that matches this boundary condition is given by the real part of (10.10) plus a constant (T_0). To decide whether we need the right-propagating or left-propagating solution or both, we need to consider another boundary condition. In the case of a thick wall extending from $x = 0$ to some large positive x, we may take it that the other condition is $T \to T_0$ as $x \to \infty$, therefore only the right-propagating solution is needed:

$$T(x, t) = T_0 + Ae^{-\alpha x}e^{i(\alpha x - \omega t)}. \tag{10.14}$$

We find that at any given $x > 0$, the temperature oscillates about T_0, but the amplitude of the oscillation is reduced by a factor $\exp(-\alpha x)$. This explains why the temperature inside a building with thick walls, or in a cave underground, does not fluctuate as much as the temperature in the surroundings. In hot countries, such buildings feel comparatively cool during the day, and warm at night, compared to the air outside. The temperature inside still rises and falls somewhat, but the phase of this oscillation as a function of time is offset from that at $x = 0$ by $\phi = \alpha x$. For example, if the exterior surface of the wall reaches a maximum temperature at time $t = 0$, then the material at x reaches a maximum temperature at the time

$$t = \frac{\alpha x}{\omega} = \sqrt{\frac{1}{2\omega D}}\, x.$$

If the temperature at the outside of the wall is given by some other function of time, the method of Fourier analysis can be used to find the solution (see Exercises).

10.1.2.2 Heating or cooling of a spherical object

In the case of a spherical object, one proposes solutions having spherical symmetry. These are spherical wave solutions. To obtain the right form, make the substitution

$$T = \frac{P}{r}$$

in the diffusion equation, where $P(r, t)$ is a function of radius and time. We have

$$\frac{\partial T}{\partial r} = -\frac{P}{r^2} + \frac{1}{r}\frac{\partial P}{\partial r},$$

so $r^2 \partial T/\partial r = -P + r\partial P/\partial r$. Substituting this into the diffusion equation (in spherical polar coordinates), we obtain

$$\frac{\partial^2 P}{\partial r^2} = \frac{1}{D}\frac{\partial P}{\partial t}.$$

The problem has now reduced to a one-dimensional diffusion equation, and we can use the solutions we have already developed previously. (The same simplification is used in other areas of physics, such as the solution of the Schrödinger equation for a spherically symmetric case.) For example, the spherical wave solutions are given by

$$P = A e^{i(kr-\omega t)} \quad \Rightarrow \quad T = A \frac{e^{i(kr-\omega t)}}{r},$$

with $k = (1+i)\sqrt{\omega/2D}$, and the solutions that decay with time are given by

$$P = A e^{ikr} e^{-Dk^2 t} \quad \Rightarrow \quad T = A \frac{e^{ikr} e^{-Dk^2 t}}{r}$$

(cf. equation (10.13)).

Consider now the problem of cooking a plum pudding. This is a roughly spherical object composed of suet, plums, and other fruit, which we shall approximate as homogenous. The pudding, initially at temperature T_0 throughout, is placed in an oven maintained at temperature T_1. We would like to find the temperature anywhere in the pudding as a function of time.

The boundary conditions are

$$T(r, 0) = T_0 \quad \text{for } r < a.$$
$$T(a, t) = T_1.$$

The first condition says that the pudding is cold and of uniform temperature T_0 throughout at $t = 0$. The second condition says the surface of the pudding is permanently hot at constant temperature T_1. At $t = 0$ there is a discontinuity in the temperature as a function of radius at $r = a$. In practice the outermost layer of the pudding will very quickly reach the temperature T_1; we proceed on the assumption that the solution will exhibit this property.

Any given function can be written down in more than one way. For example, any function can be Fourier analysed and written as a sum of sines and cosines. Part of the skill in solving a problem of this kind is in adopting a convenient way of breaking down the solution into a set of simple functions that can be summed. For cooking the pudding, we propose the form

$$T(r, t) = T_1 + \frac{P(r, t)}{r},$$

with the function P to be discovered. Here P describes the amount of departure from the *oven* temperature, so it has the boundary conditions

$$P(r, 0) = (T_0 - T_1) r \quad \text{for } r < a.$$
$$P(a, t) = 0.$$
$$P(0, t) = 0.$$

The third condition arises because we must have finite T as $r \to 0$, therefore P cannot be non-zero at the origin.

Since the oven temperature is steady, and we expect the pudding eventually to reach that temperature, we propose for P not the wave-like type of solution (10.10) but the exponentially decaying type (10.13), or to be precise, a sum of terms of the form

$$\sin(kr)e^{-k^2 Dt}. \tag{10.15}$$

Here we have picked sine functions rather than cosines or complex exponentials in order to satisfy the third boundary condition on P. This condition implies that there can be no $\cos(kr)$ functions in the solution, because if there were then they could not cancel each other out at all times (since they would each have a different time dependence) so they would give rise to $P(0, t) \neq 0$ at some or all times t, which is ruled out.

The second boundary condition on P is that $P = 0$ at $r = a$ for all times. For contributions of the form (10.15) this is satisfied if

$$ka = n\pi.$$

Hence the solution has the form

$$P = \sum_{n=1}^{\infty} A_n \sin\left(\frac{n\pi r}{a}\right) e^{-D(n\pi/a)^2 t}.$$

So far we have matched the second and third boundary conditions but not the first, except at $r = 0$. The first boundary condition is a condition on $P(r, 0)$. It gives

$$(T_0 - T_1)r = \sum_{n=1}^{\infty} A_n \sin\left(\frac{n\pi r}{a}\right).$$

This is the problem of finding the Fourier sine series for the function r. Multiply both sides by $\sin(m\pi r/a)$ and integrate from 0 and a. All the terms on the right-hand side vanish except one, which evaluates to $(a/2)A_m$. Hence

$$A_m = \frac{2(T_0 - T_1)}{a} \int_0^a r \sin\left(\frac{m\pi r}{a}\right) dr.$$

The integral can be evaluated by parts, yielding

$$A_m = \frac{2a}{m\pi}(T_1 - T_0)(-1)^m.$$

The overall solution for the temperature of the pudding is therefore

$$T(r, t) = T_1 + \frac{2a(T_1 - T_0)}{\pi r} \sum_{n=1}^{\infty} \frac{(-1)^n}{n} \sin\left(\frac{n\pi r}{a}\right) e^{-D(n\pi/a)^2 t}. \tag{10.16}$$

This concludes the derivation: we now have an explicit expression for the temperature at all times and places.

We shall examine the main features of this result.

At $r \to 0$ equation (10.16) evaluates to

$$T(0,t) = T_1 + 2(T_1 - T_0) \sum_{n=1}^{\infty} (-1)^n e^{-D(n\pi/a)^2 t}.$$

At $t = 0$ the sum oscillates between -1 and 0 as more and more terms are added, and therefore is undefined. This is merely a mathematical anomaly, however. For any $t > 0$ the series has a well-defined sum, which for small t is $-1/2$ and for large t is 0. The result is shown in Figure 10.3.

The outside of the pudding (near $r = a$) quickly reaches the final temperature. The inside (at $r = 0$) at first has a constant temperature, then rises towards T_1, approaching it exponentially when $t \gg \tau$, where $\tau = a^2/\pi^2 D$. During cooking the temperature inside the pudding is approximately a constant minus a sinc function of r. The outside is cooked more thoroughly than the inside, but this is not much of a problem because the chemical reactions involved are a strong function of temperature, so that it is more important to reach the desired temperature than it is to stay for a long time at that temperature. Indeed, a modest variation of result across the pudding improves the pleasure of eating it.

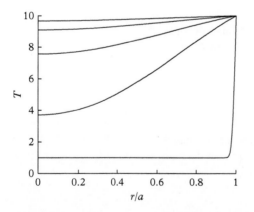
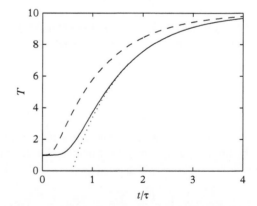

Fig. 10.3 *Temperature of a spherical pudding as a function of position and time. In this example the pudding was initially at temperature $T_0 = 1$ throughout, and it was placed in an oven maintained at temperature $T_1 = 10$. The figure shows the predicted behaviour from equation (10.16). The left-hand graph shows the temperature throughout the pudding at various times. The lowest curve is at $t = 10^{-3}\tau$, where $\tau = a^2/\pi^2 D$. The other curves are at $t/\tau = \{1, 2, 3, 4\}$, respectively. The right-hand graph shows the temperature at $r = a/2$ (dashed curve) and at the centre of the pudding ($r = 0$, full curve) as a function of time. The dotted curve shows just the first term in the series in equation (10.16), which in this example is $10 - 18e^{-t/\tau}$.*

The series on the right-hand side of (10.16) is dominated by the first term when $t \gg a^2/D\pi^2$, so for large t the solution has the form

$$T \simeq T_1 - 2(T_1 - T_0)\frac{\sin(\pi r/a)}{\pi r/a} e^{-D(\pi/a)^2 t}. \tag{10.17}$$

Thus the pudding relaxes exponentially towards its final temperature with a time constant $a^2/D\pi^2 = a^2 c/\kappa\pi^2$. It makes sense that larger puddings with a low conductivity and high heat capacity per unit volume take longer to cook.

A final comment. In the two examples treated above—the flat wall and the spherical pudding—we wished to exhibit two types of spatial solution (rectangular and spherical) and also two types of temporal solution (oscillatory and exponentially decaying). It happened that the oscillatory temporal behaviour occurred with the rectangular case, and the exponential decay with the spherical case, but this was merely because of the choice of example problems treated. There is nothing to prevent other scenarios from occurring, such as a spherical problem involving an oscillation in time, or a rectangular problem involving exponential decay in time, or a case where the answer is most simply expressed as a sum involving all types of solution.

10.2 Relaxation time

The treatment of heat flow permits an estimate of the all-important relaxation time which determines how long a body takes to reach an equilibrium state, and thus how slow a process needs to be in order to be treated as quasistatic. The two examples above looked at a temperature oscillation penetrating into a wall, or along a rod, equation (10.14), and a step change in temperature penetrating a homogenous sphere of radius a, equation (10.16). The first case gave a characteristic decay length called the skin depth:

$$\delta = \sqrt{\frac{2D}{\omega}} \tag{10.18}$$

for waves of angular frequency ω. The second case gave a characteristic time

$$\tau = \frac{a^2}{\pi^2 D}. \tag{10.19}$$

In both cases, suppose we are interested in treating the system (wall, rod, pudding) using equilibrium thermodynamics. We regard the change in external temperature as a perturbation, and we wish to know either how slow the perturbation must be in order that the system can follow it isothermally (first case), or how long it takes for our system to reach a uniform temperature (second case). In both cases, let a be the system size, such as the radius of a sphere or the distance from the wall to the centre of a cube. For an oscillating perturbation, the temperature at a is only modestly affected when $a = \delta/2$. Putting this into (10.18) and defining the relaxation time by $\tau_r = 1/\omega$, we find

$$\tau_r \simeq \frac{2a^2}{D}. \tag{10.20}$$

For the relaxation time of the sphere subject to a step change, we take as the relaxation time $\tau_r = 2\tau$, with τ as given in (10.19), since it takes a time 2τ for the temperature at the centre of the sphere to be well on the way to its final value (see Figure 10.3 and note the factor 2 in equation (10.17)). Hence

$$\tau_r \simeq \frac{2a^2}{\pi^2 D} \simeq \frac{a^2}{5D}. \tag{10.21}$$

This is an order of magnitude smaller than (10.20), in part owing to the different geometry.

The kinetic theory of gases gives $D \simeq \lambda \bar{v}$, where λ is the mean free path and $\bar{v}$ is the mean velocity (see box on page 138). In terms of the collision time $\tau_c = \lambda/\bar{v}$, equation (10.21) can be written

$$\frac{\tau_r}{\tau_c} = \frac{1}{5}\left(\frac{a}{\lambda}\right)^2. \tag{10.22}$$

10.3 Speed of sound

Sound waves are pressure waves. The speed of sound in a given material is therefore related to the compressibility. Since the sound involves acceleration of the matter and hence inertia, one should expect the density of the material to be relevant also. It is shown in the box that, for longitudinal waves in a fluid, the speed of sound is given by

$$v_s = \sqrt{\frac{\partial p}{\partial \rho}} = \sqrt{\frac{B}{\rho}}, \tag{10.23}$$

where ρ is the density; and B is the bulk modulus,

$$B = \frac{1}{\kappa} = -V\frac{\partial p}{\partial V}, \tag{10.24}$$

and the partial derivatives must be evaluated under whatever conditions correctly describe the expansion and compression of the fluid associated with the sound wave. For any material, equation (7.33) tells us that $B_S = \gamma B_T$, so we expect the speed of sound under adiabatic conditions to be faster than that under isothermal conditions by a factor $\sqrt{\gamma}$.

For an ideal gas the equation of state gives $B_T = p$, so $B_S = \gamma p$ and therefore the adiabatic speed of sound is

$$v_s = \sqrt{\frac{\gamma k_B T}{m}}, \tag{10.25}$$

where m is the molecular mass. This is similar to the mean speed of the particles in the gas.

Speed of sound in a fluid

According to classical mechanics, the smooth flow of an ideal fluid is described by the *Navier–Stokes equation* (also called the Euler equation):

$$\rho\left(\frac{\partial \mathbf{v}}{\partial t} + (\mathbf{v} \cdot \nabla)\mathbf{v}\right) = -\nabla p, \tag{10.26}$$

where $\mathbf{v}(x, y, z, t)$ is the flow velocity at any point, p is the pressure, and ρ the density. A sound wave in the fluid is a case where $\mathbf{v}$, ρ, and p all undergo oscillations. For a fluid otherwise at rest, the speeds v involved are vanishing in the limit of low amplitude oscillations, so we can ignore the second term in comparison with the first, obtaining

$$\rho \frac{\partial \mathbf{v}}{\partial t} = -\nabla p. \tag{10.27}$$

The fluid density also satisfies the *continuity equation*, which describes the conservation of mass:

$$-\frac{\partial \rho}{\partial t} = \nabla \cdot \mathbf{j}, \tag{10.28}$$

where $\mathbf{j} = \rho \mathbf{v}$ is the flux of material. This gives

$$-\frac{\partial \rho}{\partial t} = (\nabla \rho) \cdot \mathbf{v} + \rho \nabla \cdot \mathbf{v}. \tag{10.29}$$

Define s by $\rho = \rho_0(1+s)$, where ρ_0 is constant. Then the first term on the right is of order $\rho_0 k s v$ and the second is of order $\rho_0 k v$, where $k = 2\pi/\lambda$ is the wave vector. Hence for small amplitudes, $|s| \ll 1$, the first term is negligible, and we have

$$-\frac{\partial \rho}{\partial t} = \rho \nabla \cdot \mathbf{v}. \tag{10.30}$$

Now restrict to one dimension and focus attention on $\mathbf{v}$. Equations (10.27) and (10.30) give

$$\frac{\partial v}{\partial t} = -\frac{1}{\rho}\frac{\partial p}{\partial x}, \quad \frac{\partial v}{\partial x} = -\frac{1}{\rho}\frac{\partial \rho}{\partial t}. \tag{10.31}$$

Using $\partial^2 v/\partial x \partial t = \partial^2 v/\partial t \partial x$, one finds

$$\frac{\partial^2 p}{\partial x^2} = \frac{\partial^2 \rho}{\partial t^2}, \tag{10.32}$$

where further terms have been neglected in the limit of small oscillations, $|s| \ll 1$. Finally, use

$$\left.\frac{\partial p}{\partial x}\right|_t = \left.\frac{\partial p}{\partial \rho}\right|_t \left.\frac{\partial \rho}{\partial x}\right|_t \implies \frac{\partial^2 p}{\partial x^2} = \left(\frac{\partial p}{\partial \rho}\right)\frac{\partial^2 \rho}{\partial x^2}.$$

By substituting this into equation (10.32), one obtains a wave equation for ρ in which the wave speed is as given by (10.23).

The conditions prevailing in practice will depend on the wavelength of the waves. Naively one might expect high-frequency waves to be adiabatic by arguing that there is not enough time for significant heat flow, but in the case of wave motion this is mistaken because we must also consider the distance over which the heat must flow in order to establish a uniform temperature. This gets smaller at high frequencies: see equation (10.18). If there is a periodic temperature oscillation at a given plane, then the heat flow to other planes is only fast enough to establish a uniform temperature over distances small compared to δ. Therefore the condition for sound waves to be isothermal is $\lambda \ll \delta$, where λ is the wavelength. This gives

$$\lambda \ll \frac{D}{\pi v_s}. \tag{10.33}$$

Thus it is short wavelength (or high frequency) waves that are isothermal, owing to the small distance over which heat must be transported in order to establish a uniform temperature. Long wavelength waves are adiabatic. At intermediate frequencies the waves are neither isothermal nor adiabatic.

The kinetic theory of gases predicts that $D \simeq \lambda_{\text{mfp}} \bar{v}$, where λ_{mfp} is the mean free path, defined as the average distance travelled by the molecules between collisions, and $\bar{v}$ is the mean speed of the molecules. Using $v_s \simeq \bar{v}$ one finds that the cross-over condition occurs when the wavelength of the sound is of the order of the mean free path. Since this is a very short wavelength in ordinary circumstances, sound waves are usually adiabatic.

10.3.1 Ultra-relativistic gas

When the motion of a fluid is treated using special relativity, then in the instantaneous rest frame of any given element in the fluid, one obtains a modified Navier–Stokes equation, in which ρ is replaced by $(u+p)c^{-2}$, where u is the energy density and c is the speed of light. An ultra-relativistic gas has $p = u/3$ and then one finds for the adiabatic speed of sound,

$$v_s^2 = \frac{B_S}{(4/3)uc^{-2}} = c^2 \frac{\gamma p}{4p} = \frac{c^2}{3}, \tag{10.34}$$

where we have used that $\gamma = 4/3$ for the ultra-relativistic gas, and the equation of state is unchanged. Hence the sound waves travel at $1/\sqrt{3}$ times the speed of light. This result is important in the evolution of the early universe.

....

EXERCISES

(10.1) A church has limestone walls of thickness 30 cm. The outside air temperature varies daily between 10 °C and 25 °C, in a sinusoidal fashion with a maximum value at noon. Using a single flat wall model, estimate the temperature fluctuation inside the church, and the time of day when the interior is at its warmest. [Limestone has thermal conductivity 1.5 Wm^{-1}K^{-1}, specific heat capacity 910 JK^{-1}kg^{-1}, and density 2180 kgm^{-3}.]

(10.2) A cylindrical wire of radius a, resistivity ρ, and thermal conductivity κ carries a current I. In steady state, the temperature at the surface of the wire is T_0. Show that if the current is carried uniformly, then the temperature inside the wire at radius r is

$$T(r) = T_0 + \frac{\rho I^2}{4\pi^2 a^4 \kappa}\left(a^2 - r^2\right).$$

(10.3) *Newton's law of cooling* states that when a gas transports heat away from a solid or liquid surface by convection and conduction together, the net rate of transport is proportional to the area of contact multiplied by the temperature difference ΔT between the surface and the bulk of the gas. This amounts to the following statement about the heat flux:

$$\mathbf{J} = \mathbf{k}\Delta T,$$

where **k** is a coefficient whose magnitude depends on the conditions and whose direction is normal to the surface. This is an empirical rule, and a good approximation when ΔT is not too large. If the wire considered in question (10.2) is air-cooled, find an expression for T_0 in terms of k and properties of the wire.

(10.4) A cup of black coffee is initially at temperature 80 °C in a room of ambient temperature 25 °C. The temperature difference between the cup and the room decays exponentially with time constant 10 minutes. A small jug of milk in the same room has initial temperature 5 °C and its temperature difference with the room changes exponentially with a time constant of 10 seconds. The heat capacities of milk and coffee are 100 J/K and 1000 J/K respectively. At what moment should the milk be added to the coffee in order to obtain milky coffee at the highest temperature?

Practical heat engines

11.1 The maximum work theorem 152
11.2 Otto cycle 153
Exercises 155

Heat engines were introduced in Chapter 8, where we were mainly concerned with the Carnot cycle and the idealized case of reversible behaviour. In this chapter we shall introduce some further types of reversible cycle, and comment on the main imperfections which lead to irreversible behaviour and reduced efficiency.

The first widely used heat engines were the steam engines which provided one of the foundations of the Industrial Revolution. The early engines were used to pump water from mines and were a tremendous success, despite their poor energy efficiency. Although this inefficiency is partly owing to imperfections such as heat loss and friction, it is easy to see from basic thermodynamics that only a limited efficiency was possible. Although high temperatures are available from burning coal, it is not easy to heat steam much above the boiling point because of the high pressures involved. Steam was available in early engines somewhat above the boiling point, say 375 K, and condensed in a chamber somewhat below the boiling point, say 335 K. Even a perfect engine operating between these extremes would only be capable of an efficiency of 11%. James Watt introduced an important design improvement by providing a separate chamber connected to the piston by a pipe and valve. This separate chamber is maintained at a cold temperature, and when the valve is opened the steam condenses there. This effectively increases the temperature ratio to around 375/290, increasing the available ideal efficiency to approximately 23%.

By contrast with the steam engine, which involves an external heat source and an operating fluid, the *internal combustion engine* burns fuel such as petrol right inside the piston chamber. Very high temperatures are involved in the burning of petrol and in principle the thermal efficiency could approach one. It does not because of limitations on the attainable pressure and various imperfections.

Internal combustion engines have been the primary type of engine in cars for over a century. However, there is now growing interest in powering cars without the use of a heat engine at all, for example by use of a fuel cell which converts chemical energy directly into electrical energy. Fuel cells are not yet commercially viable for widespread use in cars, but they are used in commercial, industrial, and residential power generation, especially in remote locations.

Another increasingly important type of engine is the Stirling engine. This cleverly designed heat engine requires no valves and runs very smoothly. Heat is supplied to one chamber and removed from another (or a single chamber with a temperature gradient can be used), see Figure 11.1. The working fluid stays

Thermodynamics: A Complete Undergraduate Course. Andrew M. Steane.
© Andrew M. Steane 2017. Published 2017 by Oxford University Press.

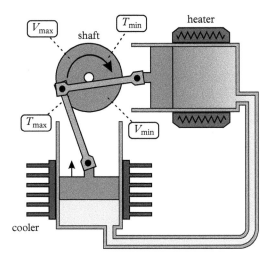

Fig. 11.1 *The Stirling engine. The engine drives the pistons which cause the shaft to rotate as shown. The labels indicate the position of the crank at four example points in the cycle. This makes it easy to see that the expansion happens mostly when the fluid is hot and the compression mostly when the fluid is cold. Hence in each cycle the fluid does more work than is done on it, the energy for which is obtained from the heater.*

permanently inside the engine. When the fluid is mostly in the hot chamber, it is heated; when it is mostly in the cold chamber it is cooled. Working the pistons a quarter cycle out of step is sufficient to arrange that the expansion to maximum volume occurs mainly when the fluid is hot, whereas compression to minimum volume occurs mainly when the fluid is cold. Hence the efficiency can in principle approach that of a Carnot engine (cf. Exercise 11.4).

Solar power is the generic name for a range of methods used to convert the energy of sunlight into useful, usually electrical, energy, without burning fossil fuels. *Photovoltaic* methods convert light into electric current using the photovoltaic effect. *Concentrated solar power* (CSP) methods use mirrors to focus sunlight onto a receiver, where it is used to drive a heat engine such as a steam turbine or a Stirling engine. Currently CSP offers the better overall efficiency in a commercial setting (around 25%), but photovoltaic methods are improving rapidly and likely to be very important in the future.

Heat pumps

Both heat engines and heat pumps typically make use of a *working fluid*. This is some fluid substance that can be easily compressed or expanded, and that exhibits a high heat capacity. A good way to obtain a high heat capacity is to make use of a phase transition between the liquid and vapour forms. During the transition the heat capacity at constant pressure goes to infinity. For example, when the fluid is in liquid form at a temperature near its boiling point, then if the pressure is reduced the fluid will begin to boil. Heat passes from the surroundings (e.g. the air inside a refrigerator) to the fluid; the amount of heat absorbed as a unit mass of fluid changes from liquid to vapour is the *specific latent heat*. The vapour is then transported through pipes to a heat exchanger in contact with the air outside the refrigerator, and there it is compressed, undergoing a phase transition back to liquid form and giving off the latent heat. An example is the *Rankine cycle* which is described in detail in Chapter 16.

11.1 The maximum work theorem

It is a common experience that friction and damping tend to place limits on what can be achieved by machines. This general intuition is made more precise by the *maximum work theorem*, which states that *for all processes leading a system from a given initial to a given final state, the work output is maximum (and the heat output minimum) for a reversible process*. To prove this, note that when moving between given states, the internal energy of the system in question changes by a given amount ΔU, so any increase in the heat output must be associated with a reduction in the work output. Therefore for maximum work, we require a process in which a minimum of heat flows out of the system in question.

To accomplish the process, the system must reach whatever is the entropy of its final state, so it must undergo some entropy change ΔS. At any given T, we have $dS \geq đQ/T$ (where $đQ$ is the heat entering the system), with equality for the reversible case. Therefore the maximum heat *input* that can happen, for the given total ΔS, occurs when the process is reversible. This condition corresponds to the minimum heat *output*, and therefore the maximum work output.

Note, in the above argument, if $\Delta S > 0$ then this does not necessarily imply that something irreversible must happen; it implies rather that the system must take in heat if the process is reversible. Also, the argument applies for either sign of ΔS, and for $\Delta S = 0$.

11.1.1 Imperfections

There are two types of issue that limit the efficiency of heat engines and heat pumps: limits on the attainable pressures and temperatures owing to material properties such as tensile strength and melting point, and irreversible processes. The latter may be deemed to be 'imperfections'. The most important ones are

(1) Friction in the moving parts.
(2) Turbulence in the working fluid.
(3) Heat conduction directly between the reservoirs.
(4) Temperature gradients at the heat exchangers.

The last item refers to the difference between the temperature of the working fluid and the temperature of the heat source or heat sink. For example, the air near the vanes on the back of a household refrigerator is normally warmer than the air in the rest of the room because it has been warmed by heat pumped out of the refrigerator. This increases the temperature difference the heat pump has to handle.

11.2 Otto cycle

The internal combustion engine usually works on a cycle involving four movements of the piston (i.e. two complete rotations of the drive shaft). Such an engine is called a 'four-stroke engine'. A two-stroke design (i.e. a single rotation per engine cycle) is also possible. The four-stroke cycle is affectionately termed 'suck, squeeze, bang, blow'.[1] In the following description we suppose the piston is mounted vertically with valves at the top, as is usually the case (see Figure 11.2).

1. *Intake*. The intake valve opens, then the piston moves down the chamber, drawing fuel and air into the cylinder.
2. *Compression*. The intake valve closes, then the piston moves up the chamber, compressing the fuel and air.
3. *Power stroke*. A spark or other mechanism initiates explosion of the fuel. The explosion is fast, i.e. a high pressure and temperature is reached before there is time for significant movement of the piston. The piston then moves down the chamber, driven by the high pressure gases.
4. *Exhaust*. The exhaust valve is opened and the combustion products, which are still at much higher pressure than the outside atmosphere, rapidly leave the chamber. The piston subsequently moves up the chamber and forces out any remaining gas.

[1] And mechanics may be familiar with even more expressive terms.

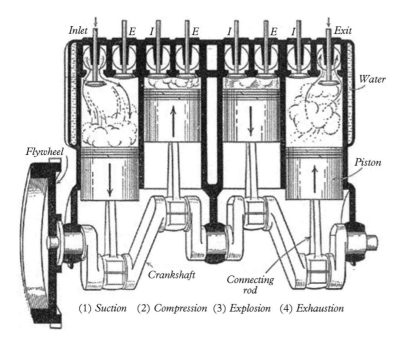

Fig. 11.2 *The internal combustion engine. The figure shows a four-cylinder example, and thus illustrates also the four stages of the four-stroke cycle. (From http://boomeria.org/physicslectures/engines/engines.html).*

154 Practical heat engines

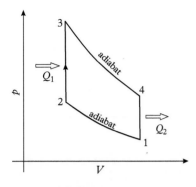

Fig. 11.3 *The Otto cycle.*

The analysis of engines such as the internal combustion engine proceeds in the first instance by inventing some reasonable representation of the main features of the engine cycle. Increasingly sophisticated models can then be used to gain further insight. A first rough approximation is to suppose that the working fluid can be treated as a single pure substance. For the internal combustion engine this is far from the truth (the fluid combines petrol gas, combustion products, and air and its composition changes through the cycle), but gives some useful basic information. Usually air is used to represent the fluid. Cycles based on air are known as 'air standard cycles.' The main features of the four-stroke cycle are modelled by the *Otto cycle*, shown in Figure 11.3. The intake and exhaust strokes are not shown since during this part of the cycle the system is not closed. For the purposes of this analysis, however, they are not important since no net work is done; they merely serve to replace the fluid. However the replacement of hot high-pressure fluid by cold low-pressure fluid has to be represented somehow, and this is handled in the Otto cycle by representing it as a drop in pressure at constant volume (4 → 1 in Figure 11.3). The rest of the cycle involves a closed system. The compression and power strokes are taken to be adiabatic; the explosion is represented as an injection of heat at constant volume.

The efficiency of the Otto cycle is calculated as follows. We model the fluid as an ideal gas and treat the heat capacities as constant. The diagram makes it easy to see that $T_1 < T_4, T_2 < T_3$, so T_3 and T_1 are the maximum and minimum temperatures. The amounts of heat absorbed and emitted are

$$Q_1 = \int_{T_2}^{T_3} C_v dT = C_v(T_3 - T_2),$$

$$Q_2 = -\int_{T_4}^{T_1} C_v dT = C_v(T_4 - T_1),$$

therefore the efficiency is

$$\eta = \frac{W}{Q_1} = \frac{Q_1 - Q_2}{Q_1} = 1 - \frac{T_4 - T_1}{T_3 - T_2}. \tag{11.1}$$

Now 1 → 2 and 3 → 4 are adiabatic so

$$T_1 V_1^{\gamma-1} = T_2 V_2^{\gamma-1},$$
$$T_3 V_2^{\gamma-1} = T_4 V_1^{\gamma-1},$$

therefore

$$(T_4 - T_1) V_1^{\gamma-1} = (T_3 - T_2) V_2^{\gamma-1}.$$

Substituting this into (11.1), we find

$$\eta = 1 - r^{1-\gamma}, \tag{11.2}$$

where $r \equiv V_1/V_2$ is the *compression ratio*.

For good efficiency a high compression ratio is desirable. However if it is made too large then regions in the fuel mixture begin to detonate rather than burn smoothly. The resulting *pinking* or *knocking* is bad for the engine. A compression ratio of about 10 can be used. Taking $\gamma \simeq 1.4$, equation (11.2) gives 60%. More realistic models suggest the thermodynamic limit for most engines made of steel is nearer 40% and real engines typically achieve half of this.

EXERCISES

(11.1) The temperature inside the engine of a crane is 2000 °C, the temperature of the exhaust gases is 900 °C. The heat of combustion of petrol is 47 MJ/kg, and the density of petrol is 0.8 g/cm³. What is the maximum height through which the crane can raise a 10000 kg load by burning 0.1 litre of petrol?

(11.2) The operation of a diesel engine can be modelled approximately by the following cycle: (i) adiabatic compression from (p_1, V_1) to (p_2, V_2), (ii) heating at constant pressure to (p_2, V_3), (iii) adiabatic expansion to (p_4, V_1), (iv) cooling at constant volume back to (p_1, V_1). Find the efficiency in terms of the two compression ratios V_1/V_2 and V_3/V_2.

(11.3) The *Stirling cycle* consists of two isothermal stages and two isochoric (constant volume) stages; see Figure 11.4. (i) Show that if the heat capacity of the working fluid is a function of temperature alone, then the amount of heat entering the fluid in one of the isochoric processes is equal to the amount leaving in the other. (ii) Hence show that the thermal efficiency is $1 - T_2/T_1$, the same as that of a Carnot engine.

It follows from (i) that the energy extracted when the fluid cools from T_1 to T_2 can be stored and used to provide the energy needed to heat the fluid from T_2 to T_1. Such an energy store is called a *regenerator* and is a common feature of Stirling cycle engines.

(11.4) The Stirling *cycle* should not be confused with the Stirling *engine*. A model cycle which approximates the action of the Stirling engine, and which also approximates a regenerative gas turbine, is the *Ericsson cycle*. This consists of two isothermal and two isobaric stages. Find the thermal efficiency if the working fluid obeys Joule's law ($U = U(T)$).

(11.5) A cycle consists of isothermal stages at temperatures T_1, T_2 and two adiathermal stages, the whole being irreversible to the extent that an entropy ΔS is generated inside the system for each cycle. Find the efficiency $\eta = W/Q_1$ in terms of $T_1, T_2, \Delta S$, and Q_1, where W is the work obtained per cycle and Q_1 is the heat extracted from the hotter reservoir (T_1) in each cycle.

(11.6) An Otto cycle uses a volume compression ratio of 9.5, with a pressure and temperature before compression of 100 kPa and 40 °C respectively.

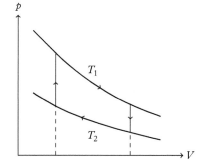

Fig. 11.4 *The Stirling cycle.*

The cycle is performed 3000 times per minute, using a mass of 11.5 g of air per cycle, and heat input 600 kJ/kg. Determine the thermal efficiency and the power output.

[*Ans.* $\eta = 0.594$, power = 205 kW.]

(11.7) The Newcomen engine was the first working steam engine with a piston; it was constructed by Thomas Newcomen through incremental improvements between 1705 and 1711 and was ultimately widely used. A large cylinder (e.g. radius 1 metre, height several metres) was closed by a piston attached to one end of a heavy balanced beam. As the beam rocked to and fro, the other end provided work, for example to drive a water pump. The working principle of the Newcomen engine is as follows. A coal fire heats a quantity of water in a boiler. During the down-stroke on the pump, which is the up-stroke on the piston, steam fills the cylinder, but no useful work is done because the steam pressure is roughly 1 atm; the weight of the pump rod provides the force required to lift the piston. At the top of the piston stroke, a valve to the boiler is closed, and another valve to a supply of cold water under pressure is briefly opened. Some cold water sprays into the chamber. Since this water is much colder than the steam, heat flows from the latter to the former, and the steam condenses. The pressure in the chamber therefore drops considerably, 'sucking' the piston downwards—which is to say, the atmospheric pressure above the piston pushes it down. This is why such engines are called 'atmospheric' engines. This is the power stroke. When the piston reaches the bottom, the valve to the boiler is re-opened and the incoming steam pushes the cold water out of the chamber.

(i) If the cylinder radius is 1 m, and the working stroke of the piston is 3 m, calculate the mass of steam required to fill the chamber at 100 °C, 100 kPa. Determine the heat input required to produce this amount of steam from water initially at 15 °C.

(ii) Determine the amount of water at 15 °C that will suffice to lower the pressure to 50 kPa.

(iii) Find the work output per cycle, and hence the average power output if the machine executes 15 cycles per minute. How does this compare to the heat input in part (i)?

For a modern steam engine, see Exercise 16.14 of Chapter 16.

Introducing chemical potential

12

12.1 Chemical potential of an ideal gas	161
12.2 Saha equation	165
Exercises	167

Chapters 6 to 8 have treated *closed* systems, which do not exchange particles with their surroundings. In order to understand open systems, it is necessary to introduce the idea of *chemical potential*. Chemical potential, denoted μ, is a property (a function of state) that any system must possess. It is intensive, it has dimensions of energy, and it is involved in any process involving transfer of material from one body to another or from one form to another (such as chemical reactions and phase changes). μ plays a role in the transfer of particles similar to the role played by pressure for mechanical work and by temperature for exchange of heat. However, whereas pressure and temperature are reasonably familiar to most of us, chemical potential is new and students (... and physicists in general ...) find it hard to get an intuitive idea of what it means.

To get some intuition, we will introduce a few facts about chemical potential without derivation. We will return to these facts and derive them in later chapters.

The name 'chemical potential' can be misleading, because it suggests that the property is something to do with chemical reactions, which is not necessarily true. The name has stuck however because it is an important and much used quantity in chemistry. The essential idea is that if two systems can exchange particles, then, other things being equal, the system whose chemical potential is higher will tend to relinquish particles, and the one whose chemical potential is lower will tend to acquire particles.[1] So the chemical potential could equally be named the 'particle potential' or 'substance potential.'

For example, in the absence of external fields, particles in a dense gas have higher chemical potential than particles in a similar but less dense gas. If there is a density gradient, particles will tend to move from the more dense to the less dense region. However, if there is a significant gravitational field present, then it will tend to pull the particles to the bottom of the chamber. The gas will then become more dense at the bottom, less dense at the top. The gradient of the density will promote particle movement upwards, while the gradient of gravitational potential promotes particle movement downwards. The chemical potential takes both tendencies into account, and its gradient gives the net force for particle movement. When equilibrium is established, the chemical potential will be uniform throughout the gas.

Similar statements may be made about the movement of charged particles in the presence of an electric field. Thus chemical potential plays a central role in

[1] Speaking of number of particles is convenient because many systems are in fact composed of small entities such as atoms or molecules which are indivisible during the processes under discussion, but we don't require this 'atomic' assumption. By a 'particle' we mean simply a small amount of material that obeys a conservation law, such as conservation of mass, so that we can keep track of how much material a system has lost or gained.

understanding electronic circuit elements such as transistors, and ion exchange across cell membranes in biology. An informal name which gives a good intuition about chemical potential is 'unlivability'. A place or a system with large 'unlivability' is a place that particles will flee from. The exact reason why they want to flee might not be easy to describe; a number of factors come into play, but as μ rises the conditions are becoming more and more 'unlivable' and particles don't want to stay there.

For a system consisting of a single type of particle species, the chemical potential can be defined as

$$\mu \equiv \left.\frac{\partial U}{\partial N}\right|_{S,V}, \qquad (12.1)$$

where N is the number of particles. When there are many different types of species in the system, then we say the system has several *components*, and the i'th component has a chemical potential

$$\mu_i \equiv \left.\frac{\partial U}{\partial N_i}\right|_{S,V,\{N_j\}}, \qquad (12.2)$$

where $\{N_j\}$ indicates that the numbers of particles in other components are all constant. The following statement (due to Gibbs) explains the meaning of the symbols: *If to any homogeneous mass in a state of hydrostatic stress we suppose an infinitesimal quantity of any substance to be added, the mass remaining homogeneous and its entropy and volume remaining unchanged, the increase of the energy of the mass divided by the quantity of the substance added is the chemical potential for that substance in the mass considered.*

The chemical potential of any substance depends on the local environment in which the substance finds itself. In other words, it depends on the intensive variables such as pressure, temperature, density, electric field, etc. In the absence of applied fields this reduces to p, T and the density. The equation of state gives a relation among the intensive variables, so this reduces the number of independent quantities by one. Therefore in the case of a pure substance (just one component) in equilibrium, the chemical potential can be expressed as a function of pressure and temperature alone,

$$\mu = \mu(p, T). \qquad (12.3)$$

Other things being equal, chemical potential typically increases with density or concentration of a substance. For example, ethanol has a higher chemical potential in a solution of scotch whisky than it has in a solution of beer. However, this is not always the case if phase changes can take place. Also, if there is more than one substance present, then the chemical potential of each can depend on the densities of all. To be precise, if there are N substances mixed together, with individual densities n_i, adding up to total density $n = \sum_i n_i$, then the chemical potential of any given substance j in the mix can depend on any $N + 1$ variables from the set $p, T, \{n_i\}$.

For example, for two substances A and B mixed together (say, salt in water), we could have

$$\mu_A = \mu_A(T, n_A, n_B), \quad \mu_B = \mu_B(T, n_A, n_B);$$
or $$\mu_A = \mu_A(p, T, n_A), \quad \mu_B = \mu_B(p, T, n_A);$$
or $$\mu_A = \mu_A(p, T, n_A), \quad \mu_B = \mu_B(p, T, n_B);$$
etc.

In the limit where the molecules of the substances are mostly kept apart (e.g. chemicals at low concentration) this mutual dependence can vanish. This can happen for a mixture of gases: if they can be treated as ideal gases then they each occupy the available volume independently and each has a chemical potential depending only on its own temperature and density.

Equation (12.1) is a useful definition, and it conveys the idea that μ should be thought of as a type of 'energy per particle'. However the conditions of constant entropy and constant volume are hard to realize in practice. An exception is when the system is maintained at a temperature close to absolute zero, so the entropy is fixed (near zero). Then μ can be regarded simply as the energy you would need to provide in order to add one more particle to the system. In the context of solid-state physics this is called the Fermi energy.

A more instructive expression is

$$\mu = \left.\frac{\partial U}{\partial N}\right|_{T,V} - T \left.\frac{\partial S}{\partial N}\right|_{T,V}, \tag{12.4}$$

which we shall derive in Chapter 13. This is useful because it is easier to think about a change taking place at constant volume and temperature, rather than constant volume and entropy. We see that at constant T, V, the chemical potential is a competition between the energy needed to add a particle and the associated change in entropy. At low temperature the energy term dominates (unless $\left.\frac{\partial U}{\partial N}\right| \to 0$); at high temperature the entropy term dominates, and therefore one expects a large negative μ in the high-temperature limit. The chemical potential of a gas, for example, is found to be large and negative at ordinary temperatures, if we don't include the rest-mass energy when calculating U.

Sometimes (12.1) is used to justify loose statements of the form 'chemical potential is the change in internal energy per particle', but it is clear from (12.4) that this is poor physics: the change in U depends on the constraints. However, we will find later that chemical potential can be equated to a related energy called the Gibbs function, divided by the number of particles.

Finally, the role of μ is brought out by the expression

$$\mu = -T \left.\frac{\partial S}{\partial N}\right|_{U,V}. \tag{12.5}$$

Fig. 12.1 *Inside a conductor, at the atomic scale the electric potential is far from uniform, but the chemical potential of the electrons is uniform. The diagram shows the typical situation at the junction of two different metals.*

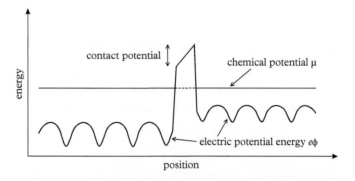

This expression enables us to see immediately an important property: *for an isolated system in thermal equilibrium, the chemical potential is uniform throughout any region of space where the movement of particles is not obstructed.* Proof: if μ were larger at A than at B, then by moving a particle from A to B the system could increase its entropy (since $\Delta S = \Delta S_B - \Delta S_A = (1/T)(\mu_A - \mu_B) > 0$). However, for an isolated system in equilibrium the entropy is already maximized, so this is not possible. From this we obtain the most common role of chemical potential in practice. *Chemical potential is the quantity that determines the movement of particles from place to another.*

If there is no physical obstruction, then particles move from regions of high chemical potential to regions of low chemical potential, and they continue to move until the chemical potential is uniform. For example, it is commonly stated that the electric potential ϕ is uniform in a good conductor in steady state and consequently the electric field is zero. In fact, this is not true; it is the chemical potential μ of the free electrons that is uniform (Figure 12.1). The electrons are like a high-pressure gas (Exercise 6.5 of Chapter 6) and their chemical potential μ includes contributions from their kinetic energy K as well as from the electric potential: $\mu = K + q\phi$, where q is the charge on an electron.[2] This simple formula applies because for this electron gas the first term in (12.4) dominates; this is not the case for an ideal gas where the second term dominates (see Eq. 12.11). At a junction between one metal and another, K changes but μ does not, therefore ϕ must change. In consequence there is a change in ϕ by an amount called the *contact potential* which can be substantial, of order several volts, in a distance of order a few atomic spacings. This is not detectable by an ordinary voltmeter, which in fact measures μ not ϕ, but it can be detected by measurement of the electric field just outside the surface of the metal near the junction.

It is also μ not ϕ which determines the flow of electrons when a current flows. Ohm's law should properly be written, not as $\mathbf{j} = \sigma \mathbf{E} = -\sigma \nabla \phi$ (where σ is the conductivity), but as $\mathbf{j} = -\sigma \nabla (\mu/q)$.

---

[2] K here is owing to the kinetic energy in the electron system in the absence of any net electric current. The electron wavefunctions then have the form of standing waves.

12.1 Chemical potential of an ideal gas

We will show in Chapter 13 that in the absence of external fields, the chemical potential of the particles in an ideal gas at given temperature and pressure is (see Figure 12.2)

$$\mu = k_B T \ln\left[\frac{\alpha p}{(k_B T)^{c_p}}\right], \tag{12.6}$$

where α is a constant and $c_p = C_p/(Nk_B) = \gamma/(\gamma - 1)$ is the heat capacity at constant pressure per particle, in units of k_B, and we have made the simplifying assumption that the heat capacity is independent of temperature. A useful way to remember this formula is

$$n\lambda_T^3 = Z_{\text{int}} e^{\mu/k_B T}, \tag{12.7}$$

where $n = N/V$ is the number density, λ_T is a characteristic distance related to temperature, and Z_{int} is a dimensionless quantity associated with internal degrees of freedom such as rotation and vibration of molecules. For a monatomic gas of atoms, Z_{int} is the number of thermally occupied electronic states, which is typically the number of states in the ground state hyperfine manifold, $Z_{\text{int}} = (2I + 1)(2J + 1)$, where I and J are the nuclear spin and the electronic total angular momentum, respectively. For a gas of fundamental particles of spin s, $Z_{\text{int}} = (2s + 1)$. Hence, in the simplest case ($s = 0$),

$$n\lambda_T^3 = e^{\mu/k_B T}, \tag{12.8}$$

and more generally Z_{int} is, in all but the most extreme conditions, a function of temperature alone. λ_T is also a function of T alone (the subscript acts as a reminder of this). By ensuring that (12.7) agrees with (12.6), we find

$$\frac{\lambda_T^3}{Z_{\text{int}}} = \frac{\alpha}{(k_B T)^{1/(\gamma-1)}}. \tag{12.9}$$

For example, for a monatomic gas at ordinary temperatures, $\gamma = 5/3$ and Z_{int} is constant, so $\lambda_T \propto T^{-1/2}$. The constant of proportionality cannot be obtained from this information alone; it is obtained either by inference from experiments, or by a quantum treatment of the motion of the particles. One finds

$$\lambda_T = \sqrt{\frac{2\pi \hbar^2}{mk_B T}}, \tag{12.10}$$

where m is the mass of the particles. This is called the *thermal de Broglie wavelength* because it is a typical de Broglie wavelength of the particles in the gas (a particle whose de Broglie wavelength is λ_T has kinetic energy $\pi k_B T$.)

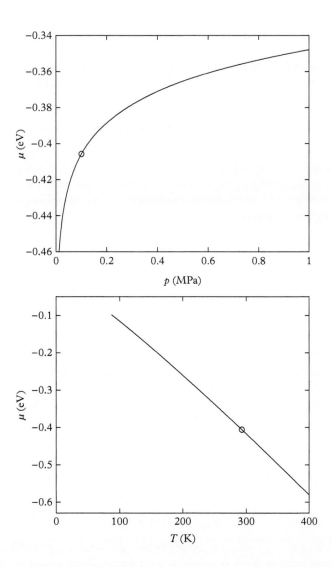

Fig. 12.2 *Chemical potential of an ideal gas as a function of p and T. The values are given for the case of argon gas, in electron volts. The first graph is at T = 293 K, the second is at p = 101 kPa; the dots indicate standard temperature and pressure.*

The condition for the ideal gas equation of state to be valid is $n\lambda_T^3 \ll 1$ (i.e. low density and/or high temperature), so we can deduce that the absolute value of the chemical potential is large and negative:

$$\mu < 0, \qquad |\mu| \gg k_B T. \quad \text{[for an ideal gas]} \qquad (12.11)$$

This implies that μ is dominated by the entropy rather than the energy term in equation (12.4).

Returning now to equation (12.6), note that μ increases with pressure (and therefore also with density) at fixed temperature. The constant α depends on the particular gas: it would be different for hydrogen, neon, carbon dioxide, etc. When we are interested primarily in the pressure dependence, it is convenient to write (12.6) in the form

$$\mu = \mu^0(T) + k_B T \ln \frac{p}{p^0}, \qquad (12.12)$$

where p^0 is some standard pressure such as 1 bar (= 10^5 Pa) and μ^0 is the value of the chemical potential at p^0 and temperature T.

12.1.1 Example: the isothermal atmosphere

Now consider an ideal gas sitting in a region of uniform gravitational field. This is a good model of the Earth's atmosphere, for example. Owing to the downwards gravitational force, we should expect that the pressure is higher at the bottom of the atmosphere than at the top. We will calculate the pressure distribution in two ways, first by a more familiar route involving pressure forces, then by using the chemical potential.

Consider a layer of gas of area A and thickness dz at some height z. The downwards gravitational force on the layer is $(\rho A dz)g$, where ρ is the mass density. The upwards pressure force from the layer below is $p(z)A$ and the downwards force from the layer above is $p(z + dz)A$. If the gas is in equilibrium, then

$$(p(z + dz) - p(z))A + \rho A dz = 0 \qquad (12.13)$$

$$\Rightarrow \quad \frac{dp}{dz} = -\rho g \qquad (12.14)$$

(by using the Taylor expansion of $p(z + dz)$). Now $\rho = mn = mp/k_B T$, where m is the mass of one molecule of the gas, so

$$\frac{dp}{dz} = -\frac{m}{k_B T} p. \qquad (12.15)$$

We shall treat the case of an *isothermal atmosphere* where T is independent of height (recall that the case of the adiabatic atmosphere was treated in Section 7.5.3). We then deduce

$$p = p_0 e^{-z/z_0}, \qquad \text{where } z_0 = \frac{k_B T}{mg}. \qquad (12.16)$$

Earth's atmosphere is not isothermal, but nevertheless this equation gives a good approximate guide to the distribution of pressure (and hence density) with height. For the case of air, averaging over the composition, we find $z_0 \simeq 8.4$ km for $g = 9.8 \text{ ms}^{-2}$; this confirms that the assumption of constant g is valid since the atmosphere's 'isothermal scale-height' z_0 is small compared to the radius of the Earth. It happens that the height of Mount Everest is roughly equal to z_0, so we should expect the pressure at the top of Everest to be roughly $(1/e)$ atmospheres.

Now let us use the chemical potential to obtain the same result. Equation (12.6) gives the chemical potential for an ideal gas in a standard situation, without further influences such as gravity. If you examine the definition of μ, equation (12.1), you should be able to convince yourself that to add a particle to the gas at height z, without changing the entropy or volume, you would need to supply the energy given by (12.6) plus a further mgz of gravitational potential energy. Therefore the expression of the chemical potential after taking gravity into account is

$$\mu = \mu^0(T) + k_B T \ln \frac{p}{p^0} + mgz. \tag{12.17}$$

Now we model the atmosphere as an isolated system. By using the argument following equation (12.5) we may argue that μ must be independent of height:

$$\mu^0 + k_B T \ln \frac{p}{p^0} + mgz = \text{const.} \tag{12.18}$$

If $p = p^0$ at $z = 0$ then the constant on the right-hand side must be μ^0, so the equation is easily solved for p (assuming again that T is independent of height):

$$p = p^0 e^{-z/z_0} \quad \text{with} \quad z_0 = k_B T/mg. \tag{12.19}$$

Thus we obtain the same result as before. The pressure p^0 in the second calculation is equal to p_0 in the first calculation, since both methods must agree on the pressure at height zero.

In this example, the first argument from pressure is arguably more simple, but the second argument serves as an introduction to the method of chemical potential. In more complicated examples it often proves easiest to use the chemical potential, and sometimes there does not exist a clear alternative.

Example 12.1

Show that, if a pressure gradient exists in an ideal gas with uniform temperature, then

$$\nabla p = n \nabla \mu, \tag{12.20}$$

and interpret this result by considering the force on a thin parcel of gas of thickness s and area A.

Solution.
From (12.12), when T is a constant we have $\nabla \mu = k_B T (1/p) \nabla p$, then by using the equation of state ($p = n k_B T$) the result follows immediately. To

> interpret the result, let the z axis be along the pressure gradient and consider a thin parcel of gas at given z. The forces on the two sides of the parcel are $p(z)A$ and $p(z+s)A \simeq (p(z) + s dp/dz)A$ (in opposite directions), where s is the thickness, so the net force is $As dp/dz$, or $\mathbf{f} = -As\nabla p$, which takes care of the direction also. Using the chemical potential this is
>
> $$\mathbf{f} = -As n \nabla \mu = -N \nabla \mu, \qquad (12.21)$$
>
> where N is the number of particles in the parcel of gas. Hence we find that for an isothermal gas, $-\nabla \mu$ can be interpreted as the force per particle.

12.2 Saha equation

An example of the role of chemical potential that is important to astrophysics is the *Saha equation*. This describes the degree of ionization of a gas of hydrogen atoms in thermal equilibrium. The ionization process is

$$\mathrm{H} \to \mathrm{p}^+ + \mathrm{e}^-.$$

The reverse process, when an electron binds to a proton to form a hydrogen atom, is called *recombination*. We are interested in the case where the process is free to take place in either direction inside a gas which is in internal equilibrium. An example is the material in the outer shell of an ordinary star, called the *photosphere*.

The energy required to ionize a given hydrogen atom has to be supplied by the rest of the gas, either from its kinetic energy or by recombining another atom. Also, if the process can proceed in both directions then it must be that there is no net entropy change of the system when any given hydrogen atom is ionized or recombined. The condition required to satisfy both constraints is

$$\mu_\mathrm{H} = \mu_\mathrm{p} + \mu_\mathrm{e}. \qquad (12.22)$$

We can prove this for the case of an isolated system by using the definition of chemical potential, equation (12.2). As we have just remarked, if the process can proceed freely in either direction at equilibrium in an isolated system, then it must be that the system entropy does not change during the process. But for a constant entropy process, we have from equation (12.2) that the energy required to add one particle of substance i is μ_i, and the energy liberated when one particle of substance i disappears (e.g. because an atom forms or breaks up) is $-\mu_i$. But for an isolated system, the total internal energy cannot change, so condition (12.22) immediately follows. It is a statement of energy conservation in adiabatic conditions for an isolated system. However, the result immediately applies more generally, no matter what constraints the system may be under, such as constant pressure or temperature or both. This is because the chemical potential is a function of state, and the reaction rates only depend on what is going on locally at the locations of the particles concerned. If an isolated system is at whatever combination of temperature and densities that makes the reaction balanced, so that ionization and recombination proceed at the same rates, then the rates will still agree if the system is not isolated but remains at that same temperature and densities.

The equilibrium condition (12.22) can be expressed in terms of density and temperature by treating the protons, electrons, and hydrogen atoms as a mixture of ideal gases, each described by equation (12.8), but we must include also the binding energy E_R. The effect of the latter can also be understood from the definition of chemical potential (12.2). The energy required to remove one hydrogen atom and provide one proton and one electron (Figure 12.3), without changing the system entropy, is

$$-\mu_{H,g} + \mu_p + \mu_e + E_R, \tag{12.23}$$

where the symbol $\mu_{H,g}$ refers to the energy liberated when a hydrogen atom in its ground state is removed from the mix without first exciting it with the energy E_R that would be required for ionization. This is given by equation (12.8):

$$\mu_{H,g} = k_B T \ln(n\lambda_H^3/4), \tag{12.24}$$

where the factor 4 is the total number of internal spin states for a bound system of two spin-1/2 particles. The binding energy E_R is normally associated with the hydrogen atom in equation (12.23), so one defines

$$\mu_H = \mu_{H,g} - E_R, \tag{12.25}$$

but this is not the only way to understand the situation: the energy E_R is an interaction energy so it really belongs to the electromagnetic field. For the purposes of book-keeping, it could be assigned to any one of the particles, or shared among them. The important point is to write down correctly what is the total energy conservation condition, allowing for all the energies involved. This is done by asserting that, in equilibrium, the chemical potentials are so arranged that the sum in (12.23) is zero. That is,

$$\mu_p + \mu_e + E_R - \mu_{H,g} = 0. \tag{12.26}$$

By using equation (12.8) three times in this equation, one obtains

$$k_B T \left(\ln(n_p \lambda_p^3/2) + \ln(n_e \lambda_e^3/2) - \ln(n_H \lambda_H^3/4) \right) + E_R = 0. \tag{12.27}$$

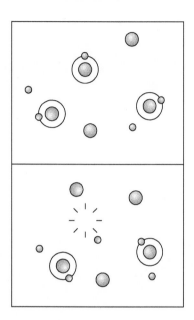

Fig. 12.3 *When a hydrogen atom is ionized, one can track the energy change by imagining that one hydrogen atom disappears from the plasma, and one proton plus one electron appears.*

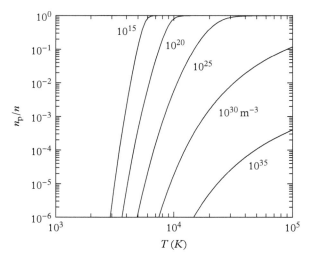

Fig. 12.4 *Degree of ionization of a neutral hydrogen plasma as a function of temperature, as predicted by the Saha equation, for five values of the number density $n \equiv n_p + n_H$ in m^{-3}.*

Therefore,

Saha equation

$$\frac{n_p n_e}{n_H} = \frac{\lambda_H^3}{\lambda_p^3 \lambda_e^3} e^{-E_R/k_B T} = \left(\frac{m_e k_B T}{2\pi \hbar^2}\right)^{3/2} e^{-E_R/k_B T}, \qquad (12.28)$$

---

where in the final step we have employed equation (12.10) and we have used the approximation that the mass of the hydrogen atom is the same as the mass of the proton.

The binding energy E_R is, to good approximation, the Rydberg energy $E_R \simeq 13.6\,\text{eV}$, except at extreme temperatures where a thermal average should be used. If the gas is electrically neutral, then $n_p = n_e$, and since whenever an atom disappears a proton appears, $n \equiv n_p + n_H$ is constant under changing degrees of ionization. Define the degree of ionization as $y = n_p/n$, then $n_p n_e/n_H = y^2 n/(1-y)$. Hence

$$\frac{n_p}{n} = \frac{-K + \sqrt{K^2 + 4nK}}{2n},$$

where $K(T)$ is equal to the right-hand side of the Saha equation. This result is plotted in Figure 12.4. It is interesting to note that significant ionization can occur at temperatures far below $E_R/k_B \simeq 158\,000\,\text{K}$.

....

EXERCISES

(12.1) A gas is contained in a harmonic potential well $V(r) = \alpha r^2$, where α is a constant and r is the distance from the centre of the well. Show that under isothermal conditions, the number density profile is $n = n_0 \exp(-\alpha r^2/k_B T)$.

(12.2) The chemical potentials for free electrons in metals A and B differ by 1.1 eV when the metals are not in contact. When cubic blocks of these metals are brought into contact, electrons move from A to B until the chemical potential is uniform. How many electrons are transferred, if the contact area is 1 cm² and the junction can be modelled as a parallel plate capacitor with gap 10^{-10} m? Is this a significant fraction of the total number of free electrons in the metal blocks? [Ans. 6×10^{13}; no.]

(12.3) (i) Use the Saha equation to find the degree of ionization of the photosphere of a hot star having $T = 12\,000$ K, $n \simeq 10^{23}$ m^{-3}. (ii) In the limit $n \to 0$ the Saha equation predicts that a hydrogen cloud in thermal equilibrium would be completely ionized even at $T \ll E_R/k_B$. This seems to go against our ordinary intuition that an excited atom will quickly decay to its ground state and stay there. What is going on?

(12.4) Generalize the argument for the Saha equation, so as to obtain the following equation for the ratio of densities of ions in a plasma:

$$\frac{n_{i+1}}{n_i} = \frac{2g_{i+1}}{g_i n_e} \left(\frac{m_e k_B T}{2\pi \hbar^2}\right)^{3/2} e^{-\Delta E/k_B T}, \qquad (12.29)$$

where i is the degree of ionization and $g_i = Z_{\text{int},i}$.

Functions and methods

13

In this chapter we will first extend the fundamental relation of the subject to the case of open systems. Then we shall introduce several new thermodynamic functions, and methods for treating thermodynamic behaviour.

The new functions (enthalpy, Helmholtz function, Gibbs function) are not as fundamental as energy and entropy, but they are of great use in understanding thermodynamic behaviour. This may be compared to introducing concepts such as energy and angular momentum in classical mechanics. In principle, Newton's laws of motion describe everything, but in order to understand what is going on, further concepts are extremely useful.

13.1 The fundamental relation	169
13.2 Thermodynamic potentials	172
13.3 Basic results for closed systems	177
Exercises	186

13.1 The fundamental relation

For an open system, energy can be exchanged with other systems by moving material between systems as well as by transfer of heat or by doing work. We have already introduced the chemical potential in equation (12.1), which we shall repeat here:

$$\mu \equiv \left.\frac{\partial U}{\partial N}\right|_{S,V}. \tag{13.1}$$

With this and our prior knowledge of heat and work, we can now present:

> **The fundamental relation of thermodynamics for a single-component simple system**
> $$dU = TdS - pdV + \mu dN. \tag{13.2}$$

It is worth revisiting the reasoning behind this, as follows. For an open simple system we know the state depends on three independent generalized coordinates, such as p, V, N. Therefore any function of state can be considered to be a function of three independent variables. In particular, we can, if we choose, express U as a function of S, V, N. Those variables are chosen because they are all extensive. Then we have $U = U(S, V, N)$, so

$$dU = \left.\frac{\partial U}{\partial S}\right|_{V,N} dS + \left.\frac{\partial U}{\partial V}\right|_{S,N} dV + \left.\frac{\partial U}{\partial N}\right|_{S,V} dN.$$

Thermodynamics: A Complete Undergraduate Course. Andrew M. Steane.
© Andrew M. Steane 2017. Published 2017 by Oxford University Press.

We have previously established that $(\partial U/\partial S)_{V,N} = T$ and $(\partial U/\partial V)_{S,N} = -p$, so, using also (13.1), we have (13.2).

An important way to 'read' equation (13.2) is to see that it tells us that there exists, for any single-component simple system, a mathematical formula for U in terms of S, V, and N, such that T, $-p$, and μ are its partial derivatives. Equally, there exists a mathematical formula for S in terms of U, V, and N (and also one for V in terms of U, S, N, and for N in terms of U, S, V). An example is the Sackur–Tetrode equation (13.42). We will find out in this chapter why such a formula is sufficiently useful to be called fundamental. The phrase 'fundamental relation' was originally introduced by Gibbs as a way to refer to the specific expression for U or S, which will vary from one system to another. The phrase 'fundamental relation' is now mostly used to refer to the relationship among the differentials, as in equation (13.2), because this relationship implies the existence of the formula noted by Gibbs. There are expressions in terms of other functions, to be defined in this chapter, that are equivalent to (13.2), and which may therefore equally be called 'the fundamental relation'; examples are given in equations (13.9)–(13.15).

13.1.1 Euler relation, Gibbs–Duhem relation

Suppose we take a system of given p, V, N and divide it in two, without changing the pressure or temperature. Then we should expect to obtain two systems, each with half the volume and half the particles of the original. Other extensive properties such as internal energy and entropy should also be shared equally. Therefore the intensive properties, such as density, internal energy per particle, etc., are not changed by halving the system. More generally, for given p, T, such properties will not depend on N. The properties of one particular drop of water in the Pacific Ocean do not depend on the size of the rest of the ocean. This is a useful idea because it implies that although the total energy of a simple system is a function of three variables (such as S, V, N) for an open system, the internal energy per particle can be written as a function of just two intensive variables, such as $s \equiv S/N$ and $v \equiv V/N$. This suggests that it should be possible to write the fundamental relation in the form

$$\mathrm{d}u = T\mathrm{d}s - p\mathrm{d}v, \tag{13.3}$$

where $u \equiv U/N$. The term involving μ has disappeared because no process can change the number of particles per particle—it is always 1.

In the following we shall prove (13.3), after introducing some related ideas.

Consider a system which consists of λ copies of a given system of properties U, S, V, p, T, μ. The new system has properties $\lambda U, \lambda S, \lambda V, p, T, \mu$, and therefore the fundamental relation for the new system is

$$\mathrm{d}(\lambda U) = T\mathrm{d}(\lambda S) - p\mathrm{d}(\lambda V) + \mu \mathrm{d}(\lambda N)$$
$$\Rightarrow \lambda \mathrm{d}U + U\mathrm{d}\lambda = \lambda(T\mathrm{d}S - p\mathrm{d}V + \mu \mathrm{d}N) + \mathrm{d}\lambda(TS - pV + \mu N).$$

Now use the fundamental relation to cancel the terms multiplied by λ, and divide out the dλ, to obtain

Euler relation
$$U = TS - pV + \mu N. \tag{13.4}$$

This is called the Euler relation. Its form is reminiscent of the fundamental relation, but it is a statement about total quantities, not small changes, so it is a new (and interesting!) piece of information.

The derivative of (13.4) gives

$$dU = TdS - pdV + \mu dN + SdT - Vdp + Nd\mu.$$

Therefore (after using the fundamental relation again),

Gibbs–Duhem relation (see also equation (21.3))
$$0 = SdT - Vdp + Nd\mu. \tag{13.5}$$

This is also commonly written

$$d\mu = -sdT + vdp, \tag{13.6}$$

where $s \equiv S/N$, $v \equiv V/N$ are the entropy and volume per particle.

The Gibbs–Duhem relation shows that changes in temperature, pressure, and chemical potential are not independent. In fact we could have anticipated that some such relationship must exist, from the existence of the equation of state. For, there is a unique temperature associated with each equilibrium state, therefore a change in temperature can't be an independent quantity, it must be determined by the changes in all the other parameters. The essential idea is that the *intensive* variables can be written as functions of one another: $p = p(T, \mu)$, or $T = T(p, \mu)$, or $\mu = \mu(T, p)$, without the need to introduce a third variable in each case. This extends to other intensive variables also. For example, for an ideal gas we can write $n = p/k_B T$, where n is number density.

This is essentially the same idea as that we proposed in the argument leading to equation (13.3), which we shall now prove. Starting from $U = uN$, we have

$$dU = Ndu + udN. \tag{13.7}$$

Therefore,

$$du + \frac{udN}{N} = \frac{dU}{N} = T\frac{dS}{N} - p\frac{dV}{N} + \mu\frac{dN}{N}$$

by using the fundamental relation (13.2). Writing $dS = Nds + sdN$ and similarly for dV, we obtain

$$du + \frac{udN}{N} = Tds - pdv + (Ts - pv + \mu)\frac{dN}{N}$$
$$\Rightarrow \quad du = Tds - pdv,$$

since the other terms cancel (by using the Euler relation (13.4) divided by N). This proof shows that (13.3) amounts to, or can be regarded as, a statement of the Gibbs–Duhem relation.

An argument similar to the above, but starting from $U = \tilde{u}V$, leads to

$$d\tilde{u} = Td\tilde{s} + \mu dn, \tag{13.8}$$

where $\tilde{u}, \tilde{s}$, and n are the energy density, entropy density, and number density. Now the dV term has disappeared because the volume per unit volume is always 1.

13.2 Thermodynamic potentials

Consider the quantity: $(U - TS)$. Its derivative is $d(U - TS) = dU - TdS - SdT = -SdT - pdV - \mu dN$, a sum of three terms. By contrast the quantity $(U + TS)$ has derivative $dU + TdS + SdT = 2TdS + SdT - pdV + \mu dN$, a sum of four terms. Owing to the simpler form of its derivative, the former quantity $(U - TS)$ is more useful than the latter $(U + TS)$, and we give it a name:

$$F \equiv U - TS. \qquad \text{Helmholtz function}$$

We can generate further useful quantities, by adding pV or subtracting TS, with or without subtraction of μN. There are a total of eight ways to do this, starting with U as the first:

internal energy U	$dU = TdS - pdV + \mu dN$	(13.9)
enthalpy $H = U + pV$	$dH = TdS + Vdp + \mu dN$	(13.10)
Helmholtz function $F = U - TS$	$dF = -SdT - pdV + \mu dN$	(13.11)
Gibbs function $G = U + pV - TS$	$dG = -SdT + Vdp + \mu dN$	(13.12)
$\tilde{U} = U - \mu N$	$d\tilde{U} = TdS - pdV - Nd\mu$	(13.13)
$\tilde{H} = H - \mu N$	$d\tilde{H} = TdS + Vdp - Nd\mu$	(13.14)
Grand potential $\Omega = F - \mu N$	$d\Omega = -SdT - pdV - Nd\mu$	(13.15)
$G - \mu N = 0$		(13.16)

The eighth form, $G - \mu N$, is zero by the Euler relation (13.4). The general mathematical idea of developing the expression $dg = -xdy$ from the expression $df = ydx$ by introducing $g \equiv f - xy$ is called a *Legendre transformation*.

We now have seven energy-related functions. The first four of these are the most useful for closed systems, since for closed systems the dN term on the right is zero. The next two ($\tilde{U}$ and $\tilde{H}$) are little used and have no standard name or symbol. The last one (grand potential) is useful in the study of open systems.

Notice that $G = \mu N$ means the chemical potential can be thought of as the Gibbs function per particle.

All of these new functions are functions of state. They are called 'thermodynamic potentials' because they have dimensions of energy, and they behave in some respects like potential energy, as we shall see.

Enthalpy, Helmholtz function, and Gibbs function are all examples of something called 'free energy', under different external conditions; this is explained in Chapter 17. Sometimes the term 'free energy' on its own is understood to mean the Helmholtz function F, and the Gibbs function can be called the 'Gibbs free energy'. Both are slight departures from clear terminology, unless the external conditions are right. The symbol A is also used for the Helmholtz function. When this is done, sometimes (beware!) the symbol F is then used for the Gibbs function, instead of G. This practice is *not* recommended.

The differential forms on the right-hand side of equations (13.9)–(13.15) reveal the *natural variables* of each function (see further comments in Section 13.2.2). These are listed in Table 13.1 for clarity. This table only lists the most commonly used thermodynamic potentials; see also Figure 13.1.

Table 13.1 *Thermodynamic potentials with their main use, natural variables, and the Maxwell relations. Constant N is to be taken as understood in the Maxwell relations.*

Function	Significance	Natural variables	Maxwell relation
U	Energy content	S, V, N	$\left.\dfrac{\partial T}{\partial V}\right|_S = -\left.\dfrac{\partial p}{\partial S}\right|_V$
F	Effective potential energy for system at fixed T	T, V, N	$\left.\dfrac{\partial S}{\partial V}\right|_T = \left.\dfrac{\partial p}{\partial T}\right|_V$
H	Related to energy changes at fixed pressure $\Delta H =$ process energy, latent heat, heat of reaction	S, p, N	$\left.\dfrac{\partial T}{\partial p}\right|_S = \left.\dfrac{\partial V}{\partial S}\right|_p$
G	Determines direction of phase and chemical changes	T, p, N	$-\left.\dfrac{\partial S}{\partial p}\right|_T = \left.\dfrac{\partial V}{\partial T}\right|_p$
Ω	Useful in general study of open systems	T, V, μ	

13.2.1 Free energy as a form of potential energy

The idea of 'potential energy' is that work done on a system may be thought of as converted into stored energy which is subsequently recoverable. This idea is relevant in mechanics, for example, when the energy is not dissipated by friction or radiated away. When two electric charges in vacuum are brought from far apart to a separation r, the work performed is $q_1 q_2/(4\pi\epsilon_0 r)$, and we may say that this work is stored as electrostatic potential energy. In fact it is stored in the electric field as field energy. Similarly, when a spring is compressed we may speak of potential energy $(1/2)kx^2$, where k is the spring constant. However, in the latter case the energy may be stored in the spring in a variety of forms, and the spring may meanwhile be exchanging energy with its surroundings by heat flow. Therefore when the spring subsequently expands again, it is not guaranteed to have the same spring constant, and the work may or may not be recovered. It follows that the concept of 'potential energy' has to be treated with caution when systems exchange energy in the form of heat as well as work.

It turns out that we can define a 'potential energy' for a thermodynamic system which is *kept at fixed temperature*, and this is often of practical interest. This potential energy is not the internal energy of the system. It is the Helmholtz function. The reason is that whenever work đW is performed on the system, we have

$$\text{đ}W = \text{d}U - \text{đ}Q \geq \text{d}U - T\text{d}S, \qquad (13.17)$$

where equality holds for reversible processes. But for an isothermal change, $\text{d}F = \text{d}U - T\text{d}S$, so we have

$$\text{đ}W = \text{d}F \quad \text{[In isothermal reversible processes]} \qquad (13.18)$$

and more generally, đ$W \geq \text{d}F$. It follows from this that the Helmholtz function is playing the role of 'potential energy' in the case of reversible isothermal processes. By tracking the change in F, either by calculation or by the use of tables, we can conveniently track energy movements for any given system in conditions of constant temperature. The total work done is given by the change in the 'potential energy' F. The reason why F and not U appears here is that there is almost always a heat flow when a change happens at fixed T, and this is taken into account by the term $-TS$ in F. The term 'free energy' was coined as a way of speaking of 'energy that is available', in the sense of available to perform work. Thus the Helmholtz function can be interpreted as the capacity of a system to perform work under isothermal conditions.

More generally, the Helmholtz function F is useful when considering systems in terms of their temperature and volume. Because the variables T, V, N are much easier to manipulate experimentally than S, V, N, the Helmholtz function is much easier to connect to experiments (and to theory in statistical mechanics) than the internal energy, even though the latter occupies the more central place in our thinking.

The enthalpy H keeps track of heat movements in a process at constant pressure, and is often of interest in chemical reactions taking place in a vessel open to the atmosphere, which maintains constant pressure. It is conserved in certain types of flow process, to be discussed in Chapter 16. It is readily verified that the heat capacity at constant pressure can be written

$$C_p = (\partial H/\partial T)_{p,N}.$$

The Gibbs function G is useful when considering processes at constant p and T, and more generally, because of its close relation to chemical potential, it is relevant in processes involving change of phase and chemical reactions. The sign of dG determines the direction of chemical reactions.

13.2.2 Natural variables and thermodynamic potentials

We have already noted that the equation of state (relating T to p, V, and N) does not furnish all the information about a system, because it does not allow the heat capacities to be determined. Similarly, if U is expressed as a function of, T, V, and N, then this equation also does not provide complete information. However, what if we knew the formula expressing U in terms of S, V, and N? Then something interesting happens. For, we can immediately obtain the temperature of any given state (S, V, N), by evaluating $(\partial U/\partial S)_{V,N}$ from the formula, and we can obtain the pressure by evaluating $-(\partial U/\partial V)_{S,N}$. Therefore we have the equation of state. The chemical potential can also be found from its expression (12.1). Thus we have complete information about all the states of the system, i.e. the values of all the properties (U, S, V, N, p, T, μ) for each state. From these, other things such as heat capacities and compressibilities can be obtained. For this reason, the variables (S, V, N) are said to be the *proper variables* or *natural variables* of internal energy: the formula expressing U in terms of its natural variables (S, V, N) is sufficient to fully characterize the system.

Similar statements can be made about all the other thermodynamic potentials. By examining the variables appearing as infinitesimals on the right-hand sides of equations (13.9)–(13.15), one deduces that the natural variables of the other main thermodynamic potentials are as listed in Table 13.1.

In statistical mechanics, the usual starting point is conditions of constant volume, temperature, and particle number, because these are easiest to treat theoretically. These are the natural variables of the Helmholtz function F, and therefore F plays a prominent role. The essential job of equilibrium statistical mechanics is to find an expression for $F(T, V, N)$ from an analysis of the internal structure of the system, especially the energy level structure. This is usually done via a related function Z called the partition function. For a system at fixed V, T, N it can be shown that $Z = \exp(-F/k_B T)$. Therefore once one has an expression for Z as a function of V, T, N, one can obtain $F(T, V, N)$ and thus complete thermodynamic information.

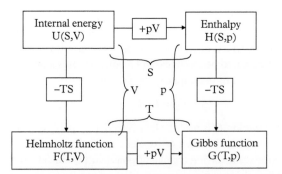

Fig. 13.1 *The four thermodynamic potentials useful for closed systems.*

13.2.3 Maxwell relations

The Maxwell relations are four very useful relationships between partial derivatives of p, V, T, and S; see Table 13.1. They are easily obtained from U, H, F, and G. We restrict attention to a closed system, so drop the dN term, and use $(\partial^2 f/\partial y \partial x) = (\partial^2 f/\partial x \partial y)$. For example, for a closed system we have

$$dH = T dS + V dp$$
$$\Rightarrow \left.\frac{\partial H}{\partial S}\right|_p = T, \quad \left.\frac{\partial H}{\partial p}\right|_S = V.$$

and $\quad \dfrac{\partial^2 H}{\partial p \partial S} = \dfrac{\partial^2 H}{\partial S \partial p} \Rightarrow \left.\dfrac{\partial T}{\partial p}\right|_S = \left.\dfrac{\partial V}{\partial S}\right|_p.$

This is a Maxwell relation. The reader should check that the results listed in Table 13.1 follow similarly from consideration of U, F, G.

Further relations can be obtained from the other thermodynamic potentials. We don't list them here because they are not so important and are easy to derive when needed.

The Maxwell relations are statements about physical behaviour, not pure mathematics; they follow from the laws of thermodynamics as expressed in the fundamental relation. For example, the relation $-(\partial S/\partial p)_T = (\partial V/\partial T)_p$ yields the prediction that if the system is in a state where the entropy does not change with pressure in an isothermal process, then the isobaric thermal expansion vanishes. This connection is not self-evident. Similar observations emerge from the other Maxwell relations, and they can be used to establish many useful connections and formulas, as this and subsequent chapters will show.

Since the Maxwell relations feature in many thermodynamic arguments, it is useful to memorize them. This can be done by noting that the numerators and denominators on the left- and right-hand sides 'cross multiply' in pairs ($dT \times dS$) and ($dp \times dV$), and the things being held constant come from the natural variables, so they never consist of p and V together or S and T together (the remaining combinations $p, S; p, T; V, S; V, T$ all occur). Finally, the minus sign appears next to a partial derivative involving S and p (mnemonic: 'Sign please').

13.2.4 Obtaining one potential function from another

Often one has an expression for one thermodynamic potential function, and one would like to obtain information about another. Suppose we have $U(V, T)$, for example. This is useful, but it is in terms of the 'wrong' variables (i.e. not the natural variables of U). We would like to obtain the Helmholtz free energy, whose natural variables are T and V. The formula $F = U - TS$ will do this, but it is not convenient because we don't yet have an expression for the entropy. A simple mathematical argument will come to the rescue. Notice that, from (13.11), $S = -(\partial F/\partial T)_V$. Therefore

$$U = F + TS = F - T \left.\frac{\partial F}{\partial T}\right|_V = -T^2 \left.\frac{\partial (F/T)}{\partial T}\right|_V. \quad (13.19)$$

This, together with other forms given below, is called a *Gibbs–Helmholtz equation*. It permits one to find the internal energy if one already knows the Helmholtz function in terms of T and V. Also, and this is more significant, this equation says that if you know U, then you know the temperature dependence of $F(T, V)$.

Similarly, one has (exercise for the reader)

$$H = -T^2 \left.\frac{\partial (G/T)}{\partial T}\right|_p, \quad (13.20)$$

so if you know H, then you know the temperature dependence of G, and vice versa. An example occurs in chemistry, where one deduces the heat of reaction (related to H) from the temperature dependence of the equilibrium concentrations of various chemical components (related to G), as shown in equation (21.19). These are the most common forms. Another is

$$G = -V^2 \left.\frac{\partial (F/V)}{\partial V}\right|_T. \quad (13.21)$$

In statistical mechanics, equation (13.19) is often written in a more convenient way by introducing $\beta \equiv 1/k_B T$. Then equation (13.19) reads

$$U = \left.\frac{\partial (\beta F)}{\partial \beta}\right|_V. \quad (13.22)$$

13.3 Basic results for closed systems

We will now apply the previously introduced methods, especially the Maxwell relations, to obtain some thermodynamic relationships valid for all simple compressible systems. In this section we will restrict attention to closed systems. All partial derivatives should be understood to be at constant N in addition to whatever other variable is being held constant.

A remark on thermodynamic manipulations

Mathematical manipulations involving partial derivatives can be confusing at first. This is not because any single step is difficult, but because there is such a wealth of possible things to try. One can easily lose one's sense of direction in a calculation of more than a few steps, so that either one argues in a loop (re-deriving the very thing one started out from), or takes a lengthy route to an answer which could have been obtained more directly.

The following guidelines may help.

(1) Begin by considering what is held constant, and what you would prefer to be held constant. To change the variable being held constant, you can use the reciprocity relation or a Maxwell relation.

(2) Recognize heat capacities. If the equation contains $(\partial S/\partial T)$ you have a heat capacity divided by T. If the equation contains $(\partial U/\partial T)$ or $(\partial H/\partial T)$ then you have a heat capacity or a combination of a heat capacity and a coefficient provided by the equation of state.

(3) Spot Maxwell relations. Be alert to the presence of any of the eight quantities that appear in the Maxwell relations, in case a Maxwell relation may be useful.

These tips should make the derivations in the rest of this chapter reasonably straightforward.

13.3.1 Relating internal energy to equation of state

First let us return to the question considered in Section 7.3.1, that of obtaining U as a function of V and T. The expression for $(\partial U/\partial V)_T$ can now be treated as follows:

$$\left.\frac{\partial U}{\partial V}\right|_T = T \left.\frac{\partial S}{\partial V}\right|_T - p \quad \text{[from } dU = TdS - pdV] \quad (13.23)$$

$$= T \left.\frac{\partial p}{\partial T}\right|_V - p. \quad \text{[using Maxwell relation]} \quad (13.24)$$

This is a notable result: it is universal, and relates $(\partial U/\partial V)_T$ directly to the equation of state. It was not possible to do this using the first law alone—compare with equation (7.25). Previously we had to use a difference of the two principal heat capacities, but now we find that that difference is in fact not an independent piece of information, it is determined by the equation of state (see equation (13.26)). The formula for internal energy change can now be written

$$\Delta U = \int C_V dT + \int \left[T \left.\frac{\partial p}{\partial T}\right|_V - p\right] dV. \quad (13.25)$$

This is simpler and more useful than (7.25). It also contributes to a very useful and fundamentally important idea, which we will expound in Section 13.3.3, after introducing some further tools.

13.3.1.1 Heat capacities again

Comparing (13.24) with (7.25), we find

$$C_p - C_V = T \left.\frac{\partial p}{\partial T}\right|_V \left.\frac{\partial V}{\partial T}\right|_p. \tag{13.26}$$

This connects the difference between the heat capacities to the equation of state. It is a useful result, so let us repeat the derivation. Having in mind $U = U(V,T)$, we first obtain (13.24), and therefore $dU = C_V dT + (T(\partial p/\partial T)_V - p)dV$. Now divide by dT at constant p (to make the left-hand side look more like a heat capacity), and recall $(\partial U/\partial T)_p = C_p - p(\partial V/\partial T)_p$ which follows directly from the fundamental relation. This completes the derivation.

Equation (13.26) is very useful in connecting theory to experiment in the study of condensed matter. For a theoretician it is much easier to calculate heat capacity at constant volume, whereas experimentally C_p is much easier to measure because solids and liquids expand so powerfully on heating, so conditions of constant volume are hard to achieve. As it stands, equation (13.26) has a constant volume term on the right-hand side, but reciprocity converts it to $-(\partial p/\partial V)_T (\partial V/\partial T)_p = \alpha/\kappa_T$, where α is isobaric expansivity and κ_T is isothermal compressibility (see Table 3.4). Hence,

$$C_p - C_V = -T \left.\frac{\partial V}{\partial T}\right|_p^2 \left.\frac{\partial p}{\partial V}\right|_T = \frac{T V \alpha^2}{\kappa_T}. \tag{13.27}$$

Several very nice insights emerge from this.

(1) Since the compressibility κ_T cannot be negative (when you squeeze a system, it must get smaller, cf. equation (17.8)), and α appears as a square, it follows that C_p *is always greater than or equal to* C_V.

(2) As $T \to 0$, $C_p \to C_V$: *the heat capacities become equal at absolute zero.* (It is possible to deduce this because it would be unphysical for a system to become strictly incompressible (i.e. zero κ_T) at any temperature, including absolute zero.)

(3) The heat capacities are equal when $\left.\frac{\partial V}{\partial T}\right|_p = 0$, i.e. when the volume as a function of temperature at fixed pressure goes through a stationary point. This happens for water at atmospheric pressure at 4 °C.

(4) It is often easier to measure C_p than C_V in the laboratory, and the quantities on the right-hand side are also measurable, so the equation yields a very good way to deduce C_V from laboratory measurements.

[1] An isotherm has $dT = 0$, and if it is horizontal on a pV diagram then it also has constant p, so it must be that $(\partial T/\partial V)_p = 0$ and therefore $(\partial V/\partial T)_p = \infty$.

(5) If the isotherms are horizontal on the indicator diagram, then (13.26) implies that C_p is infinite[1] (assuming $(\partial p/\partial T)_V \neq 0$). This is perfectly possible. It means that the system can take in heat energy at fixed pressure without raising its temperature. This happens for example during phase changes, over a finite range of volume, and for thermal radiation at all volumes.

In Chapter 7 we related the ratio of compressibilities κ_T/κ_S to the heat capacities, by applying the first law and using either Figure 7.4 or equation (7.49). We can now derive the relationship again by using the fact that we know that a function of state (namely S) stays constant in adiabatic processes. For this the Maxwell relations are not needed,[2] we simply use the reciprocity theorem in the numerator and denominator:

[2] Eqn (13.29) was derived in chapter 7 from energy conservation, without using the second law, and this is why no Maxwell relation is needed here. However, the present derivation is more general. By appealing to entropy it implicitly evokes the second law, since it assumes the existence of adiabatic surfaces in the state space.

$$\frac{\kappa_T}{\kappa_S} = \frac{\left.\frac{\partial V}{\partial p}\right|_T}{\left.\frac{\partial V}{\partial p}\right|_S} = \frac{-\left.\frac{\partial V}{\partial T}\right|_p \left.\frac{\partial T}{\partial p}\right|_V}{-\left.\frac{\partial V}{\partial S}\right|_p \left.\frac{\partial S}{\partial p}\right|_V} \quad (13.28)$$

and then bring the constant p and constant V terms together and use the chain rule:

$$\frac{\kappa_T}{\kappa_S} = \frac{\left.\frac{\partial S}{\partial V}\right|_p \left.\frac{\partial V}{\partial T}\right|_p}{\left.\frac{\partial S}{\partial p}\right|_V \left.\frac{\partial p}{\partial T}\right|_V} = \frac{\left.\frac{\partial S}{\partial T}\right|_p}{\left.\frac{\partial S}{\partial T}\right|_V} = \frac{C_p}{C_V}. \quad (13.29)$$

By using (13.27) and (13.29), measurement of α, κ_T, and κ_S at given T, V enables both the difference and the ratio of the heat capacities to be deduced, and therefore the heat capacities themselves.

13.3.1.2 Extending knowledge of C_V or C_p

Continuing the theme of finding out more from limited initial information, we discover that the equation of state can serve to extend limited knowledge of either heat capacity. Consider

$$\left.\frac{\partial C_V}{\partial V}\right|_T = T \left.\frac{\partial}{\partial V}\right|_T \left.\frac{\partial S}{\partial T}\right|_V = T \left.\frac{\partial}{\partial T}\right|_V \left.\frac{\partial S}{\partial V}\right|_T = T \left.\frac{\partial^2 p}{\partial T^2}\right|_V, \quad (13.30)$$

using a Maxwell relation. The last quantity is given by the equation of state. If the latter is known, then this result allows us to obtain the value of C_V for all volumes at some given temperature, if its value is measured at just one volume and the given temperature. For example, in the case of a gas, at any given temperature we know that in the limit $V \to \infty$ the behaviour tends to that of an ideal gas. Therefore, to find the heat capacity of a real gas at some given volume one may use

$$C_V(T, V) = C_V(T)^{(\text{ideal})} + \int_\infty^V T \left.\frac{\partial^2 p}{\partial T^2}\right|_V dV, \quad (13.31)$$

where the integral is evaluated for the real (non-ideal) gas in question. More generally, if one has the equation of state, then to obtain C_V for all states, for any system, it is sufficient to measure it at values along only a single line on a (V, T) diagram, provided that the line runs through all temperatures for which the information is needed.

Similar reasoning leads to $(\partial C_p/\partial p)_T = -T(\partial^2 V/\partial T^2)_p$, which shows we can obtain C_p from measurements along a single line on a (p, T) diagram as long as the equation of state is known.

For an ideal gas the expressions vanish, as one must expect from Joule's law. For a van der Waals gas, $(\partial C_V/\partial V)_T = 0$ but $(\partial C_p/\partial p)_T \neq 0$.

13.3.1.3 The Grüneisen parameter

Consider the parameter Γ, defined by

$$\Gamma \equiv V \left.\frac{\partial p}{\partial U}\right|_V = -\frac{V}{T} \left.\frac{\partial T}{\partial V}\right|_S. \tag{13.32}$$

You can quickly prove that the two versions are equal by use of a Maxwell relation. This parameter may be regarded either as a way of characterizing the relationship between pressure and internal energy, or as a dimensionless number that tells us about adiabatic expansion. It was originally introduced in the study of crystalline materials, because then to good approximation it is constant for a wide range of pressure and temperature, but it is applicable to any simple compressible system. It is called the Grüneisen parameter. There are two main uses. First, by integrating the equation $(\partial U/\partial p)_V = V/\Gamma$, we obtain $U = pV/\Gamma$ plus some function of volume alone, and therefore

$$p = \Gamma U/V + \phi(V), \tag{13.33}$$

where $\phi(V)$ is some function of volume. This is a useful way of writing the general relationship between pressure and energy density; we met it for the ideal gas in equation (7.28), from which one can observe that for an ideal gas with constant heat capacity,

$$\Gamma = \gamma - 1, \quad \text{[for ideal gas]} \tag{13.34}$$

where γ is the adiabatic index. Thus for monatomic gases one has $\Gamma = 2/3$, and for diatomic gases the value is lower. Using the first definition (13.32) one can interpret this as an indication that a given injection of energy raises the pressure less for a diatomic gas, because some of the energy goes to rotation of the molecules. For solids, the value of Γ is typically between 1 and 3.

More generally, Γ and γ are related as follows:

$$\gamma - 1 = \frac{TV\alpha^2}{C_V \kappa_T} = T\alpha\Gamma. \tag{13.35}$$

For the first result we have used (13.27) and the second can be proved by developing $(\partial U/\partial p)_V$ (Exercise 13.10).

The second main use of Γ is in the study of adiabatic processes. By using the second statement in (13.32), one can show that for a system where Γ is constant, the entropy per particle takes the form

$$s = s(y), \qquad (13.36)$$

where $y = TV^\Gamma$.

> **Example 13.1**
>
> Consider the sodium metal body described in Example 7.1. Its heat capacity is $C_p = 120\,\mathrm{JK^{-1}}$. Find the changes in volume and temperature of the body if the pressure increase is performed adiabatically.
>
> *Solution.*
> To find the volume change, we need $(\partial V/\partial p)_S$, which can be obtained from $\kappa_T = 1/B_T$ and γ, using equation (13.29):
>
> $$-\frac{1}{V}\left.\frac{\partial V}{\partial p}\right|_S = \kappa_S = \frac{\kappa_T}{\gamma}.$$
>
> To find γ, use equation (13.27), which gives $1 - (1/\gamma) = TV\alpha^2 B_T/C_p$, hence $\gamma = 1.065$ and so $\kappa_S = (1/\gamma B_T) = 1.52 \times 10^{-10}\,\mathrm{Pa^{-1}}$. Therefore the volume change is
>
> $$\Delta V = -V\kappa_S \Delta p = -1.5 \times 10^{-7}\,\mathrm{m^3}.$$
>
> Since this is small, it is valid to treat V as constant in the integral, and the change in κ_S is also small for this pressure change. To find the temperature change, first we obtain the Grüneisen parameter from equation (13.35), which gives $\Gamma = 1.097$. Then, from (13.32), we have $\Delta T = -\Gamma T \Delta V/V = 0.49\,\mathrm{K}$.

13.3.2 Sackur–Tetrode equation

To find an expression for entropy, recall that $C_V = T(\partial S/\partial T)_V$, so the further information we need is $(\partial S/\partial V)_T = (\partial p/\partial T)_V$, by a Maxwell relation. Therefore for a closed system,

$$\mathrm{d}S = \frac{C_V}{T}\mathrm{d}T + \left.\frac{\partial p}{\partial T}\right|_V \mathrm{d}V. \qquad (13.37)$$

For an ideal gas, this is

$$dS = \frac{C_V}{T}dT + \frac{Nk_B}{V}dV \qquad (13.38)$$

$$\Rightarrow S = \int \frac{C_V}{T}dT + Nk_B \ln V. \qquad (13.39)$$

Taking the case of constant C_V for simplicity, and using $C_V = Nk_B/(\gamma - 1)$, we obtain[3]

$$S = S_0 + \frac{Nk_B}{\gamma - 1} \ln T + Nk_B \ln V. \qquad (13.40)$$

The last term was discussed previously in Section 9.2.1.

Equation (13.40) is tantalizingly close to an expression which would be very useful if we could get it: an expression for the entropy of an ideal gas in terms of U, V, and N. This would be a prize because U, V, N are the natural variables of S (because $dS = (1/T)dU + (p/T)dV - (\mu/T)dN$). Therefore a formula for $S(U, V, N)$ would provide complete thermodynamic information about the system. Let's see if we can get it.

We have already assumed constant C_V, so we have $U = C_V T + \text{const.}$, and the constant here can be set to zero by choosing the zero of energy so as to bring this about. Note, this does *not* mean we think the equation $U = C_V T$ applies all the way to absolute zero temperature (and in fact it does not). We only assume a linear relation between U and T over the range of temperatures of interest. Under this assumption, the reference to T in (13.40) can be converted into one to U. The integrations were carried out at constant N. To get the dependence on N, we must return to the full fundamental relation for open systems, (13.2). We get precisely the same result, except that now S_0 can be a function of N. To find the dependence, we appeal to the fact that S is extensive, so $S(\lambda U, \lambda V, \lambda N) = \lambda S(U, V, N)$. This is true for any λ, so let's write it for the case $\lambda = N_0/N$:

$$S(U, V, N) = \frac{1}{\lambda} S(\lambda U, \lambda V, \lambda N) = \frac{N}{N_0} S\left(\frac{N_0}{N} U, \frac{N_0}{N} V, N_0\right)$$

$$= \frac{N}{N_0}\left[S_0(N_0) + \frac{N_0 k_B}{\gamma - 1} \ln\left(\frac{\gamma - 1}{k_B}\frac{U}{N}\right) + N_0 k_B \ln \frac{N_0 V}{N}\right], \qquad (13.41)$$

where for the energy term we have used $T(U, N) = ((\gamma - 1)/k_B)U/N$ and so for any λ, $T(\lambda U, \lambda N) = T(U, N)$. Note that $S_0(N_0)$ is here a constant. Gathering this and various other constants together, the result can be expressed:

The Sackur–Tetrode equation

$$S = Nk_B \ln\left[a\frac{V}{N}\left(\frac{U}{N}\right)^{\frac{1}{\gamma - 1}}\right], \qquad (13.42)$$

---

[3] In equation (13.40) we allow the practice of writing $\ln x$ where x is a dimensional quantity. This breaks the rule that the argument of a function ought always to be dimensionless, but it is valid as long as the equation contains a further term such as $-\ln x_0$, so that the two together make $\ln x - \ln x_0 = \ln(x/x_0)$, for which the argument is dimensionless. In (13.40) the further (constant) terms involving $\ln T_0$ and $\ln V_0$ have been bundled into S_0.

where a is a constant for any given gas. This is the equation giving the entropy of an ideal gas of fixed heat capacity, in terms of the natural variables of S, from which, the complete thermodynamic behaviour can be derived.

This equation is associated with the names of Sackur and Tetrode when the constant a is expressed in terms of fundamental quantities (see after equation (13.50)), since their contribution was to obtain such an expression, based on a microscopic model of a gas using quantum theory. Here we are using the equation in strictly thermodynamic terms, so we don't require any model other than Boyle's law and Joule's law. In this case the constant a has to be regarded as an unknown quantity which can in principle be obtained by measurement for any particular gas that approximates an ideal gas.

The constant a can be expressed as

$$a = \left[\frac{V_0}{N_0}\left(\frac{U_0}{N_0}\right)^{\frac{1}{\gamma-1}}\right]^{-1} \exp\left(\frac{S_0}{N_0 k_B}\right), \tag{13.43}$$

where S_0 is the entropy of some reference state (U_0, V_0, N_0). Then

$$S = S_0 \frac{N}{N_0} + N k_B \left(\ln \frac{V}{V_0} + \tilde{c}_V \ln \frac{U}{U_0} - (\tilde{c}_V + 1) \ln \frac{N}{N_0}\right), \tag{13.44}$$

where we have written $\tilde{c}_V$ for $1/(\gamma - 1)$: this is the constant-volume heat capacity per particle in units of k_B.

According to the Sackur–Tetrode equation, the entropy of an ideal gas tends to $-\infty$ as $T \to 0$, which is unphysical. This means that the ideal gas is itself an impossibility in the limit $T \to 0$: no real substance can satisfy both Boyle's and Joule's laws as the temperature tends to absolute zero. In practice both laws break down. This is one reason why it is important not to rely on the ideal gas for the definition of absolute temperature and entropy.

On the other hand, (13.44) is perfectly acceptable, and in fact highly accurate, at high temperature.

To illustrate the value of having a general formula for a state function expressed in terms of its natural variables, let us now extract from the Sackur–Tetrode equation some information about the other variables, T, p, and μ.

Recalling the fundamental relation (13.2), we have

$$\frac{1}{T} = \left.\frac{\partial S}{\partial U}\right|_{V,N} = \frac{N k_B}{(\gamma - 1)U}, \tag{13.45}$$

which gives the energy equation (7.27). Next,

$$\frac{p}{T} = \left.\frac{\partial S}{\partial V}\right|_{U,N} = \frac{N k_B}{V}, \tag{13.46}$$

which is the equation of state. Finally,

$$\frac{\mu}{T} = -\frac{\partial S}{\partial N}\bigg|_{U,V} = -\left(\frac{S}{N} - \frac{k_B \gamma}{\gamma - 1}\right), \qquad (13.47)$$

which tells us the chemical potential:

$$\mu = k_B T \left(\frac{\gamma}{\gamma - 1} + \ln\left[\frac{N}{aV}\left(\frac{N}{U}\right)^{\frac{1}{\gamma-1}}\right]\right) = k_B T \ln\left[\frac{nb}{(k_B T)^{\tilde{c}_V}}\right], \qquad (13.48)$$

where $n = N/V$ is the number density and b is a constant. (The second form bundles all the constants into b in order to bring out the dependence on number density and temperature.)

Equation (13.47) can also be written

$$\mu = \gamma \frac{U}{N} - T \frac{S}{N}, \qquad (13.49)$$

which is an example of (12.4).

To find the value of the unknown constant a, one may adopt either an experimental or a theoretical approach. Experimentally, one could discover S_0 at some given (U_0, V_0, N_0) by studying the heat capacity and latent heat and using the integral and sum given in equation (9.4). Theoretically, one models the gas as a collection of particles or quantum field modes. One then finds that in the case of a monatomic gas at low pressure (where particle interactions can be neglected), the entropy is

$$S = Nk_B \left(\frac{5}{2} - \ln n\lambda_T^3\right), \qquad (13.50)$$

where λ_T is the thermal de Broglie wavelength given in (12.10). This implies that for the monatomic gas, the constant a in (13.42) is $a = e^{5/2}(3\pi \hbar^2/m)^{-3/2}$, where m is the mass of one atom. Note that (13.50) gives the entropy at high temperature perfectly accurately, even though it is not valid at low temperature; see Figure 13.2.

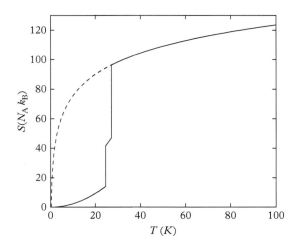

Fig. 13.2 *The entropy of one mole of neon at one atmosphere, as a function of temperature. The full line is the experimental result, the dashed line is the prediction of equation (13.50), which describes the gaseous behaviour very well. The dashed line crosses the T axis at some positive but small T; this is unphysical and shows that the ideal gas model cannot describe any system accurately at very low temperatures.*

13.3.3 Complete thermodynamic information

The previous discussion of entropy was applied to an ideal gas, but the concept can be generalized. We are now ready to show that if one has the equation of state and a knowledge of C_V for all temperatures at one volume V_0, then one has complete thermodynamic information. We show this by constructing the Helmholtz function in terms of its natural variables.

Let U_0 and S_0 be the values of internal energy and entropy at some fixed state (T_0, V_0). Then equations (13.25) and (13.37) give

$$U(T, V) = U_0 + \int_{T_0}^{T} C_V(T', V_0) dT' + \int_{V_0}^{V} \left. T \frac{\partial p}{\partial T}\right|_V - p \, dV', \quad (13.51)$$

$$S(T, V) = S_0 + \int_{T_0}^{T} \frac{C_V(T', V_0)}{T'} dT' + \int_{V_0}^{V} \left.\frac{\partial p}{\partial T}\right|_V dV'. \quad (13.52)$$

Therefore, using $F = U - TS$, we obtain

$$F(T, V) = U_0 - TS_0 + \int_{T_0}^{T} \left(1 - \frac{T}{T'}\right) C_V(T', V_0) dT' - \int_{V_0}^{V} p(T, V') dV'. \quad (13.53)$$

The dependence on N can then be obtained by using the fact that F is extensive (Exercise 13.7).

....

EXERCISES

(13.1) Use the Gibbs–Duhem relationship to show that $(\partial \mu / \partial N)_{p,T} = 0$.

(13.2) A closed simple compressible system has work ΔW done on it in a reversible process without heat exchange. (i) By how much does the internal energy change? (ii) Is it possible to tell, from the given information, the change in the Helmholtz function, enthalpy, and Gibbs function? (iii) Suppose the work were done in conditions of constant temperature. Now what can you say about U, F, H, G?

(13.3) Derive all the Maxwell relations.

(13.4) Sketch an isotherm and an adiabat passing through a given point p, V on the indicator diagram for a gas. Hence show that more work energy can be extracted from a gas in an isothermal expansion than in an adiabatic one. Where has the energy for this extra work come from?

(13.5) Give an example of a physical process which can take place at constant pressure and temperature. What thermodynamic potential is unchanged in such a process?

(13.6) A certain simple compressible system has Helmholtz function given by $F = -k_B T \ln(z_1^N/N!)$, where $z_1(T, V)$ is independent of N. (i) Assuming N is large so that the Stirling approximation $\ln N! \simeq N \ln N - N$ is valid, show that the equation of state is $pV = Nk_B T$. [Hint: find μ and use $G = \mu N$.] (ii) If $z_1(T, V)$ has the form $z_1 = f(T)g(V)$, where f and g are unspecified functions, show that the internal energy per particle is a function of T alone. [Hint: Gibbs–Helmholtz equation.]

(13.7) Prove the following *homogeneity relations*:

$$F(T, \lambda V, \lambda N) = \lambda F(T, V, N)$$
$$H(\lambda S, p, \lambda N) = \lambda H(S, p, N)$$
$$G(T, p, \lambda N) = \lambda G(T, p, N). \quad (13.54)$$

Hence show that $F(T, V, N) = (N/N_0)F(T, N_0 V/N, N_0)$.

(13.8) (a) Show that, if only the equation of state of a system is known, then it is possible to calculate the work done in an isothermal process, but not in an adiabatic one.

(b) If both the equation of state and C_V are known, show how the work done in an adiabatic process could be obtained.

(13.9) Show that the rate of change of p with T in an adiabatic compression may be expressed as $(\partial P/\partial T)_S = (C_p/T)(\partial T/\partial V)_p$.

(13.10) Obtain equation (13.35).

(13.11) Compare some of the measured values in Table 9.1 with the prediction from equation (13.50).

(13.12) Confirm that (13.42) may equally be written

$$U = \frac{N^\gamma}{(aV)^{\gamma-1}} e^{(\gamma-1)S/Nk_B} \quad (13.55)$$

and use this to obtain the equation of state.

(13.13) Show that the Helmholtz function of an ideal gas of constant heat capacity is given by

$$F(T, V, N) = \frac{Nk_B T}{\gamma - 1}\left(1 - \ln \frac{k_B T}{\gamma - 1}\right) - Nk_B T \ln a \frac{V}{N}. \quad (13.56)$$

(13.14) Starting from the Sackur–Tetrode equation, obtain an expression for the pressure of an ideal gas as a function of temperature and chemical potential. Hence confirm that $(\partial p/\partial T)_\mu = S/V$ for this system (an example of the Gibbs–Duhem relation).

(13.15) A piece of rubber of length L is subject to work by hydrostatic pressure and a tensional force f.

(i) Construct an expression for dU.

(ii) Generate the potentials which have as proper variables (S,V,f) and (S,p,f).

(iii) Derive the Maxwell relation (first developing any potential you may need)

$$\left.\frac{\partial S}{\partial L}\right|_{T,p} = -\left.\frac{\partial f}{\partial T}\right|_{p,L}.$$

14 Elastic bands, rods, bubbles, magnets

14.1 Expressions for work 188
14.2 Rods, wires, elastic bands 188
14.3 Surface tension 190
14.4 Paramagnetism 192
14.5 Electric and magnetic work 200
14.6 Introduction to the partition function 210
Exercises 212

Table 14.1 *State variables and work done in a reversible change for various simple systems.*

Fluid	p, V	$-p\,dV$
Elastic rod	f, L	$f\,dL$
Liquid surface	σ, A	$\sigma\,dA$
Dielectric sample	$\mathbf{E}, \mathbf{p}$	$-\mathbf{p}\cdot d\mathbf{E}_0$
Magnetic sample	$\mathbf{m}, \mathbf{B}$	$-\mathbf{m}\cdot d\mathbf{B}_0$

Double, double toil and trouble
Fire burn and cauldron bubble

William Shakespeare, *Macbeth*, Act 4, Scene 1.

14.1 Expressions for work

Table 14.1 gives expressions for the work done on various simple systems in a reversible change.

We have already discussed a simple compressible fluid. An elastic rod is described to good approximation by its length L and the tension f. To stretch the rod by dL against this force, the work done is clearly $f\,dL$. The surface tension in a liquid surface is, by definition, the parameter σ such that the work required to increase the area of the surface by dA is $\sigma\,dA$. These cases are all straightforward. The expressions for electric and magnetic systems are discussed at the end of the chapter (Section 14.5).

14.2 Rods, wires, elastic bands

When a long, straight object such as a rod or wire is stretched, most of the work is done against the tension in the rod or wire. There is also a small contribution coming from pressure forces acting on the curved surface, but these can be neglected to good approximation under many circumstances. We can therefore treat a rod of length L as a closed thermodynamic system with fundamental relation

$$dU = T\,dS + f\,dL. \tag{14.1}$$

The natural length L_0 is the length when the tension falls to zero. The *strain* is defined as the fractional extension, dL/L_0 when the rod is put into tension. The *stress* is defined as the force per unit area, df/A. The *Young's modulus* is defined as the ratio of stress to strain. Hence the isothermal Young's modulus is

$$E_T = \frac{L}{A}\left.\frac{\partial f}{\partial L}\right|_T. \tag{14.2}$$

Thermodynamics: A Complete Undergraduate Course. Andrew M. Steane.
© Andrew M. Steane 2017. Published 2017 by Oxford University Press.

This is a measure of how stiff the material of the rod is. (The corresponding property for a simple compressible system is the isothermal bulk modulus). The parameter k appearing in Hooke's law (6.16) is $k = AE_T/L$.

The other commonly considered quantity is the *linear expansivity at constant tension*,

$$\alpha_f = \frac{1}{L} \left.\frac{\partial L}{\partial T}\right|_f. \tag{14.3}$$

This is a measure of how readily the rod stretches or shrinks when its temperature changes. (The corresponding property for a simple compressible system is the isobaric expansivity.) Both α_f and E_T are determined by the equation of state. E_T must be positive in stable equilibrium (if you pull on something, it must get longer). α_f may have either sign, depending on the material (if you heat something, it may expand or it may shrink). A constant tension can be simply arranged by hanging a weight on the end of a vertical rod or wire. For most materials, such as metals and ceramics, heating such a wire will result in expansion, so $\alpha_f > 0$. For some materials, such as rubber, heating such a wire will often result in contraction, so $\alpha_f < 0$. We would like to employ thermodynamic arguments to connect these observations to entropy and to other properties.

One simple experiment is to first stretch a wire, then hold it at constant length (think of the strings on a guitar, for example), and enquire into what will happen when the wire is heated. Will the tension go up or down? The intuition is that a metal wire will 'try to stretch' and thus have reduced tension as its temperature rises at fixed length, and a rubber one will do the opposite. This is correct, since

$$\left.\frac{\partial f}{\partial T}\right|_L = -\left.\frac{\partial f}{\partial L}\right|_T \left.\frac{\partial L}{\partial T}\right|_f = -AE_T\alpha_f, \tag{14.4}$$

by reciprocity (5.11).

Now consider our fundamental relation, (14.1). By taking derivatives, and introducing further functions such as $U - TS$ and $U - TS - fL$, one can find a set of Maxwell relations precisely analogous to those we found before for a simple compressible system. In fact, it is not hard to see that one may simply make the replacements $p \to -f$, $V \to L$ in the Maxwell relations in Table 13.1. For example,

$$\left.\frac{\partial S}{\partial L}\right|_T = -\left.\frac{\partial f}{\partial T}\right|_L. \tag{14.5}$$

Using (14.4), this gives

$$\left.\frac{\partial S}{\partial L}\right|_T = AE_T\alpha_f. \tag{14.6}$$

Therefore, when α_f is positive, stretching the rod at fixed temperature increases its entropy; and when α_f is negative, stretching the rod at fixed temperature decreases its entropy. In the first case, heat is drawn into the rod. In the second case, heat leaves. These facts can be understood in terms of the structure of the material in each case. For a crystalline or polycrystalline material, stretching the material gives the atoms more space to move in and also makes the crystallites less ordered. This is a higher entropy situation. For a material made of long chain molecules, on the other hand, stretching the material does not cause the links in each chain to stretch in the first instance, but rather each chain straightens out, which is a lower entropy situation. This continues until the molecules are almost straight, after which further stretching may result in the sign of α_f becoming positive. See Exercise 14.2 for further information.

Try the following experiment. Take an ordinary elastic band and first stretch it as much as you dare, and hold it steady until it has attained ambient temperature in its stretched condition. At this point release the tension quickly and then immediately place the band against your lips in order to sense its temperature before it has time to reach equilibrium with the surroundings. It will feel cold. This is in contrast to most materials. Exercise: confirm this prediction by connecting the adiabatic process to α_f, and suggest the microscopic mechanism.

14.3 Surface tension

Surface tension is the tendency of the surface of a body to oppose being stretched. It can be exhibited by solids and gases, but it is most readily observed at the interface between a liquid and a gas, because such surfaces can be easily deformed. The work required to increase the area of the surface by a small amount dA is given by

$$\mathrm{d}W = \sigma \mathrm{d}A, \tag{14.7}$$

where σ is the surface tension. Surface tension is typically a strong function of temperature, but almost independent of pressure. It must vanish at the critical point (where the liquid and vapour become indistinguishable; see Chapters 15 and 22). A reasonable approximation for many liquids (e.g. water, see Figure 14.1) is Eötvös's rule:

$$\sigma \simeq k V_\mathrm{M}^{-2/3}(T_\mathrm{c} - T), \tag{14.8}$$

where V_M is the molar volume, T_c is the critical temperature, and k is a constant which has approximately the same value for many liquids, $k \simeq 2.1 \times 10^{-7}$ JK^{-1}mol$^{-2/3}$. For water, $V_\mathrm{M} = 18.0$ cm^3 and $T_\mathrm{c} = 647.096$ K.

Consider the situation shown in Figure 14.2. Let $V = (4/3)\pi r^2$ be the volume of the spherical drop, p be the pressure inside the drop, and p_0 the ambient pressure. When the plunger is pushed down by a small amount, the volume of the spherical drop gets larger by dV and the plunger does work pdV if we assume the liquid is incompressible. The net work done on the system by all the external forces is therefore

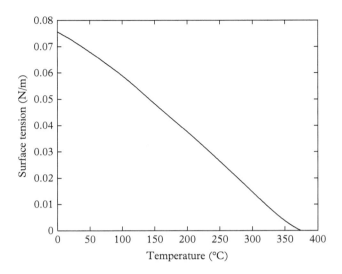

Fig. 14.1 *The surface tension of water in contact with air.*

$$\mathchar'26\mkern-12mu d W = p\,dV - p_0\,dV. \tag{14.9}$$

If the system is in mechanical equilibrium, this energy is precisely equal to the work required to increase the surface area, given by (14.7):

$$(p - p_0)\,dV = \sigma\,dA. \tag{14.10}$$

For a spherical drop of radius r, we have $A = 4\pi r^2$ and $V = (4/3)\pi r^3$, so $dA/dr = 8\pi r$ and $dV/dr = 4\pi r^2$. Therefore the pressure difference is given by

$$(p - p_0) = \frac{2\sigma}{r}. \tag{14.11}$$

In the above we have neglected heat. This is correct when we are merely interested in a balance of mechanical forces: we may use the principle of virtual work. From a thermodynamic point of view this amounts to exploring the out of equilibrium situations near to the equilibrium point; this will be explained in Chapter 17.

If we now consider instead of a liquid drop, a bubble, then there are two surfaces to account for. Assuming the bubble wall is thin, so that the inner and outer surfaces have approximately the same radii, one has

$$(p - p_0) = \frac{4\sigma}{r}. \tag{14.12}$$

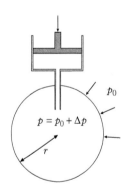

Fig. 14.2 *A spherical drop.*

Although the complete system in these examples consists of both the liquid and its surface, we may choose the surface itself as our thermodynamic system, and write the fundamental relation

$$dU = T\,dS + \sigma\,dA, \tag{14.13}$$

where all the quantities refer to properties of the surface. This is not completely obvious, because as the surface grows, more molecules are at the surface, and we have not included a chemical potential term to account for this. This is correct because there is a contribution to the entropy and energy that can be associated with the surface, without the need to assign it any volume or particles. If the surface grows, no new particles are added to the system, but some of those already present change their role, and the energy and entropy associated with this is precisely what we are tracking. This concept is discussed more fully in Section 24.1.

By taking derivatives, and introducing further functions such as $U - TS$ and $U - TS - \sigma A$, one can now find a set of Maxwell relations precisely analogous to those we found before for a simple compressible system. In fact, it is not hard to see that one may simply make the replacements $p \to -\sigma$, $V \to A$ in the Maxwell relations in Table 13.1. For example,

$$\left.\frac{\partial S}{\partial A}\right|_T = -\left.\frac{\partial \sigma}{\partial T}\right|_A. \tag{14.14}$$

Our system parameters are (σ, A) and we already have the equation relating these to temperature—equation (14.8)—so that is the equation of state. It remains, then, to consider the energy equation, and the movement of heat into and out of the surface.

We obtain the energy equation by considering $U = U(A, T)$. Then $(\partial U/\partial T)_A$ is a heat capacity and $(\partial U/\partial A)_T$ can be obtained from the fundamental relation and the equation of state via the Maxwell relation (14.14):

$$\left.\frac{\partial U}{\partial A}\right|_T = T \left.\frac{\partial S}{\partial A}\right|_T + \sigma = -T \left.\frac{\partial \sigma}{\partial T}\right|_A + \sigma \tag{14.15}$$

$$\simeq k V_{\mathrm{m}}^{-2/3} T_{\mathrm{c}}, \tag{14.16}$$

where the first line is exact and the second makes use of the approximate equation of state. Although equation (14.8) is approximate, the fact that σ is a function of temperature alone, without regard to A, is to be expected, and so is the fact that $(\partial \sigma/\partial T)_A < 0$. Hence we deduce from the Maxwell relation (14.14) that, during an isothermal change, when the area increases, heat flows *into* the surface. Therefore at given T, the entropy of the surface increases with its area.

14.4 Paramagnetism

The subject of paramagnetism was introduced in Section 6.2.4. An experimental setup is shown schematically in Figure 6.6, and with some further details in Figure 14.7. The quantities that typically we are chiefly interested in are the heat emitted or absorbed during an isothermal process, and the change in temperature during an adiabatic process. The first is quantified by

$$\frac{\mathrm{d}Q_T}{\mathrm{d}B} = T \left.\frac{\partial S}{\partial B}\right|_T, \tag{14.17}$$

the second by

$$\left.\frac{\partial T}{\partial B}\right|_S, \tag{14.18}$$

where B is the magnetic field. We would like to relate these quantities to the equation of state and the heat capacities.

Before proceeding we must note that the above expressions are ambiguous as they stand, because there are several magnetic fields in the problem. These are:

1. The field $\mathbf{B}_i$ inside the sample.
2. The field $\mathbf{H}_i = (1/\mu_0)\mathbf{B}_i - \mathbf{M}$ inside the sample, where $\mathbf{M}$ is the magnetization.
3. The fields $\mathbf{B}'$ and $\mathbf{H}'$ outside the sample.
4. The field $\mathbf{B}_0$ that would be present in the solenoid if the sample were removed while keeping the total flux Φ in the solenoid constant.
5. The field $\mathbf{B}_{0I}$ that would be present in the solenoid if the sample were removed while keeping the current I in the solenoid constant.

In the limit of small magnetization, all of these agree (apart from the constant factor μ_0 between $\mathbf{B}$ and $\mathbf{H}$), and therefore the distinction between them does not need to be made when treating weakly magnetized materials. More generally, *all* the above may feature in the treatment of magnetic systems, though $\mathbf{B}'$ and $\mathbf{H}'$ are of little interest and seldom used.

A given sample will respond to the field present in the sample, so the *magnetic susceptibility* χ is defined by

$$\chi = \frac{M}{H_i}, \tag{14.19}$$

where $M = |\mathbf{M}|$ and $H_i = |\mathbf{H}_i|$. The fields $\mathbf{B}_0$ and $\mathbf{B}_{0I}$ are useful for simplifying expressions for magnetic work, as explained in Section 14.5. In the simplest geometry they are related by

$$B_0 = \mu_r B_{0I}, \tag{14.20}$$

where $\mu_r = 1 + \chi$ is the relative permeability of the medium, and then $H_i = H_{0I}$. This is why one often sees equation (14.19) quoted with H_{0I} instead of H_i on the right-hand side. For a more general geometry, see Exercises 14.8–14.10.

Either of the fields $\mathbf{B}_0$ and $\mathbf{B}_{0I}$ may be termed 'the external field' or 'the applied field' but for a precise treatment one must specify which is under consideration. Unfortunately, both these descriptive phrases are somewhat misleading, because $\mathbf{B}_0$ and $\mathbf{B}_{0I}$ do not refer to fields anywhere in the apparatus when the sample is present. However in the remainder of this section we will limit ourselves to the case $\chi \ll 1$. Then $\mu_r \simeq 1$ and it no longer matters which field is termed 'the applied field'. The B field quantity referred to in (14.17) and (14.18) can then be taken as the applied field without further comment, and the expression for magnetic work (derived in Section 14.5) is

$$\dbar W = -m \d B, \tag{14.21}$$

where m is the total dipole moment of the sample.

We would like now to relate the temperature and the heat exchange to the other properties. Following the by now standard procedure, we first write down the fundamental relation. This is

$$dU = TdS - pdV - mdB. \tag{14.22}$$

When treating the paramagnetic behaviour of a gas, the mechanical work $-pdV$ cannot be ignored, and even for a solid it is not always negligible. However, it can be neglected to good approximation in many cases, and we will do so here, so that the relation becomes

$$dU = TdS - mdB. \tag{14.23}$$

From this and related quantities, Maxwell relations are derived. One can see that they will correspond to those for pV systems, with the replacements $p \to m$, $V \to B$.[1] Hence we find for the isothermal process,

$$T \left. \frac{\partial S}{\partial B} \right|_T = T \left. \frac{\partial m}{\partial T} \right|_B, \tag{14.24}$$

and the latter can be obtained from the equation of state. Using first reciprocity and then a Maxwell relation, we find for the adiabatic process,

$$\left. \frac{\partial T}{\partial B} \right|_S = - \left. \frac{\partial T}{\partial S} \right|_B \left. \frac{\partial S}{\partial B} \right|_T = -\frac{T}{C_B} \left. \frac{\partial m}{\partial T} \right|_B. \tag{14.25}$$

From this we deduce that when the magnetization depends on temperature, then we can expect to observe the temperature of a thermally isolated sample changing when the magnetic field does.

The equation of state is usually established in terms of the susceptibility, defined by equation (14.19). Curie's law, $\chi = a/T$ (equation 6.22) is an example, but we will not assume it here in order to keep the discussion general. Note, M/H is not in general the same as $(\partial M/\partial H)$ except in the limit $M, H \to 0$, or when χ is independent of H.

In the case $\chi \ll 1$ one has $M \ll H$, and then one can take $B \simeq \mu_0 H$. Then, using $m = VM$, where V is the volume, we have

$$m = VB\chi/\mu_0. \tag{14.26}$$

---

[1] The Maxwell relations for the magnetic system can also be obtained by the replacements $V \to m$, $p \to -B$.

χ is generally a decreasing function of temperature because when the material is hot, the small magnetic dipoles inside it are reluctant to line up. Hence $(\partial m/\partial T)_B < 0$ and so heat is given out during an isothermal magnetization (Eq. 14.24) and the temperature rises during an adiabatic magnetization (Eq. 14.25). Hence the temperature falls during an adiabatic demagnetization.

In terms of χ, equations (14.24) and (14.25) read

$$T\left.\frac{\partial S}{\partial B}\right|_T = \frac{TVB}{\mu_0}\left.\frac{\partial \chi}{\partial T}\right|_B, \qquad \left.\frac{\partial T}{\partial B}\right|_S = -\frac{TVB}{\mu_0 C_B}\left.\frac{\partial \chi}{\partial T}\right|_B. \qquad (14.27)$$

We return to these equations after Section 14.4.1.

14.4.1 Ideal paramagnet

We introduced the ideal paramagnet in Section 6.2.4, where we gave the equation of state (Eq. 6.25). This equation, and related information such as the heat capacities, can be obtained from first principles by a simple theoretical model. The model ignores the possibility of interactions between the magnetic particles which make up the paramagnet, and in consequence it does not describe the phase transition at the Curie point described in Section 14.4.2. However it does capture the main features of paramagnetic behaviour above the Curie point in a useful way.

The calculation starts out from specific details of the microscopic structure, so this part of the argument is not thermodynamic in character. Rather, it is a way of gaining some basic information about the system so that we can then get to work with thermodynamic reasoning.

The basic idea is to take an interest in the thermodynamic potential whose natural variables are temperature and another, where the 'other' is the parameter that determines the placement of the energy levels of the system. In the present case this is the magnetic field, so the function we want is the Helmholtz function $F = U - TS$, because from the fundamental relation (14.23), we have

$$dF = -SdT - mdB, \qquad (14.28)$$

hence the natural variables of F are T and B. It also follows that F is the free energy under conditions of constant T and B.

Now, an important basic result in statistical mechanics is that for a collection of distinguishable and very weakly interacting particles, the free energy is given by

$$e^{-F/Nk_BT} = \sum_{i \in \text{states}} e^{-\epsilon_i/k_BT}, \qquad (14.29)$$

where the sum runs over the quantum states (energy eigenstates) of one particle, and ϵ_i are the energies of those states. This sum is called the *partition function*, and this is a remarkably powerful formula. It makes, in one step, the connection between microscopic physical details and the macroscopic thermodynamic function we need. We will not offer a general proof here, but we provide a derivation for the two-state system in Section 14.6. For the moment we will illustrate the method by using it.

We model a simple paramagnetic system by supposing that it is composed of N small dipoles, each with dipole moment of size μ, and we suppose each dipole can be in one of two states: pointing either along $+z$ or $-z$, where the z direction is along the direction of the applied magnetic field. Therefore there are just two terms in the sum on the right-hand side of (14.29). The interaction energy of each small dipole with the magnetic field $\mathbf{B}$ is $-\boldsymbol{\mu} \cdot \mathbf{B}$, so the energies of the two states of a single dipole are $\epsilon_1 = -\mu B$ and $\epsilon_2 = +\mu B$. Hence we obtain

$$\sum_{i \in \text{states}} e^{-\epsilon_i/k_B T} = e^{\mu B/k_B T} + e^{-\mu B/k_B T} = 2\cosh(x), \tag{14.30}$$

where

$$x \equiv \frac{\mu B}{k_B T}. \tag{14.31}$$

Therefore, using (14.29),

$$F = -N k_B T \ln(2\cosh x). \tag{14.32}$$

Since we have in (14.32) an expression for the free energy in terms of its natural variables, all the thermodynamic information can immediately be obtained. One finds

$$m = -\left.\frac{\partial F}{\partial B}\right|_T \quad = N\mu \tanh x, \tag{14.33}$$

$$S = -\left.\frac{\partial F}{\partial T}\right|_B \quad = N k_B (\ln(2\cosh x) - x \tanh x), \tag{14.34}$$

$$U = F + TS \quad = -N\mu B \tanh x \tag{14.35}$$

$$\qquad\qquad\quad = -mB, \tag{14.36}$$

$$C_B = T \left.\frac{\partial S}{\partial T}\right|_B = \left.\frac{\partial U}{\partial T}\right|_B = N k_B x^2 \operatorname{sech}^2 x, \tag{14.37}$$

$$C_m = 0. \tag{14.38}$$

The first of these results is the equation of state, which we previewed in Section 6.2.4 (Figure 6.8) and we plot it again in Figure 14.3. It predicts the Curie law at low B, and also shows the value of B at which the law breaks down because the magnetization saturates. The heat capacity at constant field, C_B, was also previewed in Figure 7.3b, and some related remarks are made in Section 9.4.

Note that, at fixed N, m is a function of x alone, and so is S, so S can be written as a function of m alone. This explains the form of the illustrative Carnot cycle, which was shown as an example in Figure 8.1, and it results in $C_m = 0$.

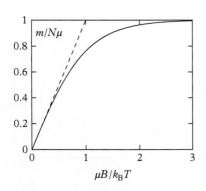

Fig. 14.3 *Magnetic dipole moment of an ideal paramagnet. The dashed line shows Curie's law.*

It means that, for the ideal paramagnet, you can't change the entropy without also changing the magnetization. This is because the only way to change the entropy is to change the distribution of the spins between the two states, and this inexorably changes the magnetization when there are only two quantum states in play. The fact that, for the ideal paramagnet, S can be expressed as a function of (B/T) alone (Figure 14.4), also enables one to see immediately that in an adiabatic process, a decrease in B is accompanied by a decrease in T.

14.4.2 Cooling by adiabatic demagnetization

Adiabatic demagnetization is a method of cooling a paramagnet to low temperatures. The magnetic degrees of freedom are cooled directly, and if there is a coupling between the magnetic dipoles of the molecules and their vibrational motion in the lattice (for a solid) then the vibrations can be cooled also. A good insight into the underlying physics is given by examining an energy level diagram, as in Figure 14.5. If each molecule has a total electronic angular momentum $\mathcal{J}$, then in a magnetic field there are $2\mathcal{J}+1$ energy levels, equally spaced, with spacing proportional to B. The temperature is a measure of how rapidly the population of these energy levels falls off as a function of energy, with low temperatures corresponding to a rapid fall-off. The population P of any given quantum state of energy ϵ is given by the *Boltzmann factor*, $P \propto \exp(-\epsilon/k_B T)$, so one finds $T = \epsilon_{1/2}/k_B \ln 2$, where $\epsilon_{1/2}$ is the energy above the ground state where the population per quantum state falls to half that of the ground state. In the initial isothermal step, T and therefore $\epsilon_{1/2}$ is fixed while the applied magnetic field is increased. A large applied B field makes the energy levels spread far apart (step (a)–(b) in the diagram); notice that the population of the highest energy state falls. After this, the sample is thermally isolated and the applied field is reduced. Therefore the second process is adiabatic, and in an adiabatic process the level populations stay fixed. As the field falls, the energy levels come close together (step (b)–(c)) and therefore $\epsilon_{1/2}$ falls. Consequently the final $\epsilon_{1/2}$, and therefore the final temperature, is low.

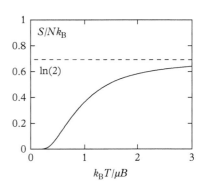

Fig. 14.4 *Entropy of an ideal paramagnet. Note that the entropy is shown here as a function of T/B, not B/T as in the previous figure.*

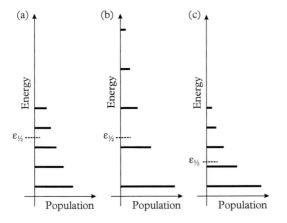

Fig. 14.5 *Energy level changes for cooling by adiabatic demagnetization (see text). The example shown here involves a change in B by only a factor 2, and consequently only a small cooling effect. In practice, B is changed by a large factor and a substantial cooling is obtained.*

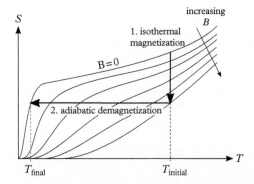

Fig. 14.6 *The entropy of a paramagnetic salt as a function of temperature, for various values of the applied magnetic field. The magnetic cooling method proceeds in two steps as shown.*

The above gives a good insight into the process. To analyse it more quantitatively, one relies on empirical measurements or a more sophisticated model. The typical behaviour is shown in Figure 14.6.

The general trends are reasonably intuitive: as B increases at given T, the entropy falls because the sample is becoming magnetized, so the many small dipoles inside it are lining up. If we then decrease B at given S, the temperature falls because the dipoles have less interaction energy with the field, and no heat has come in to replenish their energy, so the temperature falls. The two steps, isothermal magnetization, followed by adiabatic demagnetization, achieve, overall, a considerable cooling. The final temperature achievable by this method is somewhere on the $B = 0$ line, and since this line is steep at low temperatures, the final temperature does not depend strongly on the initial temperature nor on the value to which the B field is raised during the isothermal step. This is the case, that is, as long as we begin in the temperature regime where the magnetic effects influence the entropy. This is the regime where the magnetic heat capacity is non-negligible. It corresponds in practice to temperatures of order 1 kelvin, so other methods such as cooling by liquid helium are required to reach an appropriate starting point.

The magnetic dipoles in the sample are associated with angular momentum in the atoms or molecules and are said to constitute a 'spin system'. They occupy sites on a crystal lattice, and typically there is a non-negligible coupling between the dipoles and the motional degrees of freedom, i.e. vibrations of the lattice. Therefore as the spin system is cooled, so are the lattice vibrations, as long as the process proceeds slowly enough for them to remain in mutual equilibrium. This can be achieved in a few seconds. If it is desired to use the whole sample to also cool something else, then the times required may be much longer, some hours, because thermal resistance at boundaries is high at low temperatures.

There are many technical difficulties that must be overcome in practice in order to introduce both good thermal contact during the isothermal stage, and also good isolation during the adiabatic stage. A modern system may employ thermal conduction via a metallic link interrupted by a superconducting heat switch (Figure 14.7). In the electrically superconducting state the thermal conductivity is very low; it can fall by a factor 10^5 or more when the material undergoes the superconducting transition. This is brought about by removing a (modest) magnetic field in the switch.

The lower limit of the cooling method is set by the fact that the curve at $B = 0$ plunges to very low entropy before $T = 0$ is reached. This is because at low temperatures the spin system undergoes a phase transition at a temperature called the *Curie point*. The spins themselves generate a small magnetic field, and thus influence one another. They also influence one another through stronger electrostatic interactions because they are situated on charged particles. As a result, at low temperature they line up even in the absence of an applied field, and thus assume a lower entropy configuration. Therefore the lowest temperature attainable by adiabatic demagnetization is the temperature at the Curie point, called the *Curie temperature*, T_c.

There is a trade-off between low temperature and heat capacity. In order to get a low T_c the spins must have weak interactions with one another, and this can be achieved if they are at low density in the sample. But then there are fewer of them so the heat capacity of the sample is reduced. This does not matter if we only want to cool the spin system, but it is a problem if we want to use the magnetic cooling as a refrigerator in order to cool something else. The paramagnetic salt cerium magnesium nitrate has magnetic ions spaced out by the rest of the molecular structure, and with this material it is possible to reach a few millikelvins.

Lower temperatures still are available by cooling the nuclear spins, rather than the electronic spins. The dipole moments are smaller by about a factor 1000, and therefore so are the coupling to the applied field, the heat capacity, and the Curie point. Thus microkelvins can be attained for the nuclear spin system, but they do not offer any useful cooling effect on the lattice vibrations.

The use of empirical data such as in Figure 14.6 is often the most accurate way to analyse the cooling method. However, an approximate model of the equation of state and the heat capacity is useful to get further insight. The ideal paramagnet model correctly predicts that there will be cooling, but it cannot treat the behaviour near the Curie point. For this, a better approximation is the *Curie–Weiss law*:

$$\chi = \frac{a}{T - T_c}, \qquad (14.39)$$

where a is a constant depending on the sample, especially its density, and T_c is the Curie temperature, which is also a constant for a given material. (The Curie law, equation (6.22), can be written $\chi = a/T$, which agrees with the Curie–Weiss law when $T \gg T_c$.) It is notable that this model says χ is independent of B at given T. This holds quite well at low fields; it means the material has a linear response.

In reality, close to the Curie point the susceptibility becomes large but not infinite, and is no longer independent of B. Therefore the Curie–Weiss law is rather approximate at temperatures just above the Curie point (and below T_c the Curie–Weiss law does not apply). However it gives a good general insight.[2] Exercise 14.6 shows you how to use it to estimate the cooling effect, and one finds

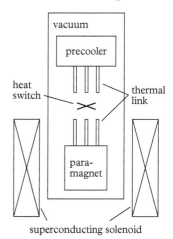

Fig. 14.7 *Apparatus for cooling by adiabatic demagnetization. The vacuum vessel may be immersed in liquid helium or otherwise cooled to around 1 K. The precooler is used during the isothermal stage while a large B field is applied. Then the heat switch breaks the thermal contact, so that the adiabatic stage can proceed.*

[2] At high temperature the behaviour is better matched by (14.39) if we replace T_c by a temperature somewhat higher than the Curie temperature. For example, for iron the ferromagnetic transition occurs at 1043 K, while the best fit to the high temperature behaviour is obtained with $T_c = $ 1093 K in the Curie–Weiss law.

that, if B_0 is the low field and B is the high field, then the final temperature T_f is related to the initial temperature T_i by

$$\frac{T_f - T_c}{T_i - T_c} \simeq \frac{B_0}{B}. \qquad (14.40)$$

Hence if $B_0 = 0$ then $T_f \simeq T_c$.

14.5 Electric and magnetic work

In this section we consider the performance of work on two types of system: a dielectric medium polarized by an electric field, and a magnetic medium magnetized by a magnetic field.[3] The subject has given rise to very much confusion in the scientific community for a long period, and there remain misleading or ambiguous treatments in some recent textbooks. The reason for this confusion will become apparent as we go along, but in the end there should be no remaining doubt as to what are the correct expressions and what are the physical circumstances to which they apply.

To begin, consider the following puzzle.

Take an ordinary parallel plate capacitor with plates of area A and separation x, with charge q on one plate, and charge $-q$ on the other (Figure 14.8). By using Gauss' law one finds that the electric field between the plates is $E = q/\epsilon_0 A$. An electric field in vacuum has an energy density $(1/2)\epsilon_0 E^2$, where ϵ_0 is the permittivity of free space, so the total energy of the field between the capacitor plates (treating it as uniform for simplicity) is

$$U = \frac{1}{2}\epsilon_0 E^2 A x = \frac{q^2}{2\epsilon_0 A} x. \qquad (14.41)$$

Suppose one of the plates is displaced by dx. The energy of the system rises, and we can use this to find the force on the plate:

$$\frac{dU}{dx} = \frac{q^2}{2\epsilon_0 A} = \frac{1}{2}qE. \qquad (14.42)$$

Since U increases with x, this is an attractive force.

Now consider the electric potential (or voltage) ϕ between the plates. It is given by $\phi = Ex$, so we can write

$$U = \frac{1}{2}\epsilon_0 \frac{\phi^2}{x^2} A x = \frac{\epsilon_0 \phi^2 A}{2x}, \qquad (14.43)$$

which gives

$$\frac{dU}{dx} = -\frac{\epsilon_0 \phi^2 A}{2x^2} = -\frac{1}{2}qE. \qquad (14.44)$$

Now it seems as if the force between the plates is repulsive, because the energy falls as x increases. But this contradicts our previous conclusion, so the puzzle is: what is really going on here?

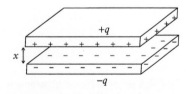

Fig. 14.8 An isolated parallel-plate capacitor.

---

[3] Our discussion is indebted to Carrington.

There is no direct mathematical contradiction between (14.42) and (14.44), although there is a poor notation, because in fact in both cases we have made an assumption about what does or does not change when x changes. The two equations should really be written

$$\left.\frac{\partial U}{\partial x}\right|_q = \frac{1}{2}qE, \qquad (14.45)$$

$$\left.\frac{\partial U}{\partial x}\right|_\phi = -\frac{1}{2}qE. \qquad (14.46)$$

Now it is clear that there is no direct mathematical contradiction, but there remains the puzzle of which, if any, of these expressions correctly gives the force between the plates. They agree on magnitude but disagree on the direction. One equation says the energy increases with x, which represents an attractive force between the plates, while the other equation says the energy decreases with x, which represents a repulsive force. Since like charges attract, we expect the plates to attract one another, so (14.45) looks plausible as a statement of the energy change, and in fact it does give the force correctly.

So what is wrong with (14.46)?

What is going on here is that we have failed to consider another source of energy which has been implicitly assumed when we consider a change at constant potential ϕ. In the first case (constant charge q), we can assume the capacitor is an electrically isolated system, because the charge on isolated metal plates will not change when the plates are moved. In the second case (constant potential ϕ) the capacitor *cannot* be an isolated electrical system, because if it were then ϕ would change when the plate separation is changed. So if in fact ϕ stays constant, it must be that an electric circuit is in place, with a voltage source (battery), such that charge can move off one plate, through the battery, and onto the other plate. But when the charge moves through the battery, electrical work is done either on or by the battery. In the present case, the amount of charge that must move, if ϕ stays constant when x increases by δx, is

$$\delta q = -\frac{q}{x}\delta x, \qquad (14.47)$$

where the sign indicates that the charge moves *off* the positive plate, and so works *against* the battery, doing work on it. The work done *by* the battery is $\phi \delta q$, so the work done *on* the battery is $-\phi \delta q = (\phi q/x)\delta x$. This energy must be supplied by whatever system provides the external force that moves the plate, so the full accounting is

$$\text{work performed by the external force} = \frac{\phi q}{x}\delta x + \left.\frac{\partial U}{\partial x}\right|_\phi \delta x$$

$$= \frac{1}{2}qE\delta x. \qquad (14.48)$$

This agrees fully with (14.45), as it must, because the force between the plates is what it is, for a given charge and field, irrespective of whether or not the plates are connected to a voltage source.

The above serves as an introduction to an issue that is often left ambiguous in the treatment of electric and magnetic problems. This is the issue of being careful to state precisely which system is under discussion, and to keep track of all the ways in which it interacts with other systems.

14.5.1 Dielectrics and polarization

We now turn to the treatment of a polarizable dielectric. We model a dielectric as a homogenous medium in which Maxwell's equations take the form

$$\nabla \cdot \mathbf{D} = q, \qquad \nabla \wedge \mathbf{E} = -\frac{\partial \mathbf{B}}{\partial t}, \tag{14.49}$$

$$\nabla \cdot \mathbf{B} = 0, \qquad \nabla \wedge \mathbf{H} = \mathbf{j} + \frac{\partial \mathbf{D}}{\partial t}, \tag{14.50}$$

where it is understood that q refers to free charge (not polarization charge) and $\mathbf{j}$ refers to free current (not magnetization current nor changing polarization). Here,

$$\mathbf{D} = \epsilon_0 \mathbf{E} + \mathbf{P}, \tag{14.51}$$

$$\mathbf{H} = (1/\mu_0)\mathbf{B} - \mathbf{M}, \tag{14.52}$$

where $\mathbf{P}$ is the polarization (electric dipole moment per unit volume) and $\mathbf{M}$ is magnetization (magnetic dipole moment per unit volume).

First consider electrostatic problems.

When the field under consideration is sourced by a charge q located on a conductor at potential ϕ, then the work required to change the charge (and hence the associated fields) by bringing in a charge δq from another conductor at zero potential is

$$\delta W = \phi \delta q. \tag{14.53}$$

It can be shown (see Appendix A) that

$$\phi \delta q = \int \mathbf{E} \cdot \delta \mathbf{D} \, dV, \tag{14.54}$$

where the integral is over all space. This result holds both in empty space, and also in the case where there is a polarizable medium of some arbitrary shape in the vicinity of the charged conductors. In general, $\mathbf{E}$ and $\mathbf{D}$ are functions of position.

For a composite system consisting of a pair of conductors with charges q and $-q$, and a dielectric nearby (for example, our system could be a capacitor with a slab of dielectric in the middle), and all their associated fields, the principle of conservation of energy (first law of thermodynamics) therefore gives

$$\delta U_{\text{tot}} = \delta Q + \int \mathbf{E} \cdot \delta \mathbf{D}\, dV, \qquad (14.55)$$

where δQ is the heat transfer.

One can study polarization phenomena perfectly well using equation (14.55). It is a correct and workable equation. However, it is often found convenient in practice to study the system another way. The idea is that the heat δQ will be distributed, in general, between the conductor and the dielectric, but if we are interested primarily in the latter then it makes sense to separate δQ into two parts. Similarly, it makes sense to try to separate out the work that is associated with changing the polarization of the dielectric from the work that would have been done in any case if the dielectric were not there. The idea is that, when q changes, some of the work is associated with changing the vacuum field, and some is associated with properties of the dielectric. It turns out that a convenient way to make the separation is to *define* a field $\mathbf{E}_0$ such that

$$\mathbf{E}_0 \equiv \text{the field that would be present if the dielectric were removed while leaving the charge on the conductor unchanged.}$$

This $\mathbf{E}_0$ is commonly called 'the applied field', but it is important to be clear that it is not 'the applied field' in any simple sense. It is not the field near the dielectric, nor the field inside the dielectric, nor any field present anywhere in the system under consideration! (though it may be numerically equal, in some cases, to one of those fields). It is, *by definition*, not a field in the system we are discussing, but the field in some other system, a fictitious one, but one related to the system under consideration. This fictitious (but useful) system is one just like the complete system we are studying, but with no dielectric, and (crucially) with the *same charge on the conductor*. Note, it is defined to have the same charge, not the same potential.

We are now ready to separate the work term in equation (14.55) into two parts. We can write

$$\int \mathbf{E} \cdot \delta \mathbf{D}\, dV = \delta W_e + \delta W_p, \qquad (14.56)$$

where the subscript 'e' indicates 'external' and 'p' indicates 'polarization'; the latter refers to the dielectric medium. We *define*

$$\delta W_e = \int \mathbf{E}_0 \cdot \delta \mathbf{D}_0\, dV \qquad (14.57)$$

and therefore

$$\delta W_p = \int \mathbf{E} \cdot \delta \mathbf{D}\, dV - \int \mathbf{E}_0 \cdot \delta \mathbf{D}_0\, dV. \qquad (14.58)$$

Now, since $\mathbf{E}_0$ and $\mathbf{D}_0$ are vacuum fields, they are related by $\mathbf{D}_0 = \epsilon_0 \mathbf{E}_0$, so $\mathbf{E}_0 \cdot \delta \mathbf{D}_0 = \mathbf{D}_0 \cdot \delta \mathbf{E}_0$. One can show (see Appendix A) that

$$q_0 \delta\phi_0 = \int \mathbf{D}_0 \cdot \delta \mathbf{E}_0 \mathrm{d}V \tag{14.59}$$

By using this and equation (14.54) in (14.58), we obtain

$$\delta W_p = \phi \delta q - q_0 \delta \phi_0. \tag{14.60}$$

But, by assumption, $q = q_0$, and therefore also $\delta q = \delta q_0$, so we have

$$\delta W_p = \phi \delta q_0 - q \delta \phi_0 = \int (\mathbf{E} \cdot \delta \mathbf{D}_0 - \mathbf{D} \cdot \delta \mathbf{E}_0) \, \mathrm{d}V = \int (\epsilon_0 \mathbf{E} - \mathbf{D}) \cdot \delta \mathbf{E}_0 \mathrm{d}V,$$

which gives

$$\delta W_p = -\int \mathbf{P} \cdot \delta \mathbf{E}_0 \, \mathrm{d}V. \tag{14.61}$$

Roughly speaking, this may be considered 'the work done on the dielectric medium'. Strictly speaking, it is the difference between the work done on the complete actual system and the work which would have been done if the dielectric were not there, under the assumption that the system would then have the same conductor with the same charge.

Equation (14.61) accurately treats both weak and strong polarization, and any geometry. In the case of a uniform field $\mathbf{E}_0$, we can take $\delta \mathbf{E}_0$ outside the integral (whether or not $\mathbf{P}$ is uniform) and thus obtain

$$\delta W_p = -\mathbf{p} \cdot \delta \mathbf{E}_0, \tag{14.62}$$

where $\mathbf{p}$ is the total electric dipole moment of the sample.

We can now make the separation of equation (14.55) into two parts. We write

$$\delta U_\mathrm{tot} = \delta U_e + \delta U_p, \tag{14.63}$$

where

$$\delta U_e = \delta Q_e + \int \mathbf{E}_0 \cdot \delta \mathbf{D}_0 \, \mathrm{d}V, \tag{14.64}$$

$$\delta U_p = T \mathrm{d}S - \int \mathbf{P} \cdot \delta \mathbf{E}_0 \, \mathrm{d}V. \tag{14.65}$$

Equation (14.65) is useful for studying polarizable media, because it has the form of a fundamental relation, in which T and S are the temperature and entropy of the dielectric slab, and the second term gives the work done in terms of further functions of state. Although the field $\mathbf{E}_0$ is not a property of the system in quite such a direct and obvious way as the volume is a property of a compressible system, it can be considered and treated as one because it is a function of the state of the dielectric in the thermodynamic sense. That is, for a given state of the dielectric (defined, for example, by a given polarization and entropy), the value of $\mathbf{E}_0$ is determined.

When $\mathbf{E}_0$ is uniform, we have

$$dU_p = TdS - \mathbf{p} \cdot d\mathbf{E}_0. \qquad (14.66)$$

This equation may be regarded as a fundamental relation in terms of which the behaviour of polarizable media can be studied, and we recommend it. However, it is not the only way to formulate the problem, because it treats part of a composite system, and there is more than one possible convention on the way the energy of the composite is assigned to the parts. The other standard choice is to consider that 'what would have been the case if the dielectric were not present' is not given by the field $\mathbf{E}_0$ but by a field $\mathbf{E}_{0\phi}$, defined as

$\mathbf{E}_{0\phi} \equiv$ the field that would be present if the dielectric were removed while leaving the electric potential on the conductor unchanged.

Using this idea, we argue in a similar way as before, but now instead of using equation (14.58), we *define*

$$\delta W_\phi \equiv \int \mathbf{E} \cdot \delta \mathbf{D} \, dV - \int \mathbf{E}_{0\phi} \cdot \delta \mathbf{D}_{0\phi} \, dV. \qquad (14.67)$$

When $\phi = \phi_0$ one has $\phi \delta q - q_0 \delta \phi_0 = \phi_0 \delta q - q_0 \delta \phi$, and so one finds

$$\delta W_\phi = \int \mathbf{E}_{0\phi} \cdot \delta \mathbf{D} \, dV - \epsilon_0 \mathbf{E}_{0\phi} \cdot \delta \mathbf{E} \, dV = \int \mathbf{E}_{0\phi} \cdot \delta \mathbf{P} \, dV. \qquad (14.68)$$

With this choice, the total energy is divided up as $\delta U_{\text{tot}} = \delta U_{e\phi} + \delta U_{p\phi}$, where

$$\delta U_{e\phi} = \delta Q_e + \int \mathbf{E}_{0\phi} \cdot \delta \mathbf{D}_{0\phi} \, dV,$$

$$\delta U_{p\phi} = TdS + \int \mathbf{E}_{0\phi} \cdot \delta \mathbf{P} \, dV. \qquad (14.69)$$

Equation (14.69) may now be interpreted as a fundamental relation for a polarized medium, but in our view this second way of dividing up the energy is confusing. Therefore, although there is nothing inaccurate or unworkable about (14.69), we do not recommend it.

To illustrate the confusion that may result, consider the following experiment. We prepare an isolated capacitor with a dielectric completely filling the space between the plates. Note, there are no electrical connections: the capacitor is prepared with charge q on one plate, $-q$ on the other, and then electrically isolated. Suppose the dielectric medium is linear and homogenous, so that the polarization is given by $\mathbf{P} = \epsilon_0 \mathbf{E}_i (\epsilon_r - 1)$, where ϵ_r is the relative permittivity and $\mathbf{E}_i$ is the field inside the dielectric. In this case one finds

$$\mathbf{P} = \frac{q}{A}\left(1 - \frac{1}{\epsilon_r}\right), \qquad \phi = \frac{qx}{\epsilon_0 \epsilon_r A}, \qquad (14.70)$$

where the plates have area A and separation x. Now suppose the dielectric is cooled, and as a result its relative permittivity increases. In this case, $\mathbf{P}$ is increased and ϕ falls. Since there is no change in q, the field $\mathbf{E}_0$ does not change, so the work W_p described by equation (14.61) is zero; which makes sense: no external source of emf, such as a battery, has done any work on the system in this example, since none was connected. The change in the dielectric is wholly attributed to the heat which left it. However, the work W_ϕ described by (14.68) is not zero! What can that mean? In this experiment, the increase of ϵ_r (which was totally owing to heat leaving the system) results in a fall in ϕ, and therefore the field $\mathbf{E}_{0\phi}$ gets smaller, and (on the view adopted in (14.69)) this is the field deemed to be the one that would be present in the absence of the medium. Therefore the energy in the applied field, on this approach, is deemed to have fallen. The 'work done on the medium', as described by W_ϕ, is here provided by the loss of energy from the 'applied field' $\mathbf{E}_{0\phi}$.

> **Example 14.1**
>
> A capacitor is filled with a linear homogenous dielectric slab with constant relative permittivity ϵ_r, area A, and thickness x. A variable voltage source is connected, and used to raise the charge on the capacitor from 0 to q in a quasistatic adiabatic process. Find the work done by the source and the work done on the dielectric material.
>
> *Solution.*
> When the charge on the capacitor is q, the relevant quantities have the following values:
>
> | | | |
> |---|---|---|
> | fields in the medium | $D = q/A,$ | $E = q/A\epsilon_0\epsilon_r$ |
> | total dipole moment | $p = qx(1 - 1/\epsilon_r)$ | |
> | voltage between the plates | $\phi = xE.$ | |
> | 'applied field' | $E_0 = q/A\epsilon_0$ | $= \epsilon_r E$ |
> | 'applied field' | $E_{0\phi} = \phi/x$ | $= E.$ |
> | work done by the source | $W = \int \phi\,dq$ | $= xq^2/2A\epsilon_0\epsilon_r = q\phi/2$ |
>
> The last line answers the first part of the question. The second part of the question is ambiguous, so has no single answer. On one model, the energy assigned to the 'applied field' is $(1/2)\epsilon_0 E_0^2 Ax = \epsilon_r q\phi/2$ and the energy accounted as work on the medium is
>
> $$W_p = \int -p\,dE_0 = \tfrac{1}{2}q\phi(1-\epsilon_r).$$
>
> On the other model, the energy assigned to the 'applied field' is $(1/2)\epsilon_0 E_{0\phi}^2 Ax = q\phi/2\epsilon_r$ and the energy accounted as work on the medium is
>
> $$W_\phi = \int E_{0\phi}\,dp = \tfrac{1}{2}q\phi(1-1/\epsilon_r).$$

For example, if $\epsilon_r = 3$, $q = 3 \times 10^{-6}$ C, and $\phi = 20$ volt then the energy is accounted as follows:

	Work done by source (μJ)	Increase in applied field energy (μJ)	Work assigned to medium (μJ)
Model 1	30	90	−60
Model 2	30	10	20

14.5.2 Magnetic work

The case of a magnetizable medium subject to a magnetic field has many similarities with the case of electric polarization described above. The puzzle of the force between the plates of an empty capacitor is closely mirrored by a similar puzzle involving a solenoid. To make the point, it is very helpful to consider, in the first instance, an isolated solenoid.

This is a solenoid which can be taken to be of cylindrical shape, with many turns of current-carrying wire, but with no current source in the circuit (Figure 14.9). The wire is simply connected to itself in a closed circuit and we idealize it as having no electrical resistance. The current then flows forever. In this case, one can easily show from the Maxwell equation (14.49ii) (Faraday's law of induction) that the magnetic flux in the solenoid stays constant. This is done by integrating the equation along a closed path lying everywhere inside the wire. Then $\mathbf{E} = 0$ all along the path, so one finds

$$\frac{d\Phi}{dt} = 0,$$

where $\Phi = \int \mathbf{B} \cdot d\mathbf{S}$ is the flux. In particular, this flux will stay constant when other things change, such as the shape of the solenoid, or when a magnetic sample is brought up.

Treating the magnetic field inside the solenoid as uniform for the sake of simplicity, the field in an empty solenoid is $B = \mu_0 n I = \Phi/A$, where n is the number of turns per unit length, μ_0 is the permeability of free space, I is the current, and A is the cross-sectional area of the solenoid. If the solenoid has length x then the total field energy is

$$U = \frac{1}{2} B H A x = \frac{\Phi^2}{2\mu_0 A} x = \frac{\mu_0 N^2 I^2}{2x},$$

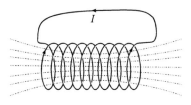

Fig. 14.9 *An isolated current-carrying solenoid. The wire is idealized as offering no electrical resistance.*

where $N = nx$ is the number of turns. Hence one finds, for a fixed number of turns,

$$\left.\frac{\partial U}{\partial x}\right|_\Phi = \frac{1}{2}\Phi H, \qquad \left.\frac{\partial U}{\partial x}\right|_I = -\frac{1}{2}\Phi H.$$

Using this, one can invent a puzzle about the force on the ends of the solenoid, similar to the one we explored for the capacitor in connection with equations (14.45) and (14.46).

This introduction helps one to familiarize oneself with the fact that for a solenoid in closed circuit, in the absence of a source of e.m.f., the variable that naturally stays constant is the *flux* Φ and not the current I. The flux plays the role analogous to the charge on the plates of a capacitor.

We can now proceed to the case of a magnetizable sample interacting with a field produced by a solenoid, where now we allow a source of e.m.f. in the circuit, so that the flux can change when the source does work. The complete composite system consists of the sample, the solenoid, and their fields. The source of e.m.f.[4] is regarded as an external device which can interact with this composite system. Changes in the total internal energy of the composite system are given by (see Appendix A)

$$\delta U_{\text{tot}} = \delta Q + \int \mathbf{H} \cdot \delta \mathbf{B} \, dV, \qquad (14.71)$$

where δQ is the heat transfer. We now follow a procedure similar to the one previously described for electric polarization. We write $\delta Q = \delta Q_e + \delta Q_m$ and $\delta W = \delta W_e + \delta W_m$, with the latter defined as

$$\delta W_m \equiv \int \mathbf{H} \cdot \delta \mathbf{B} \, dV - \int \mathbf{H}_0 \cdot \delta \mathbf{B}_0 \, dV, \qquad (14.72)$$

where

$$\mathbf{B}_0 \equiv \begin{array}{l}\text{the field that would be present if the sample were removed} \\ \text{while leaving the flux in the solenoid unchanged.}\end{array} \qquad (14.73)$$

Now $\mathbf{B}_0 = \mu_0 \mathbf{H}_0$ so $\mathbf{H}_0 \cdot \delta \mathbf{B}_0 = \mathbf{B}_0 \cdot \delta \mathbf{H}_0$ and hence we have

$$\delta W_m = \int \mathbf{H} \cdot \delta \mathbf{B} - \mathbf{B}_0 \cdot \delta \mathbf{H}_0 \, dV = I \delta \Phi - \Phi_0 \delta I_0, \qquad (14.74)$$

using equations (A.13) and (A.14). The definition of $\mathbf{B}_0$ is such that $\Phi = \Phi_0$, so we have

$$\delta W_m = I \delta \Phi_0 - \Phi \delta I_0 = \int \mathbf{H} \cdot \delta \mathbf{B}_0 - \mathbf{B} \cdot \delta \mathbf{H}_0 \, dV$$
$$= - \int \mathbf{M} \cdot \delta \mathbf{B}_0 \, dV. \qquad (14.75)$$

This justifies the equation included in Table 14.1 and used in Section 14.4. The result is valid for strong as well as weak magnetization.

The above is exact and applies to any geometry and any magnetization strength. A useful physical intuition about the result is offered by the following argument which applies in the case of weak magnetization. We model the

---

[4] The source here is often called a 'current source' because this is the standard term in circuit theory for a source of *electrical energy* that ordinarily provides a constant current. We have in mind such a source which can be adjusted to provide any desired current.

magnetic medium as a collection of permanent magnetic dipoles $\mathbf{m}_i$, and ignore the interaction energy of the dipoles with one another (the *self-energy*). Then the total interaction energy between the dipoles and the field is

$$U_{\text{int}} = -\sum_i \mathbf{m}_i \cdot \mathbf{B}_i, \tag{14.76}$$

where $\mathbf{B}_i$ is the field at any given dipole due to all sources except the dipoles themselves. If the field is uniform, then

$$U_{\text{int}} = -\mathbf{m} \cdot \mathbf{B}_i,$$

where $\mathbf{m} = \sum_i \mathbf{m}_i$ and $\mathbf{B}_i$ can be read as the 'internal' field. Since in this simple argument we are ignoring the self-energy, this field can be taken as $\mathbf{B}_0$. Changes in this interaction energy are then given by

$$dU_{\text{int}} = -\mathbf{B}_0 \cdot d\mathbf{m} - \mathbf{m} \cdot d\mathbf{B}_0 = đQ + đW. \tag{14.77}$$

Given (14.75), we can interpret the first term as $đQ$ and the second term as $đW$. It is natural to associate a change in $\mathbf{m}$ with a change in entropy (consider the ideal paramagnet for example), so this gives a further motivation to adopt (14.75) as the magnetic work, and it shows that we can interpret the sum of this heat and work as a change in the interaction energy between the magnetic sample and the field.

The other possible convention is to take

$$\delta W_{mI} \equiv \int \mathbf{H} \cdot \delta \mathbf{B} - \mathbf{B}_{0I} \cdot \delta \mathbf{H}_{0I} \, dV, \tag{14.78}$$

where

$$\mathbf{B}_{0I} \equiv \begin{array}{l} \text{the field that would be present if the sample were removed} \\ \text{while leaving the current in the solenoid unchanged.} \end{array} \tag{14.79}$$

Then one obtains

$$\delta W_{mI} = \int \mathbf{B}_{0I} \cdot \delta \mathbf{M} \, dV. \tag{14.80}$$

Thus, just as in the electrostatic case, there are two possible conventions as to how the energy is divided between the part that is attributed to the presence of the medium, and the part that is attributed to the external field. The choice between these conventions turns on what one chooses to consider would be the situation if the magnetizable sample were not present. Both conventions lead to correct equations that can be used to study magnetizable media, as long as one is clear about which field is being considered to be the 'applied' field. Equation (14.75) is often preferable because it treats the isolated solenoid in a physically intuitive way, it connects straightforwardly to calculations in statistical mechanics, and for weak magnetization it has the physical interpretation offered by equation (14.77). However, in the case of strong magnetization, some results are more simply expressed in terms of $\mathbf{B}_{0I}$, and therefore this is often taken as 'the applied field'. This choice is sometimes justified by the argument that the current acts as source of the magnetic field in Maxwell's equations, and constant current sources of e.m.f. are easy to construct in practice. However, although there is nothing fundamentally wrong with this point of view, one can also choose to regard fluxes as sources of magnetic fields.

Both choices also lead to further thermodynamic potentials, such as the 'magnetic enthalpy',

$$\mathcal{H}_m = U_m + \mathbf{m} \cdot \mathbf{B}_0, \tag{14.81}$$

$$\mathcal{H}_{mI} = U_{mI} - \mathbf{m} \cdot \mathbf{B}_{0I}. \tag{14.82}$$

When $B_0 \simeq B_{0I}$, then one finds $\mathcal{H}_m \simeq U_{mI}$ and $\mathcal{H}_{mI} \simeq U_m$. Thus if we introduced a change of name, calling $\mathcal{H}_{mI}$ 'internal energy' and U_{mI} 'enthalpy', the two models would agree in all respects in the limit $\chi \to 0$.

14.6 Introduction to the partition function

Before leaving this chapter, we offer the reader a derivation of the relationship between the free energy and the sum over states in equation (14.29). For a two-state system the derivation does not require the general methods of statistical physics. However, we do require some basic assumption about the way entropy relates to microscopic properties, and for this we will adopt, without proof, the formula which we have already mentioned in equation (9.15) and Exercise 9.12 of Chapter 9:

$$S = -Nk_B \sum_i p_i \ln p_i. \tag{14.83}$$

This gives the entropy for a system of $N \gg 1$ very weakly interacting particles, each of which may be in one of a set of quantum states (energy eigenstates) i with probability p_i. This entropy formula may be derived from Boltzmann's formula (9.8), and the related ideas. We shall not try to justify it further here, but simply use it to obtain the *Boltzmann factor* (14.94) for a two-state system, and hence equation (14.29).

For N very weakly interacting particles, each having two energy states, we have

$$S = -Nk_B [p \ln p + (1-p) \ln(1-p)], \tag{14.84}$$

where the probabilities of being in the two states are $p_1 = p$ and $p_2 = 1 - p$. The two-state system is simple because, for fixed N, the entropy is a function of a

> **Alternative route for the ideal paramagnet** Equation (14.32) can be derived directly from entropy, without going via the partition function, as follows. Using the model of N small dipoles, each with two possible states, we find that the total interaction energy with the applied field is given by
>
> $$U = -\sum_{i=1}^{N} \mathbf{m}_i \cdot \mathbf{B} = -\mathbf{m} \cdot \mathbf{B} = -mB, \quad (14.85)$$
>
> where the sum is over all the small dipoles, each of dipole moment $\mathbf{m}_i$, and $\mathbf{m} = \sum_i \mathbf{m}_i$ is the total dipole moment.
>
> Let p be the fraction of the N dipoles that are aligned with z, and $1-p$ the fraction anti-aligned with z. Then
>
> $$m = pN\mu - (1-p)N\mu = (2p-1)N\mu, \quad (14.86)$$
>
> since each dipole of size $\mu = |\mathbf{m}_i|$ contributes either $+\mu$ or $-\mu$ according to its orientation.
>
> Finally, we use equation (14.84).
>
> We have in equations (14.85), (14.86), and (14.84) complete thermodynamic information about the system. For, using (14.85) we can write $m = -U/B$ and using (14.86) we can write
>
> $$p = \frac{1}{2}\left(1 + \frac{m}{N\mu}\right) = \frac{1}{2}\left(1 - \frac{U}{BN\mu}\right). \quad (14.87)$$
>
> By using this in (14.84) we have an expression for S in terms of U and B. But, from the fundamental relation (14.23), these are the natural variables of S, so this is just what we need: all the thermodynamic properties can now be found. Once again, entropy is king and energy is prime minister.
>
> However, this box is included primarily as an observation. The method of calculation via the Boltzmann factor and the partition function is to be preferred because it furnishes more generally useful methods and insights (and it involves easier algebra).

single parameter, p. Suppose the energies of the states are $\epsilon_1 = -\epsilon$ and $\epsilon_2 = +\epsilon$, respectively (for the magnetic case, $\epsilon = \mu B$). Then the energy of the system is

$$U = Np(-\epsilon) + N(1-p)\epsilon = N\epsilon(1-2p). \quad (14.88)$$

We would like to relate p to the temperature. This can be done via the fundamental relation,

$$dU = TdS + đW, \quad (14.89)$$

which gives

$$T = \left.\frac{\partial U}{\partial S}\right|_{(đW=0)} = \left.\frac{\partial U}{\partial S}\right|_{\epsilon}. \quad (14.90)$$

The energy level ϵ is constant in this expression because the work term is $đW = -mdB$, and changes in B result in changes in ϵ. Hence we have

$$T = \left.\frac{\partial U}{\partial p}\right|_{\epsilon} \frac{dp}{dS} = \frac{-2N\epsilon}{Nk_B \ln((1-p)/p)}. \quad (14.91)$$

Solving this for p gives

$$p = \frac{1}{1 + e^{-2\epsilon/k_B T}} = \frac{e^{\epsilon/k_B T}}{e^{\epsilon/k_B T} + e^{-\epsilon/k_B T}}. \tag{14.92}$$

and

$$1 - p = \frac{e^{-\epsilon/k_B T}}{e^{\epsilon/k_B T} + e^{-\epsilon/k_B T}}. \tag{14.93}$$

Recalling that the energy of the state with probability p is $\epsilon_1 = -\epsilon$, both results can be written

$$p_i \propto e^{-\epsilon_i/k_B T}, \tag{14.94}$$

where the proportionality factor (the denominator in equation (14.92)) can always be reinserted by making sure the probabilities add to 1. This formula for the relative probability of occupation of a state of energy ϵ_i in thermal equilibrium is called the *Boltzmann factor*.

Now introduce

$$Z \equiv \sum_{i \in \text{states}} e^{-\epsilon_i/k_B T}. \tag{14.95}$$

Then we have

$$p_i = \frac{1}{Z} e^{-\epsilon_i/k_B T}. \tag{14.96}$$

Using this in (14.83) gives

$$S = -Nk_B \sum_i p_i \left(\frac{-\epsilon_i}{k_B T} - \ln Z \right)$$

$$= Nk_B \ln Z + \frac{N \sum_i p_i \epsilon_i}{T} \tag{14.97}$$

$$= Nk_B \ln Z + \frac{U}{T}, \tag{14.98}$$

where we have recognized the sum in the second step as the expression for the average energy per particle, and N times this is the total energy U. We thus have

$$ST - U = Nk_B T \ln Z \tag{14.99}$$

and equation (14.29) follows immediately.

....

EXERCISES

(14.1) Show that, for an elastic rod,

$$\left. \frac{\partial C_L}{\partial L} \right|_T = -T \left. \frac{\partial^2 f}{\partial T^2} \right|_L, \tag{14.100}$$

where C_L is the heat capacity at constant length.

(14.2) Consider the 'ideal elastic', whose equation of state is given by (6.20).
(i) Show that the internal energy is a function of temperature alone (in this respect the ideal elastic substance is like an ideal gas. It is a reflection of the fact that the energy is associated with kinetic not potential energy of the molecules).
(ii) If the heat capacity C_L is independent of temperature at $L = L_0$, show that the Helmholtz function is given by

$$F = F_0 - TC_L \ln \frac{T}{T_0} + KT \left[\frac{L^2}{2L_0} + \frac{L_0^2}{L} \right], \quad (14.101)$$

where $F_0(T) = U_0 - TS_0 + (T - T_0)C_L - (3/2)L_0KT$. [Use equation (13.53), suitably modified.]
(iii) Show that the difference between the primary heat capacities is

$$C_L - C_f = -\frac{K(L^3 - L_0^3)^2}{L_0 L(L^3 + 2L_0^3)}. \quad (14.102)$$

(14.3) The surface tension of liquid argon is given by $\sigma = \sigma_0(1 - T/T_c)^{1.28}$, where $\sigma_0 = 0.038$ N/m and the critical temperature $T_c = 151$ K. Find the surface entropy per unit area at the triple point, $T = 83$ K.

(14.4) Obtain equations (14.33)–(14.38) from (14.32).

(14.5) Calculate $(\partial S/\partial U)_B$ from equations (14.34) and (14.35), and comment.

(14.6) Obtain equation (14.40) from the Curie–Weiss law, as follows. Consider the magnetic Helmholtz function $F = U - ST$. We have $dF = -SdT - mdB$ and it is useful to consider $F(T, B)$. This can be written

$$F(T, B) = F(T, 0) + \int \left. \frac{\partial F}{\partial B} \right|_T dB.$$

Use this to obtain

$$F(T, B) = F(T, 0) - \frac{aVB^2}{2\mu_0(T - T_c)}$$

and hence

$$S(T, B) = S(T, 0) - \frac{aV}{2\mu_0} \frac{B^2}{(T - T_c)^2}.$$

The first term is the entropy at $B = 0$ for any given temperature. The second term is the magnetic contribution. If we assume that the *change* in entropy is dominated by the latter contribution, then for an adiabatic process, explain why equation (14.40) is obtained. (This derivation illustrates the concept introduced in equation (13.30): an equation of state does not give the heat capacity directly, but it can be used to extend partial knowledge of the heat capacities.)

(14.7) *Alternative route for the ideal paramagnet.* Introduce $y = m/N\mu$ in equation (14.87) and show that

$$S = \frac{Nk_B}{2}\left(y\ln\frac{1-y}{1+y} - \ln\frac{1}{4}(1-y^2)\right).$$

Then introduce x defined by $y \equiv \tanh x$, and obtain equation (14.34). Then find an expression for T in terms of x, using $(1/T) = (\partial S/\partial U)_B$, and hence obtain (14.31). Finally, use (14.85) to obtain (14.35), and hence (14.32).

(14.8) *Geometric factor.* A long straight solenoid has cross-section A and contains a long straight paramagnetic sample with cross-section A_m. Show that, in the case $\chi \ll 1$, the total magnetic flux through the system is $\Phi = \mu_0 nI(A + A_m\chi)$, and hence

$$B_0 = B_{0I}(1 + \chi A_m/A).$$

Thus $B_0 = B_{0I}$ when the sample is thin and does not fill much of the solenoid, and $B_0 = (1+\chi)B_{0I}$ when the sample fills the solenoid. In general we define a *geometric factor* γ_0 such that $B_0 = (1+\gamma_0\chi)B_{0I}$ for $\chi \ll 1$. These results are modified when χ is not small (see question 14.9).

(14.9) *Demagnetizing factor.* For a strongly magnetic material, the internal B_i field may differ considerably from $\mu_0 H_i$. One has $M = \chi H_i$ and

$$H_i = H_{0I} - \alpha M, \qquad (14.103)$$

where α is called the *demagnetizing factor*; its value depends on the shape of the sample. For a long cylindrical sample, $\alpha = 0$; for a sphere, $\alpha = 1/3$.

(i) Show that $H_i = H_{0I}/(1+\alpha\chi)$ and hence

$$M = \frac{\chi}{1+\alpha\chi}H_{0I}. \qquad (14.104)$$

This means that the influence of the magnetization on the field can be accounted for in many expressions by replacing the susceptibility by $\chi/(1+\alpha\chi)$, but this simple rule is not universally true because we still have $B_i = \mu_0(1+\chi)H_i$.

(ii) Show that the generalization of (14.20) is

$$B_0 = B_{0I}\left(1 + \gamma_0\frac{(1-\alpha)\chi}{1+\alpha\chi}\right), \qquad (14.105)$$

where γ_0 is the geometric factor defined in question 14.9. Note, since α also depends on the geometry, it is common to absorb the factor $(1-\alpha)$ into the geometric factor, such that one has

$$B_0 = B_{0I}\left(1 + \gamma\frac{\chi}{1+\alpha\chi}\right), \qquad (14.106)$$

where $\gamma = (1-\alpha)\gamma_0$.

(14.10) Show that for strong as well as weak magnetization,

$$\mu_0 M = \frac{\chi}{1 + \alpha\chi} B_{0I} = \frac{\chi}{1 + (\gamma + \alpha)\chi} B_0. \qquad (14.107)$$

Hence, show that, for a sample described by the Curie–Weiss law, the equation of state can be written

$$\frac{\mu_0 M}{B} = \frac{a}{T - T'_c}, \qquad (14.108)$$

where $T'_c = T_c - \alpha a$ for $B = B_{0I}$ and $T'_c = T_c - (\gamma + \alpha)a$ for $B = B_0$. Hence for strong magnetization the Curie–Weiss law is modified simply by a shift in the parameter T_c. Note, this is not strictly the transition temperature, but a parameter in a model that is valid for $T \gg T_c$ (the actual critical point is reached at a temperature slightly below this T_c).

(14.11) Compare the elastic band experiment described at the end of Section 14.2 with the process of cooling by adiabatic demagnetization.

15 Modelling real gases

15.1 van der Waals gas · 219
15.2 Redlich–Kwong, Dieterici, and Peng–Robinson gas · 224
Exercises · 226

The ideal gas has proved to be a useful idealized system for learning about thermodynamics, but it is limited in two ways. First, it does not exhibit some important thermodynamic phenomena such as phase changes, and second it is unphysical in the limit of low temperature.

In any case we want to apply thermodynamics to the study of real systems, not just idealized ones.

Consider the isothermal compression of a real gas. For example, imagine the gas is placed in a cylinder in thermal contact with a temperature reservoir, and the piston is slowly moved in. If we start from a large volume and low pressure, then initially the isotherm is like that of an ideal gas: the pressure rises almost in inverse proportion to the volume. However, as the volume is reduced the isotherm departs more and more from Boyle's law—see Figure 15.1. If the temperature is not too high, then at some particular pressure and volume, the behaviour abruptly takes on a new character: the pressure stops changing, while the volume continues to fall: the isotherm is horizontal. At some smaller volume, the compression changes character a second time: the pressure now rises abruptly, even for small volume changes: the isotherm is very steep, the system has become almost incompressible.

The behaviour just described is an example of a phase change. At the horizontal portion of the isotherm, the material inside the cylinder is in two forms, (two *phases*): one gaseous and one liquid, and as the system is compressed the substance gradually goes over from a wholly gaseous phase to wholly liquid. Once the system is completely in liquid form, it becomes much harder to compress.

Other types of phase change also occur, such as freezing; this general concept is the subject of Chapter 22. In the present chapter we focus attention on the gaseous behaviour, but with a view to understanding it over a wider range of pressure and temperature, where the ideal gas model is no longer accurate.

At higher temperatures, the phase change from liquid to vapour takes place over a smaller range of volume—see Figure 15.2. This indicates that the density difference between the liquid and vapour phases is getting smaller. Indeed, the liquid and the vapour become increasingly hard to tell apart. Eventually, at a particular isotherm called the *critical isotherm*, the phase change vanishes. At one particular pressure and volume, called the *critical point*, this isotherm is both horizontal, $(\partial p/\partial V)_T = 0$, and has a point of inflexion, $(\partial^2 p/\partial V^2)_T = 0$. These two conditions suffice to uniquely identify the pressure and volume; the temperature is then fixed by the equation of state. Therefore there is just one such critical

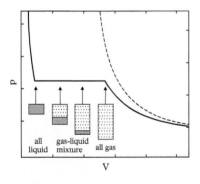

Fig. 15.1 *Isothermal compression of a real fluid. The full line shows an example observed isotherm, the dashed line shows Boyle's law.*

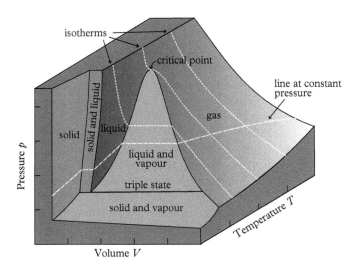

Fig. 15.2 *The equilibrium states for a typical inert substance treated as a simple compressible system. The surface in p, V, T 'space' shows that there are just two independent variables, the third being fixed by an equation of state.*

point. Its state parameters are called the critical pressure, critical volume, and critical temperature.

The complete picture of the behaviour for all values of pressure, volume, and temperature can be displayed in a diagram such as Figure 15.2. In principle all this behaviour can be captured by a single equation of state, giving p as a function of T and V, for example, but the formula has no simple mathematical expression. Rather than trying to manipulate such a formula, it is more useful to model the substance using approximate methods. We will consider a few examples. These are intended primarily to model the gaseous behaviour while taking into account the effect of non-zero forces between the particles. As a consequence, they also predict that there will be a vapour/liquid phase transition and they give some valid qualitative insights into this aspect. They do not account for the solid phase.

van der Waals' equation

$$\left(p + \frac{a}{V_m^2}\right)(V_m - b) = RT. \tag{15.1}$$

Redlich–Kwong equation

$$\left(p + \frac{a}{V_m(V_m + b)T^{1/2}}\right)(V_m - b) = RT. \tag{15.2}$$

Dieterici's equation

$$p(V_m - b)\, e^{a/(RTV_m)} = RT. \tag{15.3}$$

Virial expansion

$$PV_m = RT\left(1 + \frac{B_2}{V_m} + \frac{B_3}{V_m^2} + \frac{B_4}{V_m^3} + \ldots\right) \tag{15.4}$$

Here V_m is the molar volume and R is the molar gas constant, $R \equiv N_A k_B$. In the first three equations, a, b are constants. In the virial expansion, the coefficients $B_2, B_3, \ldots$ are functions of temperature.[1]

The gas volume V is related to V_m by $V = mV_m$, if m is the number of moles ($m = N/N_A$ for N particles). Therefore, if one wants an expression in terms of the gas volume V, one should replace V_m by V/m in the above. In this chapter we will consider one mole of gas, so that $V_m = V$ and one can drop the m; but if any doubt arises one must reinsert m as appropriate.

All these equations are modifications of the ideal gas equation. The constants a, b and the functions $B_i(T)$ are chosen so as to match the observed behaviour at some range of pressure and temperature. This means the equations are guaranteed to match the true behaviour accurately in the chosen regime, and they are valuable in so far as they also describe the behaviour correctly far from that regime. All the equations succeed in predicting the presence of a critical isotherm, and this means one can use the observed critical pressure and temperature to establish values for the parameters a, b. However, the equations then predict the critical molar volume and they are not guaranteed to get it right. When the constants are established this way, the equations predict the behaviour away from the critical point. If one wanted the most accurate model of the low pressure behaviour, i.e. far from the critical point, then one would establish empirical values for the constants by direct comparison with the real gas in the low pressure regime, and then the values of the constants would differ from those inferred from the critical point.

The standard procedure is to use the critical pressure and temperature to determine the coefficients.

Compression factor and Boyle temperature

One piece of information which can be useful to compare a real gas with an ideal gas is the value of pV/RT. This quantity is called the *compression factor* or *compressibility factor*, and is often given the symbol Z. Its value is $Z = 1$ for an ideal gas. At high pressures (e.g. 500 atmospheres) Z is greater than 1 for most gases. At low pressures its value may be above or below 1, depending on temperature, and tending to 1 as the pressure tends to zero. Figure 15.3 shows some examples. The low pressure region is where any real gas is most like an ideal gas, and for any gas there exists a temperature for which Z remains close to 1 at low pressure. This is the temperature where Z does not change as a function of pressure, to first-order approximation, i.e.

$$\lim_{p \to 0} \left.\frac{\partial Z}{\partial p}\right|_T = 0 \quad \text{at } T_B. \tag{15.5}$$

The temperature where this occurs is called the *Boyle temperature*, since at this temperature Boyle's law is obeyed especially well.

In the virial expansion, the Boyle temperature is the temperature where $B_2(T) = 0$. The general form of $B_2(T)$ is shown in Figure 15.4. The Boyle

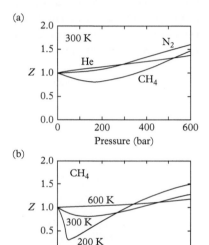

Fig. 15.3 Compression factor $Z = pV/RT$ as a function of pressure at 300 K for three gases (a), and for one gas (methane) at three example temperatures (b). The critical temperature for methane is 191 K and the Boyle temperature is 506 K.

---

[1] The virial coefficients are numbered such that B_i is the coefficient of V_m^{-i} in the expression for p/RT. This can also be written $p/k_B T = n + b_2 n^2 + b_3 n^3 + \ldots$, where $b_i = B_i/N_A^{i-1}$ and n is number density.

temperature is typically a few times higher than the critical temperature, and the ratio T_B/T_c does not vary much from one gas to another; an example of the *law of corresponding states* that we discuss in Section 15.1.2. $B_2(T)$ can be fitted to high accuracy for many gases by the form $B_2 = a_2 - b_2 \exp(c_2/T)$, where a_2, b_2, c_2 are constants.

15.1 van der Waals gas

We begin with the van der Waals equation because it is the simplest reasonably successful approximate model of a fluid. Figure 15.5 shows the isotherms on the indicator diagram.

The equation was originally obtained from theoretical considerations. First consider the b term. At a given temperature, the pressure tends to infinity, not as the volume tends to zero as would be the case for an ideal gas, but now as the volume tends to a finite value given by b. This reflects the fact that the molecules of the gas take up a small but non-zero volume, and when they are pushed close together they repel one another. This suggests the value of b should be of the order of Avogadro's number times the volume of a single molecule, and the observed values (see Table 15.1) are consistent with this.

Writing

$$p = \frac{RT}{V-b} - \frac{a}{V^2} \quad (15.6)$$

makes it clear that the a term represents a reduction in pressure compared to what would be the case for a volume $V - b$ of ideal gas at the same temperature. This reduction is associated with weak attractive forces between the molecules of the gas when they are far apart. Physical intuition suggests that if the molecules are attracted to one another, they will push less on the walls of a container, which is correct, but to make this argument precise one must make a comparison with an ideal gas under equivalent conditions. In order to make such a comparison, start from a situation of $pV^2 \gg a$, so the a term is negligible. Then a volume V of van der Waals gas has the same behaviour as a volume $V - b$ of ideal gas. Next argue that it is easier to perform a compression on the real gas (i.e. less external work is required), because the real molecules pull themselves together a little. To predict this experimental reality the equation of state must show a slightly lower pressure for the real gas compared to the ideal gas, at a given final volume (V, $V - b$ respectively) and temperature.

The $(1/V^2)$ dependence of the a term is obtained from an energy argument—see box.

Now that we have a physical interpretation for a, it is clear that we should expect a to be larger in gases with polar molecules which attract one another more readily (e.g. water, carbon dioxide) compared to inert gases such as argon. The observed values in Table 15.1 bear this out. Or, by reasoning in the

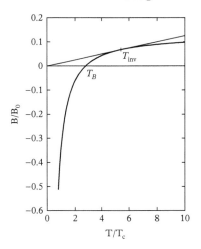

Fig. 15.4 *Generic behaviour of second virial coefficient for real gases, showing the Boyle temperature, given by $B(T_B) = 0$ and the condition $dB/dT = B/T$, which is related to the inversion temperature, discussed in Chapter 16 (see equation (16.11)). $B(T)$ is here plotted in units of $B_0 = p_c/RT_c$, and T is shown in units of the critical temperature T_c. The boiling point at any given pressure is, by definition, below T_c. The curve shown here is the function $B/B_0 = 0.599 - 0.467 \exp(0.694 T_c/T)$. This reproduces the behaviour of Ar, Kr, Xe, CH_4, N_2, O_2, and CO accurately, and provides useful but lower accuracy for other substances [Kaye and Laby online].*

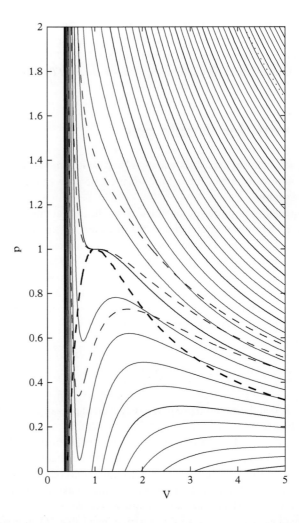

Fig. 15.5 *Isotherms given by the van der Waals equation. The thick dashed line shows the saturation curve predicted via the Maxwell construction (Figure 15.6); the dotted line at high temperature (top right) is the Boyle line (where $pV = RT$). The thin dashed lines show some isotherms for a real gas: the real critical isotherm is flatter at the critical point than the van der Waals equation predicts.*

other direction, the empirical values of a and b tell us about the intermolecular forces.

The van der Waals equation predicts that gases with more strongly attracting molecules will condense at higher temperatures, for obvious reasons.

15.1.1 Phase change

Examining the van der Waals isotherms—Figure 15.5—it is clear that the equation of state does not directly reproduce the observed horizontal isotherms during the gas–liquid phase change. However, this is not a failure of the van der Waals model. Rather, it requires interpretation. We have here an example of a phenomenon which we shall discuss at greater length in Section 17.2: an unstable region in the

Derivation of a/V^2 term We consider the effect of the intermolecular interactions when the molecules are far apart. The attraction between each pair of molecules can be described by a potential energy $\phi(r)$, where r is the distance between the molecules. The total potential energy of one molecule, owing to its interactions with all the others, is then

$$\delta U = \int_0^\infty \phi(r) n 4\pi r^2 dr \qquad (15.7)$$

$$\simeq \bar{n} \int_0^\infty \phi(r) 4\pi r^2 dr, \qquad (15.8)$$

where $\bar{n}$ is the volume-average of the number density. This type of argument is called 'mean field theory'. Since ϕ is negative for attractive interactions, we introduce

$$\epsilon \equiv \frac{1}{2} \int_0^\infty -\phi(r) 4\pi r^2 dr, \qquad (15.9)$$

so $\delta U \simeq -2\bar{n}\epsilon$. All the other molecules will also have this amount of potential energy, so the total potential of all the pairs is $N\delta U/2 = -N^2\epsilon/V$ (the factor 2 is to avoid double-counting).

This energy is internal energy in addition to that possessed by an ideal gas under the same conditions:

$$U = U_{\text{ideal}} - \frac{N^2\epsilon}{V} \qquad \text{[energy equation]}$$

$$\Rightarrow \left.\frac{\partial U}{\partial V}\right|_T = \frac{N^2\epsilon}{V^2}. \qquad (15.10)$$

Now using (13.24) this tells us about the equation of state:

$$T\left.\frac{\partial p}{\partial T}\right|_V = p + \frac{N^2\epsilon}{V^2}, \qquad (15.11)$$

which integrates to give

$$p + \frac{N^2\epsilon}{V^2} = f(V)T, \qquad (15.12)$$

where $f(V)$ is an unknown function of V. By adopting $f(V) = R(V-b)^{-1}$ one finds the van der Waals equation of state (15.6), with $a = N_A^2 \epsilon$.

---

entropy curve causes the system to separate into two phases. To understand the isotherms, we should study the Helmholtz function F (natural variables T, V, N). F can be obtained by taking the formula for an ideal gas of volume $V - b$ and adding the internal energy change calculated in the box, $-a/V$. For a monatomic gas one thus finds (for one mole)

$$F = RT\left[\ln\left(\frac{N_A \lambda_T^3}{V-b}\right) - 1\right] - \frac{a}{V}. \qquad (15.13)$$

The result is plotted in Figure 15.6. To minimize its free energy (see Chapter 17) the system follows the straight tangent line shown on the graph. Such a line has constant $(\partial F/\partial V)_T$, so it corresponds to constant p, hence a horizontal line on the indicator diagram. Thus the underlying van der Waals *theory* does successfully predict the observed qualitative behaviour, even during a phase change.

To find out where precisely the phase change begins and ends on the isotherm, one can use the free energy argument just given, or use the following argument due to Maxwell. Argue that for any horizontal part of an isotherm of a closed system we have constant p, T, N and therefore the Gibbs function is constant. Therefore, to find where the horizontal part meets the isotherms given by the van

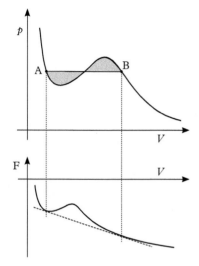

Fig. 15.6 *The van der Waals model in the phase change region. The lower graph shows the Helmholtz function F as a function of V along an isotherm, as predicted by the van der Waals model. The tangent to this curve shows the value of F actually adopted by the system between the states A and B (this is explained in Chapter 17). The upper graph illustrates the Maxwell construction. We seek a horizontal line AB such that the integral $\int_A^B p\,dV$ is equal to $p(V_B - V_A)$ (equation (15.17)). But the integral is the area under the curve, and $p(V_B - V_A)$ is the area under the line AB, so it follows that the two shaded regions must have equal area.*

der Waals equation of state, we need to find two points on a given isotherm that have the same value of G. For the isotherm shown in Figure 15.6, for example, we must have

$$G_A = G_B, \quad (15.14)$$

where the labels A, B refer to the two thermodynamic states, specified by p, T and V_A, V_B. Now, using $G = F + pV$ (the definition of G), we have

$$F_A + p_A V_A = F_B + p_B V_B \implies F_B - F_A = -p_A(V_B - V_A), \quad (15.15)$$

using $p_A = p_B$. But since $dF = -SdT - pdV$, and we are considering points on an isotherm, this difference is also given by

$$F_B - F_A = \int_A^B -p\,dV. \quad (15.16)$$

Hence, we have

$$\int_A^B p\,dV = p_A(V_B - V_A). \quad (15.17)$$

It is obvious that this equality holds for the horizontal isotherm actually followed by the system, but we are using the formula for another purpose. We follow the prediction $p(V)$ of the 'wiggly' van der Waals isotherm, just for the purpose of doing this integral, and thus, when the integral agrees with $p(V_B - V_A)$, find the places where the van der Waals theory gives $G_A = G_B$. This occurs when the shaded regions in the figure are of equal area (see figure caption). This equal-area method is called the *Maxwell construction*.

15.1.2 Critical parameters and the law of corresponding states

Expressed as a formula for V in terms of the other variables, the van der Waals equation of state is a cubic equation:

$$pV^3 - (RT + bp)V^2 + aV - ab = 0. \quad (15.18)$$

This is not readily solved for V. However, it is not hard to find the values of the critical parameters in terms of a and b by solving $(\partial p/\partial V)_T = 0$, $(\partial^2 p/\partial V^2)_T = 0$ (and see Exercise 15.3 for a neat method). One finds $p_c = a/27b^2$, $V_c = 3b$, $T_c = 8a/27Rb$. We can write the pressure, volume, and temperature in units of their critical values, by introducing $\tilde{p} = p/p_c$, $v = V/V_c$, $t = T/T_c$. This is called adopting 'critical units'. Then the equation becomes

$$(\tilde{p} + 3/v^2)(3v - 1) = 8t. \quad (15.19)$$

Table 15.1 *Critical parameters and van der Waals constants for some example gases. $a = (27/64) R^2 T_c^2/p_c$, $b = (1/8) RT_c/p_c$. The last two columns are described in the text.*

Gas	T_c (K)	p_c (MPa)	Z_c	a (J m³ mol⁻²)	b (cm³ mol⁻¹)	T_B/T_c	T_c/T_{trip}
Ne	44.4	2.76	0.311	0.0208	16.72	2.75	1.81
Ar	150.9	4.87	0.291	0.1363	32.19	2.72	1.80
Kr	209.4	5.50	0.288	0.233	39.57	2.77	1.81
Xe	289.7	5.84	0.287	0.419	51.56	2.72	1.80
N_2	126.2	3.39	0.289	0.1370	38.69	2.63	2.00
CH_4	190.6	4.60	0.286	0.230	43.06	2.65	2.10
CO_2	304.1	7.38	0.274	0.365	42.67	2.35	1.40
H_2O	647.1	22.06	0.229	0.5536	30.49	2.8	2.37
H_2	33.2	1.297	0.306	0.0248	26.60	3.5	2.40
He	5.19	0.227	0.301	0.00346	23.76	4.6	–

When the equation of state is thus expressed in critical units, the constants a and b drop out. Therefore, the van der Waals equation (and, as we will see, the other equations) predicts that all gases will behave alike if their properties are rescaled by their critical values. This is called the 'law of corresponding states' and is quite well observed in practice, see Figure 15.7. There is thus revealed a *unity in diversity*. It means that all fluids may be understood to be different examples of essentially the same sort of thing. For example, the curves in Figure 15.3b will, to good approximation, describe any gas, not just methane, in suitably rescaled pressure and temperature units.

The law of corresponding states is further illustrated by the data shown in Table 15.1. The third column of the table gives the compression factor at the critical point. The van der Waals equation gives the value $3/8 = 0.375$, which is of the correct order of magnitude, but overestimates the observed value by about 30%. However, the values observed for the various gases agree among themselves much better than this. The examples in the table all lie within 6% of 0.292, with the exception of water which has a smaller value owing to the larger interactions of polar molecules.

The penultimate column in the table gives the ratio of the Boyle temperature and the critical temperature. The former is a low density, high temperature property, whereas the latter is a property of the high density critical state, but the ratio shows little variation among the gases. The noble gases agree with each other well on this measure, and methane with its roughly spherical molecules is similar. Nitrogen differs a little, and CO_2 and H_2O more, owing to dipole forces and the shape of the molecules. Helium and hydrogen are different again, owing to their unusually light molecules which make quantum effects more important in the collisions.

The final column in the table compares the critical point with the triple point. The latter is not at all modelled by any of the equations considered in this chapter,

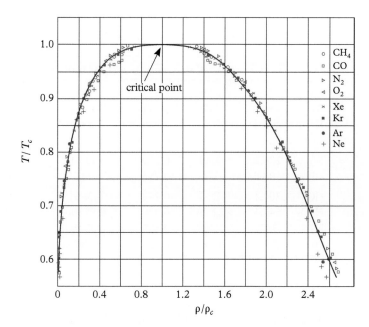

Fig. 15.7 *The temperature as a function of density for liquid and gaseous phases, for a number of substances, in critical units. [Adapted from Guggenheim (1945).]*

since it involves the solid phase (see Chapter 22). Considering the wide range of critical temperatures of the different gases, the degree of agreement of the values of T_c/T_{trip} is notable, though somewhat less than for the other ratios.

There are also further examples of universal behaviour around the critical point, extending to a rich variety of phenomena called *critical phenomena*, the study of which is introduced in Chapter 25.

15.2 Redlich–Kwong, Dieterici, and Peng–Robinson gas

The Redlich–Kwong equation is a modification of the van der Waals equation, designed to better fit empirical observations. It is one of a large number of equations of varying degrees of complexity and precision which have been proposed to model the behaviour of gases and liquids. In critical units it is

$$\tilde{p} = \frac{3t}{v - \beta} - \frac{1}{\beta v(v + \beta)\sqrt{t}}, \tag{15.20}$$

where $\beta = (1 + 2^{1/3} + 2^{2/3})^{-1} \simeq 0.259921$. See Exercise 15.4 and Table 15.2 for the critical parameters. One finds the critical compression factor predicted by this equation is $Z_c = 1/3$, which is a better approximation to observed values than that given by the van der Waals equation.

Table 15.2 *Critical parameters, critical compression factor, and inversion temperature for some example equations of state used to model real gases.* †For Redlich–Kwong, $T_B/T_c = (3\beta^2)^{-2/3}$ and $\beta = (1 + 2^{1/3} + 2^{2/3})^{-1}$.

	p_c	V_c	T_c	Z_c	T_B/T_c	T_{inv}/T_c
van der Waals	$\dfrac{a}{27b^2}$	$3b$	$\dfrac{8a}{27Rb}$	$\dfrac{3}{8} = 0.375$	$\dfrac{27}{8}$	$\dfrac{27}{4}$
Redlich–Kwong	$\left[\dfrac{\beta^7 R a^2}{3b^5}\right]^{1/3}$	$\dfrac{b}{\beta}$	$\left[\dfrac{3\beta^2 a}{Rb}\right]^{2/3}$	$\dfrac{1}{3} \simeq 0.333$	$2.898^\dagger$	$\left(\dfrac{5}{6\beta^2}\right)^{2/3}$
Dieterici	$\dfrac{a}{4e^2 b^2}$	$2b$	$\dfrac{a}{4Rb}$	$\dfrac{2}{e^2} \simeq 0.271$	4	8

The Dieterici equation is obtained from a different theoretical approach. In critical units it takes the form

$$\tilde{p}(2v - 1) = t \exp\left(2 - \frac{2}{tv}\right). \tag{15.21}$$

It makes a reasonable prediction for the critical compression factor ($2/e^2 \simeq 0.271$), and with some modifications this equation can model the liquid–gas phase transition region to within a few % for the noble gases. It yields no simple expression for the energy equation.

All the above equations are too rough for use in designing modern industrial processes. A more accurate and widely used equation is that of Peng and Robinson:

$$p = \frac{RT}{V - b} - \frac{a(1 + x(\omega, T))^2}{V^2 + 2Vb - b^2}, \tag{15.22}$$

where[2]

$$a = 0.457236 \, R^2 T_c^2 / p_c \tag{15.23}$$
$$b = 0.077796 \, RT_c / p_c$$
$$x(\omega, T) = \left(0.37464 + 1.54226\omega - 0.26992\omega^2\right)\left(1 - \sqrt{T/T_c}\right).$$

Here ω is called the *acentricity factor*; it is a constant for a given gas (for argon, $\omega = 10^{-3}$; for nitrogen, $\omega = 0.04$). The critical compression factor is $Z_c = 0.307401$. Applied to argon, at a given temperature this equation predicts the vapour pressure to 1% accuracy, the volume of liquid and gas phases to 15% and 2% respectively below 140 K, and with 10% accuracy at the critical point (150.9 K). It can be used to obtain the molar entropy to an accuracy of 1% on the saturation curve below 140 K, and 2% at the critical point.

[2] The constants here are $(y^2 - 2)(z - 2/z)^2/4y^4 \simeq 0.457$ and $(y^2 - 2)/2y^2 z \simeq 0.078$; where $y = (4 - 2\sqrt{2})^{1/3} + (4 + 2\sqrt{2})^{1/3}$ and $z = y + 2$. The critical volume is $V_c = (1 + y)b$.

EXERCISES

(15.1) Show that for a van der Waals gas,

$$\left.\frac{\partial C_V}{\partial V}\right|_T = 0, \quad C_p - C_V = \frac{R}{1 - 2a(V-b)^2/(RTV^3)}.$$

(15.2) (i) Show that for both a van der Waals gas and a Dieterici gas, the B_2 coefficient in the virial expansion is given by

$$B_2 = b - \frac{a}{RT}. \tag{15.24}$$

Hence show that the Boyle temperature is 3.375 times the critical temperature for the van der Waals gas, and 4 times for the Dieterici gas.
(ii) Find the next virial coefficient for both these equations of state.

(15.3) The van der Waals equation is a cubic equation in V, having either three real roots or just one real root at any given p, T. For $T < T_c$ and $p < p_c$ there are three real roots. As T increases, these roots approach one another and merge at $T = T_c$. Therefore at the critical point the equation of state must take the form $(V - V_c)^3 = 0$. Use this to find the critical parameters in terms of a and b, by expanding the bracket and comparing with the van der Waals equation.

(15.4) Using the method introduced in question 15.3, or otherwise, show that for the Redlich–Kwong equation, the critical compression factor $(p_c V_c / RT_c)$ is $Z_c = 1/3$, and the critical volume is $V_c = xb$, where x is a solution of the cubic equation

$$x^3 - 3x^2 - 3x - 1 = 0. \tag{15.25}$$

Confirm (e.g. by plotting the function) that this function has one real root, at $x = 1 + 2^{1/3} + 2^{2/3} \simeq 3.84732$. Show that the critical temperature is $T_c = (3a/x^2 Rb)^{2/3}$ and obtain equation (15.20).

(15.5) Using equation (13.31), show that the constant volume heat capacity of a Redlich–Kwong gas is given by

$$C_V(T, V) = C_V(T)^{(\text{ideal})} - \frac{R}{4\beta^2} t^{-3/2} \ln \frac{v}{v + \beta},$$

where $v = V/V_c$, $t = T/T_c$, and $\beta = (1 + 2^{1/3} + 2^{2/3})^{-1}$. Hence find the fractional reduction in C_V, compared to the ideal gas, at $t = 1$, $v = 2$, when $C_V^{(\text{ideal})} = 3R/2$.

(15.6) (i) Show that if the molar heat capacity in the ideal gas limit is $C_V = (3/2)R$, then the molar heat capacity of a Peng–Robinson gas is

$$C_V = \frac{3}{2}R + \frac{y(y+1)\,(I(V)-I(V_0))}{2(TT_c)^{1/2}}, \quad (15.26)$$

where $y = 0.37464 + 1.54226\omega - 0.26992\omega^2$ and

$$I(V) \equiv -\int \frac{a\,dV}{V^2 + 2Vb - b^2}$$

$$= \frac{a}{2\sqrt{2}\,b} \ln\left[\frac{V + (1+\sqrt{2})b}{V + (1-\sqrt{2})b}\right]. \quad (15.27)$$

(ii) Show that the molar Helmholtz function is given by

$$F(T,V) = U_0 - TS_0 + \frac{3}{2}R(T-T_0)$$
$$- (I(V) - I(V_0))\,(1 + x(\omega,T))^2$$
$$- RT \ln\left[\frac{V-b}{V_0-b}\left(\frac{T}{T_0}\right)^{3/2}\right]. \quad (15.28)$$

(15.7) In *Journal of Chemical Physics* vol. 115 (2001), R. Sadus proposes the two-parameter equation of state

$$pV = RT\frac{1 + y + y^2 - y^3}{(1-y)^3} e^{-a/RTV},$$

where $y = b/4V$ and a, b are parameters. Considering that it has only two empirically determined parameters, this equation of state shows a good ability to predict the density of both liquid and vapour phases and to locate the phase boundary. Show that $y = 0.357057$ at the critical point and find the critical compression factor.

16 Expansion and flow processes

16.1 Expansion coefficients 228
16.2 U: free expansion 229
16.3 H: throttle process: Joule–Kelvin expansion 230
16.4 General flow process 236
Exercises 239

In this chapter we discuss three types of expansion process, and the more general issue of thermodynamic processes involving a continuously flowing fluid.

The free expansion (or Joule expansion) was described in Section 7.5.1, where its connection to Joule's law and the ideal gas was examined. Here we elaborate that connection and also briefly consider its effect for real gases and other substances. We then discuss at length another process called the Joule–Kelvin expansion, also called the throttling process or Joule–Thomson expansion. This is important in refrigeration applications, including the liquification of gases. Finally, we briefly discuss the thermodynamics of devices such as a compressor or a gas turbine.

16.1 Expansion coefficients

If one is interested in the temperature change during an expansion, then the quantity of interest is $(\partial T/\partial V)_x$ or $(\partial T/\partial p)_x$, where x indicates the type of process, i.e. the conditions under which the expansion takes place. We will consider the cases of constant U (free expansion), constant H (Joule–Kelvin process), and constant S (adiabatic expansion). The reader is invited to derive the following formulae for the relevant coefficients (the guidelines in Section 13.3 may help):

$$\left.\frac{\partial T}{\partial V}\right|_U = -\frac{1}{C_V}\left(T\left.\frac{\partial p}{\partial T}\right|_V - p\right). \quad \text{[Joule expansion]} \quad (16.1)$$

$$\left.\frac{\partial T}{\partial p}\right|_H = \frac{1}{C_p}\left(T\left.\frac{\partial V}{\partial T}\right|_p - V\right). \quad \text{[Joule–Kelvin expansion]} \quad (16.2)$$

$$\left.\frac{\partial T}{\partial V}\right|_S = -\frac{T}{C_V}\left.\frac{\partial p}{\partial T}\right|_V. \quad \text{[adiabatic expansion]} \quad (16.3)$$

$(\partial T/\partial V)_U$ is called the 'Joule coefficient', sometimes written μ_U. $(\partial T/\partial p)_H$ is called the 'Joule–Kelvin' (or 'Joule–Thomson' or 'throttling') coefficient, sometimes written μ_H.

Thermodynamics: A Complete Undergraduate Course. Andrew M. Steane.
© Andrew M. Steane 2017. Published 2017 by Oxford University Press.

16.2 U: free expansion

$$\left.\frac{\partial T}{\partial V}\right|_U = -\frac{T \left.\frac{\partial p}{\partial T}\right|_V - p}{C_V}. \tag{16.4}$$

The free expansion is an expansion into a vacuum, in conditions of thermal isolation; recall Section 7.5.1. Since no heat is exchanged or work done, and the system is closed, the process maintains U constant.

A free expansion does not have to be sudden. It could be a slow leak of gas through a tiny hole, but it is always thermodynamically irreversible.

The reader should check that the equation of state predicts zero temperature change in a free expansion for an ideal gas. For a van der Waals gas, one finds $(\partial T/\partial V)_U = -a/V^2$: the gas cools as it freely expands. This can be understood as follows. Owing to the long-range attraction between the molecules, they gain potential energy as they move apart, which they pay for in kinetic energy. Since this kinetic energy is wholly owing to thermal motion, its reduction must represent a drop in temperature. Therefore the temperature of the gas falls.

We should expect that the random part of the potential energy (owing to the fact that the molecules have separations randomly distributed about a mean) is also characterized by a temperature, which in thermal equilibrium is equal to the temperature associated with the kinetic energy. Therefore in a free expansion of a van der Waals gas, the mean intermolecular potential energy rises but the thermal spread about the mean falls.

16.2.1 Deriving the equation of state of an ideal gas

Students sometimes assume (and books sometimes imply) that to define an ideal gas it is sufficient to assert that the equation of state is $pV = Nk_B T$. This is wrong because it omits to mention Joule's law ($U = U(T)$). If we assert Joule's law then we do not need to give the full equation of state to complete the definition of the ideal gas, because it is sufficient to assert Boyle's law ($p \propto 1/V$ at constant T). The equation of state is then a *derived* property. We will prove this through an application of equation (16.1).

Boyle's law enables us to infer that the equation of state must have the form $pV = f(T)$ for some function of T, but it does not in itself yield the function. However, Joule's law states that $(\partial T/\partial V)_U = 0$, and using (16.1), we conclude that[1]

$$\left.\frac{\partial p}{\partial T}\right|_V = \frac{p}{T} \quad \text{[for any system obeying Joule's law]}$$

$$\Rightarrow \quad \ln p = \ln T + f(V) \quad \text{[integrating at constant volume]}$$

$$\Rightarrow \quad p \propto T \quad \text{at fixed } V.$$

---

[1] We argue that C_V is a well-behaved quantity which cannot be infinite except possibly for some special case.

If $p \propto T$ at fixed V, then the only possible form for the function $f(T)$ is $f = aT$ for some constant a, therefore the equation of state is $pV = aT$. Thermodynamics can't give the value of the proportionality constant a, but, significantly, it does enable us to prove that the product pV is proportional to the *absolute* temperature T. Therefore we have proved that the ideal gas *empirical* temperature, measured by definition by pV, must equal the absolute temperature, up to a proportionality constant. (An alternative proof has already been presented in Section 8.2.2.)

16.3 *H*: throttle process: Joule–Kelvin expansion

The process of squeezing a thermally isolated fluid slowly through a constriction is called the Joule–Kelvin process, or throttle process. It is also called the Joule–Thomson process. Whenever we squeeze toothpaste slowly out of a tube, we are carrying out such a process (as long as no heat transfer takes place). Another example is the use of an aerosol can: the gas emerges from the nozzle slowly enough for the process to be quasistatic, but fast enough for heat flow from the surroundings to be negligible.

The strict definition of the Joule–Kelvin process is a quasistatic flow process, in which a fluid is moved into some region at well-defined pressure, and moves out of that region at a lower well-defined pressure, the whole taking place in conditions of thermal isolation; see Figure 16.1. The process is also called a *throttling process* because the pressure drop appears across some sort of constriction or porous plug which restricts the free motion of the fluid. The speed of the flow is fast enough for a non-negligible pressure difference to exist across the constriction, but slow enough to be quasistatic. That is, the fluid is in thermal equilibrium on either side of the constriction, with, on each side, a well-defined pressure and volume per mole. Within the constriction, on the other hand, the flow is turbulent and the fluid is out of equilibrium (and its entropy is increasing).

This is an irreversible process. The irreversible component is the restricted movement of the fluid through the constriction. After the fluid emerges from the constriction, the turbulent part of its motion dies down, the fluid re-thermalizes, and overall the entropy has increased.

In the Joule–Kelvin process work is done at both the input and output side. Consider the work required to move a given quantity (e.g. unit mass) of fluid through the apparatus. Let the pressures on the two sides be p_1, p_2 and the volume of unit mass be V_1 at the input side and V_2 at the output side. Then work $p_1 V_1$ is done on the fluid at the input, and the fluid does work $p_2 V_2$ at the output. Therefore by the first law, the change in internal energy is

$$U_2 - U_1 = p_1 V_1 - p_2 V_2, \tag{16.5}$$

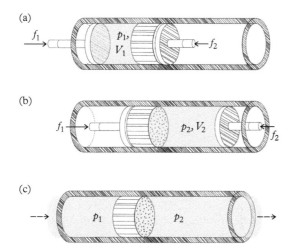

Fig. 16.1 *Joule–Kelvin process. (a),(b) Two cylinders are fitted with frictionless pistons, with some fluid enclosed on the left and none initially on the right, the whole being thermally insulated. The two regions are joined by a constriction or porous plug which greatly restricts the flow of fluid. The surroundings maintain a fixed pressure p_2 on the right-hand piston. The left-hand piston is set in motion, at a slow but not infinitesimal speed. This increases the pressure on the left to $p_1 > p_2$. The fluid then moves across the constriction while the left piston is pushed home and the right one moves out. Overall a change of state from p_1, V_1 to p_2, V_2 takes place, with no heat entering or leaving the system. For given p_1, V_1, and p_2 the final volume is that which maintains the enthalpy (see text). (c) More generally, the expansion is an example of a flow process in which the fluid behind any given body of fluid does work on the left, and work is done on whatever system receives the fluid on the right. The conditions for the Joule–Kelvin process are constrained initial and final pressures, and thermal isolation.*

where U_1, U_2 is the specific internal energy of the fluid before and after the expansion. We have $U_1 + p_1 V_1 = U_2 + p_2 V_2$, or in other words $H_1 = H_2$, that is, the process leaves the specific enthalpy unchanged.

This energy conservation argument remains valid even if notable things happen in the course of the expansion, such as a phase change, or chemical reactions.

The process is called an expansion because the fluid goes from a state of high pressure and low volume to one of low pressure and high volume. However, it is the pressure not the volume which is controlled by outside agencies, therefore if we take an interest in the accompanying temperature change, then the expansion coefficient is $(\partial T/\partial p)_H$. This is given by (16.2).

16.3.1 Bernoulli equation

In the case of a fluid flowing continuously, the internal energy is made up of a contribution $(1/2)mv^2$ from the kinetic energy of the flow, plus other contributions (thermal motion, gravitational potential energy, chemical energy, etc.). It is useful to separate out this contribution, and the one associated with potential energy, such as gravitational potential energy. Then when mass m passes through the system, the energy conservation can be written

$$u_1 m + p_1 V_1 + \frac{1}{2}mv_1^2 + mgz_1 = u_2 m + p_2 V_2 + \frac{1}{2}mv_2^2 + mgz_2,$$

where u is the internal energy per unit mass when the fluid is not moving and is at zero height. Hence

$$u_1 + \frac{p_1}{\rho_1} + \frac{1}{2}v_1^2 + \phi_1 = u_2 + \frac{p_2}{\rho_2} + \frac{1}{2}v_2^2 + \phi_2, \tag{16.6}$$

where ρ is density and ϕ is gravitational potential.

Equation (16.6) can be applied to a tube of fluid in conditions of streamline (non-turbulent) flow, as long as heat conduction and work done against viscous forces is negligible. Then one may deduce that, along any given streamline,

$$u + \frac{p}{\rho} + \frac{1}{2}v^2 + \phi = \text{const.}$$

This is called *Bernoulli's theorem*, it is a basic result in hydrodynamics. If the fluid is incompressible it further simplifies to

$$\frac{p}{\rho} + \frac{1}{2}v^2 + \phi = \text{const.}$$

The main observation following from this is that as the velocity increases the pressure falls. This fact is important in the design of industrial fluid processing plants, and aerofoils for aircraft wings. It also enters into the physics of music in musical instruments such as the clarinet and the oboe. When the performer blows between the reeds of an oboe, the reeds move *together*, not apart, owing to the drop in pressure. As they approach one another a more complicated flow pattern sets in and their own elastic properties provide restoring forces. The resulting oscillation produces the sound we hear.

16.3.2 Cooling and liquification of gases

An important application of the Joule–Kelvin process is to cool a gas, and in particular to cool it so far that it liquifies. Since the pressure falls, if we want the temperature to fall also than we need a *positive* value for the coefficient μ_H, given by equation (16.2):

$$\mu_H \equiv \left.\frac{\partial T}{\partial p}\right|_H = \frac{1}{C_p}\left(T\left.\frac{\partial V}{\partial T}\right|_p - V\right). \tag{16.7}$$

In practice the coefficient can be positive or negative, depending on the conditions.

For an ideal gas we have $(\partial V/\partial T)_p = V/T$, so $\mu_H = 0$, i.e. the temperature does not change in a Joule–Kelvin process.

Plotting the isenthalps for a real gas we typically find a situation as indicated in Figure 16.2. To get cooling we need the region where the isenthalps have positive slope. This happens if the pressure and temperature are low enough, so that the state is within the region bounded by the dashed curve. This curve marks the states where μ_H vanishes. The set of such states, where $\mu_H = 0$, defines the *inversion curve*. The inversion curve can easily be obtained from the equation of state, since it is the set of states where

$$\left.\frac{\partial V}{\partial T}\right|_p = \frac{V}{T}. \quad \text{[at inversion]} \tag{16.8}$$

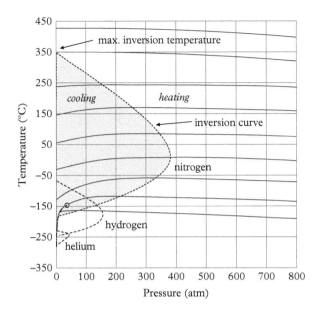

Fig. 16.2 Isenthalps (continuous lines) and inversion curve (dashed) for nitrogen. The small dash-dotted line finishing in a circle is the liquid–vapour coexistence line; the circle is the critical point. The inversion curves for hydrogen and helium are also shown.

One finds that a state on the inversion curve is at a local *maximum* of temperature as a function of pressure during a change at constant enthalpy. It follows that the greatest temperature change, for given initial temperature and final pressure, is obtained when the initial pressure is on the inversion curve. Typically this pressure is high by everyday standards (many atmospheres), an indication of the fact that we need to be well away from the ideal gas regime if we are to get a useful temperature change.

The highest temperature on the inversion curve is called the *maximum inversion temperature*, T_{inv}. This is often abbreviated to simply *inversion temperature*. Clearly, if one wants to cool a gas from an initial temperature above T_{inv}, one must first use some other process (such as cooling another gas) to get the temperature below T_{inv}, before the Joule–Kelvin method will be useful.

Consider the virial expansion, equation (15.4), keeping only the first two terms:

$$\frac{pV}{RT} = 1 + \frac{B(T)}{V}. \tag{16.9}$$

This gives

$$\mu_H = \frac{1}{C_p} \frac{T\frac{dB}{dT} - B}{1 + 2B/V}. \tag{16.10}$$

Therefore the coefficient goes to zero when

$$\frac{dB}{dT} = \frac{B}{T}, \tag{16.11}$$

Table 16.1 *Maximum inversion temperature and critical temperature, in kelvin, for various gases, and their ratio. The maximum inversion temperature was calculated from the second virial coefficient; see Exercise 16.1. The maximum T_{inv} is at $p \to 0$.*

	Ar	CO_2	He	H_2	Ne	N_2	O_2
T_{inv}	794	1372	44	224	247	641	789
T_c	150.9	304.1	5.19	33.2	44.4	126.2	154.6
T_{inv}/T_c	5.3	4.5	8.5	6.8	5.6	5.1	5.1

c.f. Figure 15.4. Note, this is not a differential equation for $B(T)$. Rather, $B(T)$ is a property of the gas in question, and the equation tells us that if there is a temperature where the gradient of $B(T)$ is equal to B/T, then at that temperature, $\mu_H = 0$. In practice the shape of the $B(T)$ curve is as shown in Figure 15.4, so there is such a temperature. This gives the temperature of a point on the inversion curve at high molar volume V, since this is the limit where the other terms in the virial expansion are negligible. High V at any given T means low p, so this is the point where the inversion curve meets the T axis on a pT diagram. This is typically the highest temperature on the inversion curve. Therefore the temperature given by (16.11) is the maximum inversion temperature. See Table 16.1 for values for some common gases.

To get the full inversion curve, we need either more terms in the virial expansion, or else a complete equation of state. For a van der Waals gas, one has

$$\left.\frac{\partial T}{\partial p}\right|_H = \frac{V}{C_p}\left(\frac{2a(V-b)^2 - RTV^2 b}{RTV^3 - 2a(V-b)^2}\right) \qquad (16.12)$$

$$\simeq \frac{1}{C_p}\left(\frac{2a}{RT} - b\right) \qquad (16.13)$$

(to obtain the exact result, use (15.6) and extract $(\partial T/\partial V)_p$; to obtain the approximate result either use this or, more simply, calculate from the equation of state, neglecting terms of order a^2, b^2 and ab).

Equation (16.13) shows that the a term, from attractive long-range intermolecular forces, leads to cooling, while the b term, from repulsive short-range forces, leads to heating. The reason for the cooling is the same as was given in Section 16.2 for the free expansion. The heating when repulsive terms dominate then follows from the same argument, since the change in potential energy then has the opposite sign.

The inversion curve, when $\mu_H = 0$, is easily obtained from (16.12). One finds

$$T = \frac{2a}{Rb}\left(\frac{V-b}{V}\right)^2. \qquad (16.14)$$

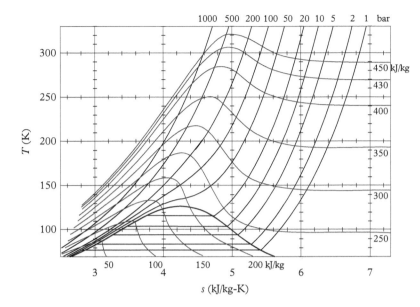

Fig. 16.3 Isobars and isenthalps for nitrogen.

By using this together with the equation of state one has an implicit equation for $p(T)$, which yields the inversion curve. The isenthalps can be obtained from $H = (3/2)RT + pV - a/V$. (For the inversion curve of the Dieterici gas, see Exercise 16.4.)

In practice one would use data such as that shown in Figures 16.2 and 16.3 to calculate the cooling effect, rather than relying on the van der Waals model. Figure 16.4 shows the main features of an apparatus used to liquify a gas using the Joule–Kelvin expansion. Gases can also be liquified by other means, such as adiabatic expansion, but the Joule–Kelvin process has the useful property that there are no mechanical moving parts at the low temperature part of the process. Karl von Linde developed the idea of using the cooled expanded gas to pre-cool the gas approaching the nozzle by means of a *counter-current heat exchanger*. All parts of the apparatus are important to the overall cycle, which consists of compression at constant temperature at around 300 K, then cooling at approximately constant pressure as the gas passes down through the counter-current heat exchanger, followed by cooling by Joule–Kelvin expansion, then heating as the gas passes back up through the heat exchanger. Of course, when the process first starts there is no cooling or heating effect in the heat exchanger, but as soon as some gas has undergone some Joule–Kelvin cooling, the heat exchanger begins to contribute. Hence although the temperature drop in the Joule–Kelvin expansion may be small, it can eventually result in a large effect as the gas passes several times around the cycle (see Exercise 16.8).

236 Expansion and flow processes

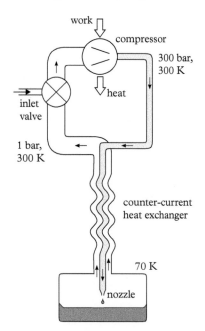

Fig. 16.4 *The Linde process for liquification of a gas using the Joule–Kelvin effect and a counter-current heat exchanger. The compressor compresses the gas to a high pressure (many atmospheres) at some easily available temperature. The gas then passes down through the counter-current heat exchanger where it is pre-cooled by heat exchange with returning gas. Then it reaches the expansion nozzle where it is cooled by the Joule–Kelvin effect. Some may turn to liquid. The rest returns through the heat exchanger, and is re-compressed and recirculated. Example numbers are given for nitrogen.*

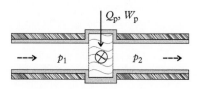

Fig. 16.5 *Generic flow process.*

Suppose a fraction λ of the gas is turned to liquid during the expansion. Then, by using the fact that the enthalpy per unit mass does not change,

$$h_i = \lambda h_L + (1-\lambda) h_f,$$

where h_i, h_f are the specific enthalpy of the gas before and after the expansion, and h_L is the specific enthalpy of the liquid. Solving for λ, we find

$$\lambda = \frac{h_f - h_i}{h_f - h_L}.$$

The specific enthalpies depend on the intensive variables. If we assume the final pressure is a fixed quantity (typically 1 atmosphere), and the final temperature is fixed because the liquid is in equilibrium with its vapour, then h_L and h_f are fixed. Therefore to maximize the efficiency λ we must minimize h_i, i.e. the enthalpy should begin at a stationary point as a function of pressure. This implies that the liquified fraction is largest when the process starts right on the inversion curve.

16.4 General flow process

The Joule–Kelvin expansion is an example of a general class of process called flow process, illustrated in Figure 16.5. In a flow process one or more substances flow steadily into a chamber or machine where some process takes place, and one or more substances flow steadily out. Turbines, compressors, jet engines, and many chemical plants are of this type. In the general case, we allow that the chamber can receive a further injection of heat and work, called *process energy*, per unit amount of the substance (or substances) passing through. You can readily show (Exercise 16.11) that the conservation of energy yields

$$Q_p + W_p = H_{\text{out}} - H_{\text{in}} \equiv \Delta H, \tag{16.15}$$

where Q_p and W_p are the process heat and work per unit mass of fluid passing through, and ΔH is the change in specific enthalpy of the fluid or fluids.

In the case of a pump or compressor, one might have $Q_p = 0$ and $W_p > 0$. In the case of a turbine where high pressure gas is used to drive a rotating shaft, one might have $Q_p = 0$ and $W_p < 0$; here one obtains output work called *shaft work* from the process:

$$W_{\text{shaft}} = -W_p = Q_p + H_{\text{in}} - H_{\text{out}}. \tag{16.16}$$

In the case of a chemical reaction vessel, one usually has $Q_p \neq 0$ and $W_p = 0$, so here the enthalpy change is equal to the heat emitted or absorbed during the reaction.

Example 16.1

A compressor draws in air at 300 K and 1 bar, compresses it adiabatically to 2 bar, and delivers the compressed air to a pipe. Find the process work per mole, assuming that $\gamma = 1.4$.

Solution.
We have
$$W_p = \Delta H = \int V \, dp$$
since $dH = T dS + V dp$ and $dS = 0$. During the adiabatic expansion, pV^γ is constant, so
$$W_p = V_1 \int_{p_1}^{p_2} \left(\frac{p_1}{p}\right)^{1/\gamma} dp = \frac{\gamma V_1 p_1}{\gamma - 1}\left[\left(\frac{p_2}{p_1}\right)^{1-1/\gamma} - 1\right].$$

For one mole, $p_1 V_1 = RT_1 = 2494$ J, so we find $W_p = 1.91$ kJ/mole.

16.4.1 S and H: the gas turbine

The gas turbine is a device used to extract work from a high-pressure gas in a continuous process. Gas is introduced into a chamber containing a rotating shaft with blades shaped so as to rotate the shaft when there is a pressure difference across them. We would like to investigate the possible efficiency of such a device. The movement of the gas over the blades is complicated, and we have no guarantee that the gas will be in any well-defined equilibrium state inside the chamber. However, just as for the Joule–Kelvin process, this is a flow process in which, in steady state, a certain mass of gas at some initial temperature and pressure T_1, p_1 is received and that same mass is passed back out at a final temperature and pressure T_2, p_2. Therefore we can apply thermodynamic reasoning to the process as a whole.

A gas turbine is modelled to good approximation by the assumption that no heat enters or leaves the gas in the turbine, but now, unlike the Joule–Kelvin expansion, we allow that work W may be done on the shaft. The conservation of energy (Eq. 16.16) reads

$$H_{\text{in}} - H_{\text{out}} = W. \tag{16.17}$$

The internal working of the gas turbine may or may not be reversible. We have said nothing about that as yet. In practice one tries to design the blades to approach as closely as possible to adiabatic (that is, isentropic) conditions, but this is not possible to achieve perfectly and we would like to assess the impact of this. To this end, construct an *HS* indicator diagram, as in Figure 16.6 (such a diagram is known as a *Mollier diagram*). It is useful to plot lines of constant pressure (isobars) on the diagram, because pressure is the controlled property in this case. The isobars are found from tables of entropy and enthalpy (in the absence of computers) or, in the modern period, by using widely available data managed by computer software.

First we measure p_1 and T_1, and use them to find the enthalpy (using the tables or software) of a unit mass of gas. Thus we locate point *A* on the diagram. We take the final pressure as a given (for example, 1 atmosphere), so we know the state of the gas leaving the turbine is somewhere on the p_2 isobar. First we consider what would happen if the functioning of the turbine were adiabatic. We draw a line vertically down from *A* to meet the p_2 isobar at *B*. Thus we find H_S, which is the final enthalpy in the case of an adiabatic expansion. The shaft work available in this case is

$$W_S = H_{in} - H_S.$$

Next we measure the final temperature T_2 actually observed for our turbine. From this, and p_2, we obtain H_{out} and thus locate point *C*. Note that point *C* lies to the right of and above *B*. This means it represents an increase of entropy, and a reduction of available shaft work. The overall performance measure (roughly speaking, efficiency) of the turbine can be defined by comparing how well it performs in practice with the best possible performance, which we can see must be given by W_S, since the turbine cannot decrease the entropy of the gas in conditions of thermal isolation. Therefore

$$\text{performance measure} = \frac{H_{out} - H_{in}}{H_S - H_{in}}. \tag{16.18}$$

It is notable how simple this whole argument is. What might have been regarded as a very complicated process has been reduced to its essential details in a few simple steps. That is the power of thermodynamics.

The performance measure we have obtained is not the only possible one, however. The enthalpy difference $H_{out} - H_S$ is called the *reheat enthalpy*, and it is associated with a temperature difference which could in principle be used to drive a heat engine. Therefore one might argue that the performance measure given does not give a fair assessment. Another measure is obtained by comparing the turbine with another process having the same input and output states (*A* and *C* on Figure 16.6) but which moves between those states in a reversible manner, the entropy being provided by reversible heat transfer from a heat reservoir. A performance measure which compares the output work with the maximum that could be achieved for a given initial and final state, in the presence of a given reservoir, is called an *exergetic* measure.

If our engineer had measured the final state and found it to lie at point *D* on Figure 16.6, then he would have been disappointed because then the turbine is

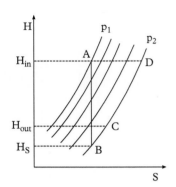

Fig. 16.6 *Enthalpy–entropy indicator diagram with isobars, for the analysis of a gas turbine (see text).*

doing no work at all! This is the case of constant enthalpy: the flow inside the turbine is so turbulent that the process has become the Joule–Kelvin process. The only hope for the engineer is that perhaps he found the final gas temperature was quite low, so he can market his machine as a gas cooling plant.

....

EXERCISES

(16.1) Use equation (16.3) to derive the relation between pressure and volume for an isentropic expansion of an ideal gas with γ independent of temperature.

(16.2) Derive equations (16.1) and (16.2). Verify that for an ideal gas $\left.\frac{\partial T}{\partial V}\right|_U = 0$ and $\left.\frac{\partial T}{\partial p}\right|_H = 0$.

(16.3) A gas obeys the equation $p(V-b) = RT$ and has C_V independent of temperature. Show that (a) the internal energy is a function of temperature only, (b) the ratio $\gamma = C_p/C_V$ is independent of temperature and pressure, (c) the equation of an adiabatic change has the form $p(V-b)^\gamma = $ constant.

(16.4) Show that the inversion curve for the Dieterici gas is given by $V = 4/(8-T)$ in critical units [Hint: differentiate the equation of state and then replace $(\partial V/\partial T)_p$ by V/T]. Hence obtain the inversion curve

$$p = (8-T)\exp\left(\frac{5}{2} - \frac{4}{T}\right). \qquad (16.19)$$

(16.5) Show that as $p \to 0$, the van der Waals inversion curve intercepts the T axis at $T = 2a/Rb$ and $T = 2a/9Rb$ (i.e. 6.75 and 0.75 in critical units).

(16.6) Show that, for a gas with second virial coefficient $B = a - be^{c/T}$, where a, b, c are constants, the maximum inversion temperature is a solution of the equation $aT = b(c+T)e^{c/T}$. Hence find the maximum inversion temperatures of helium and argon, for which $[a, b, c] = $ [114.1, 98.7, 3.245 K] and [154.2, 119.3, 105.1 K], respectively.

(16.7) Use equation (16.13) to roughly estimate the temperature change in a Joule–Kelvin expansion, as follows. First, assume we are comfortably in the region where cooling occurs, so the b term can be neglected. Then the equation is easily integrated. Show that the ratio of final to initial temperature, after a pressure change Δp, is

$$\frac{T_f}{T_i} = \sqrt{1 + \frac{4a\Delta p}{RC_p T_i^2}},$$

where $\Delta p < 0$ so $T_f < T_i$. Putting in values for a mole of nitrogen ($a = 0.137\,\text{Jm}^3$, $C_p = 3.5R$), find T_f/T_i for a starting temperature of 300 K and a pressure drop of 100 atmospheres. (For a more accurate estimate, consult Figure 16.2.) [Ans. 0.87]

(16.8) Consider the Linde process for nitrogen, with parameters as in Figure 16.4. Assuming the gas leaves the cold end of the heat exchanger at 160 K, trace the path of a small fixed mass of gas on a TS diagram such as Figure 16.3 as it passes once around the process. Estimate the heat leaving a unit mass of gas in (i) the compressor and (ii) the heat exchanger, and estimate the fraction that liquifies in the expansion.

(16.9) Roughly trace the path on a TS diagram of a small fixed mass of nitrogen passing several times around the Linde process, with the whole apparatus starting at 300 K.

(16.10) At modest pressures the equation of state of a gas can be written as $pV = RT + Bp$, where B is a function of T only. (i) Find an expression for the Joule–Kelvin inversion temperature in terms of B and dB/dT. (ii) For helium gas between 5 K and 60 K, B can be represented as $B = m - (n/T)$, where $m = 15.3 \times 10^{-6}$ m^3 mole^{-1} and $n = 352 \times 10^{-6}$ m^3 K mole^{-1}. From this information, estimate the inversion temperature for helium. For an expansion starting below the inversion temperature, comment on the efficiency of the process (i.e. cooling per unit pressure change) as the temperature falls. (iii) In a helium liquifier, compressed helium gas at 14 K is fed to the expansion valve, where a fraction x liquifies and the remaining fraction $(1-x)$ is rejected as gas at 14 K and atmospheric pressure. If the specific enthalpy of helium gas at 14 K is given by

$$h = a + b(p - p_0)^2,$$

where $a = 71.5$ kJ kg^{-1}, $b = 1.3 \times 10^{-12}$ kJ kg^{-1} Pa^{-2}, and $p_0 = 3.34$ MPa ($= 33.4$ atm), and the specific enthalpy of liquid helium at atmospheric pressure is 10 kJ kg^{-1}, determine what input pressure will allow x to achieve a maximum value, and show that this maximum value is approximately 0.18.

(16.11) Derive equation (16.15), first for a flow process with a single entrance and exit pipe, and then, more generally, for a chamber with several entrance and exit pipes containing possibly different fluids at different pressures.

(16.12) The working of many gas turbines can be modelled by the *Brayton cycle*, which consists of two adiabatic and two isobaric processes (cf. Figure 16.8). Show that if the working fluid is an ideal gas, then the thermal efficiency is $1-(p_1/p_2)^{1-1/\gamma}$. (Compare this with the Otto cycle, Eq. 11.2.)

(16.13) Consider the Mollier or hs chart shown in Figure 16.7.

 (i) Why do the isobars get steeper at high T? Is this a universal property of Mollier charts?

 (ii) If the chart only showed one of T or p as a function of h and s, then which would it be better to show?

 (iii) Use the chart to find the heat capacity at constant pressure of water at 1 bar, 50° C and of steam at 1 bar, 200° C.

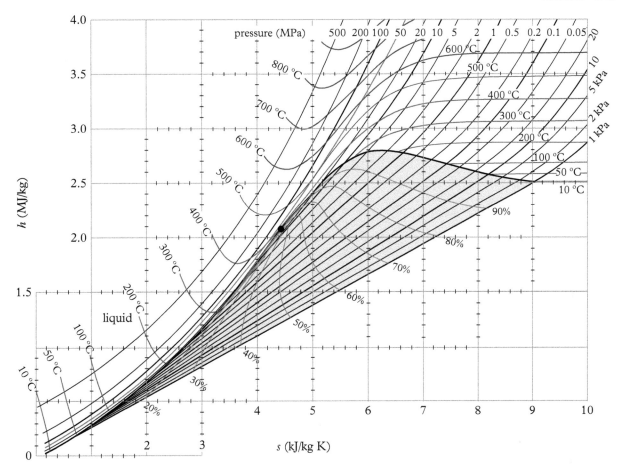

Fig. 16.7 *Enthalpy–entropy diagram (also known as h–s chart or Mollier diagram) for H_2O. In the mixed-phase region (shaded) the isotherms are not shown, in order to reduce clutter; they are parallel to the isobars. The % lines show the proportion of vapour in the mix. The saturation curve is also shown in Figure 22.9b.*

(16.14) The *Rankine cycle* (Figure 16.9) is a model suitable for the heat engines widely used in power stations which employ a steam turbine. The cycle involves a working fluid which undergoes a phase change, and consequently allows large amounts of heat transfer without great changes in pressure, and for this reason it is also useful in refrigerator applications. There are four stages: compressor, boiler, turbine, condenser. The compressor raises the pressure of a liquid at approximately constant volume. Because not much volume change takes place, not much work needs to be supplied. This also raises the temperature slightly. Then the liquid is heated in a boiler at constant pressure. Its temperature rises to the boiling point, and then stays constant as large amounts of heat

Fig. 16.8 (a) A turbojet engine, which is an example of a device employing a gas turbine. (b) The Brayton cycle: a simple model which captures the main elements of the working principle. Gas is introduced into the compressor where the pressure is raised in, ideally, an adiabatic compression. Fuel burning in the combustion chamber raises the temperature of the gas at roughly constant pressure. The gas drives the turbine in, ideally, an adiabatic expansion, cooling as it does so. Further cooling of the exhaust gas takes place in the surrounding atmosphere at constant pressure. Some of the shaft work of the turbine can be used to drive the compressor; the rest can be used to drive a generator, or to impart momentum and kinetic energy to the jet of exhaust gas.

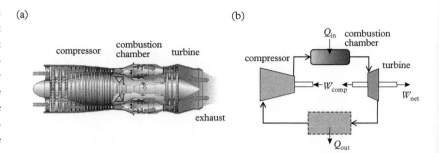

are supplied and there is a large increase in volume. Usually a complete phase change takes place. After this the fluid (now vapour) passes through a turbine which it drives in an approximately adiabatic expansion. Finally, the fluid is cooled at constant pressure by heat exchange so that its volume falls again. In early devices, condensation began in the turbine, but because water droplets can damage the turbine blades, modern power stations avoid this by using a high operating temperature at the input to the turbine, and also a further stage called reheating.

(i) Consider a Rankine cycle with water as the working fluid, as shown in Figure 16.9. Show that the work extracted per unit mass of water is $w = (h_c - h_d) - (h_b - h_a)$ and the heat input is $q_{in} = h_c - h_b$, where h is specific enthalpy.

(ii) Given that the pressure limits are 10 kPa and 5 MPa, and the temperature limits are 45 °C and 400 °C, find the thermal efficiency defined by $\eta = w/q_{in}$. [Use the hs chart shown in Figure 16.7]

(iii) Calculate the flow rate required for an output power of 1 MW.

(iv) Calculate the volume of water vapour entering the turbine per second. [*Ans.* –; 0.28; 1180 kg/s; 73 m³]

(16.15) In a refrigerator based on the Rankine cycle, the working fluid goes around the cycle in the opposite direction to the one appropriate for power generation. Briefly describe the main components of such a refrigerator. Where is the work input? How is a net transfer of heat from a colder to a hotter body achieved?

Fig. 16.9 The Rankine cycle. Example values are given for a power station. Modern power stations divide the turbine stage (cd) into two or more parts with reheating (not shown), to avoid condensation in the turbine. $s - s_0$ is the amount by which the specific entropy exceeds a reference value at the triple point.

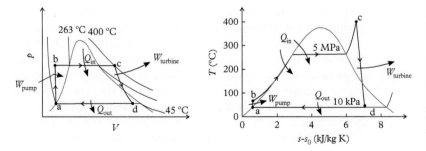

Stability and free energy

17

In this chapter we shall introduce an important type of thermodynamic reasoning, and make our first excursion into the quantitative treatment of non-equilibrium states. We shall also discuss the central concept of 'free energy' as a way to take care of entropy maximization when a system is not isolated.

Suppose a system is disturbed in some way, so that it is not in a thermal equilibrium state. For example, we could shake a bottle of liquid, or send some sound waves through a gas. After the disturbance is removed, the system will take some non-negligible period of time to settle back to a thermal equilibrium state. We now know enough about entropy to be able to claim that, if the system is isolated, then during this relaxation process the system's entropy must be increasing. We should like to study that increase, but how can we?—up to now we have only ever considered thermal equilibrium states. Out of equilibrium we don't have an equation of state; we lack even a definition of temperature; it is not clear what pressure might mean either.

One way to proceed is to imagine dividing the system up into many pieces (Figure 17.1). Each piece should be large enough that it can be treated by thermodynamic reasoning, but small enough so that, even though the system as a whole is out of equilibrium, each small piece has a well-defined temperature, pressure, etc. We shall pursue this policy in several settings, making use of the fundamental relation in the form

$$dS = \frac{1}{T}dU + \frac{p}{T}dV - \frac{\mu}{T}dN.$$

17.1 Isolated system: maximum entropy	243
17.2 Phase change	249
17.3 Free energy and availability	250
Exercises	257

17.1 Isolated system: maximum entropy

Consider first a fixed quantity of gas in a rigid, thermally insulated chamber. Suppose the gas is not in equilibrium overall, but it can be treated as a collection of small cells of gas, each of which has well-defined properties such as pressure, volume, temperature, energy, etc. These properties may change with time because each cell is busy exchanging heat, work, and particles with its neighbouring cells, but at any time each cell is in internal equilibrium.

In this situation we may claim that each cell of gas has a well-defined equilibrium entropy S_i, and the entropy of the whole system is *by definition*

$$\tilde{S} \equiv \sum_i S_i. \qquad (17.1)$$

Thermodynamics: A Complete Undergraduate Course. Andrew M. Steane.
© Andrew M. Steane 2017. Published 2017 by Oxford University Press.

244 Stability and free energy

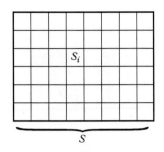

Fig. 17.1 *To reason about a thermodynamic system that is not in equilibrium, we divide it into many parts such that each part is in internal equilibrium but the different parts are not all in equilibrium with one another.*

We have attached a tilde (˜) to the total entropy defined by this equation, as a reminder that this is not necessarily an equilibrium entropy. It is the entropy of a system that may or may not be in a thermal equilibrium state.

To begin with, let us treat the simple case where just two cells are needed to describe the non-equilibrium state:

$$\tilde{S} = S_1 + S_2.$$

We suppose a process takes place in which a small amount of energy dU_1 moves from cell 2 to cell 1, accompanied by a small change in volume dV_1 and a transfer of dN_1 particles (see Figure 17.2). Then the total change in entropy is

$$\begin{aligned}
d\tilde{S} &= dS_1 + dS_2 \\
&= \frac{1}{T_1}dU_1 + \frac{p_1}{T_1}dV_1 - \frac{\mu_1}{T_1}dN_1 + \frac{1}{T_2}dU_2 + \frac{p_2}{T_2}dV_2 - \frac{\mu_2}{T_2}dN_2 \\
&= \left(\frac{1}{T_1} - \frac{1}{T_2}\right)dU_1 + \left(\frac{p_1}{T_1} - \frac{p_2}{T_2}\right)dV_1 - \left(\frac{\mu_1}{T_1} - \frac{\mu_2}{T_2}\right)dN_1. \quad (17.2)
\end{aligned}$$

If the system is relaxing towards equilibrium, then $d\tilde{S}$ must be positive:

$$\left(\frac{1}{T_1} - \frac{1}{T_2}\right)dU_1 + \left(\frac{p_1}{T_1} - \frac{p_2}{T_2}\right)dV_1 - \left(\frac{\mu_1}{T_1} - \frac{\mu_2}{T_2}\right)dN_1 > 0, \quad (17.3)$$

and when equilibrium is reached, $\tilde{S}$ is maximized, so $d\tilde{S} = 0$.

Now, the quantities dU_1, dV_1, dN_1 can all be independent (dU_1 can be exchanged without influencing V_1, N_1 by a heat transport; dV_1 can be exchanged without influencing U_1, N_1 by an expansion combined with some heat transport; etc.). Therefore we may consider them one at a time.

(1) dU. Setting $dV_1 = 0$, $dN_1 = 0$, we observe that to obtain $d\tilde{S} > 0$ it must be that dU_1 is positive if $T_1 < T_2$, and negative if $T_1 > T_2$. This is none other than the Clausius statement: heat flows from the hotter to the colder body. At equilibrium, we deduce $T_1 = T_2$.

(2) dV. Setting $dU_1 = 0$, $dN_1 = 0$, we observe that to obtain $d\tilde{S} > 0$ it must be that dV_1 is positive if $p_1/T_1 > p_2/T_2$, and negative if $p_1/T_1 < p_2/T_2$. When the two temperatures are equal, this corresponds to the expected property that the cell at higher pressure 'wins' and expands. When equilibrium is reached the two pressures must be equal.

(3) dN. Setting $dU_1 = 0$, $dV_1 = 0$, we observe that to obtain $d\tilde{S} > 0$ it must be that dN_1 is positive if $\mu_1/T_1 < \mu_2/T_2$. When the temperatures are equal, therefore, particles move from the region of higher chemical potential to the region of lower chemical potential. This is consistent with what we learned about chemical potential in Chapter 12. In equilibrium, $\mu_1 = \mu_2$.

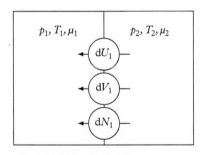

Fig. 17.2 *An out-of-equilibrium system consisting of two parts, each part in internal equilibrium at well-defined p, T, μ, but the two parts not in equilibrium with each other. A process involving transfer of dU_1, dV_1, dN_1 from one part to the other is considered. If the system is relaxing towards equilibrium, this process must achieve a net increase of entropy.*

The two cells of gas we have just considered could be any two of many regions in a large sample. Therefore we have shown that *at equilibrium a gas must be uniform*: it has uniform temperature, density, and chemical potential.

In fact the argument is more general: you should be able to see immediately that the statement about temperature would apply whenever two parts of an isolated system are free to exchange heat at fixed volume and particle number. The equality of (p/T) at equilibrium will obtain whenever two parts of an isolated system can compete for volume independently of each other's internal energy or particle number. This happens when heat flow is unrestricted, so that both the pressures and the temperatures equalize (we consider the case of restricted heat flow in a moment). The equality of μ/T will obtain whenever two parts of an isolated system can exchange particles independently of each other's internal energy. This independence is usually achieved by the situation that the systems can exchange heat as well as particles, so both μ and T equalize. This happens for example when two phases of a substance (e.g. water and steam) are in equilibrium together.

We have now placed on a firmer footing the claim made after equation (12.5), where we first met chemical potential.

17.1.1 Equilibrium condition with internal restrictions

Consider now a fluid contained in a cylinder, with a freely movable partition dividing the cylinder into two volumes V_1, V_2. Suppose this partition is both impervious to particle movement through it, and also thermally insulating. In this case the two parts of the cylinder can be at different temperatures without any heat flow between them, so the equilibrium state might have $T_1 \neq T_2$. It might appear from (17.3) that in this situation the equilibrium condition is $p_1/T_1 = p_2/T_2$, but that is a false impression. The case is described by Figure 17.2 and equation (17.3), but now dU_1 and dV_1 are not independent (and $dN_1 = 0$). In the absence of heat flow and particle flow, the only way subsystem 2 can pass energy dU_1 to subsystem 1 is by doing work on it, so we have

$$dU_1 = -p_1 dV_1, \qquad dU_2 = -p_2 dV_2, \qquad (17.4)$$

and by conservation of energy and volume, $dU_2 = -dU_1$ and $dV_2 = -dV_1$, as before. Therefore we have $p_1 dV_1 = -p_2 dV_2$, which gives

$$p_1 = p_2. \qquad (17.5)$$

Hence the equilibrium condition is equal pressures, as one should expect, even when other properties such as temperature and chemical potential may differ on the two sides of the partition. This case does involve a further subtlety, however, in that, as stated, we have not described a way for the system to change its entropy through movement of the partition, since we have only discussed reversible movement of the partition. This point is treated in Appendix B.

One can also have particle movement between regions separated by a permeable membrane. When the membrane is fixed in place, or otherwise restricted in its ability to move, there can be a pressure difference across it in thermal equilibrium. This occurs in the phenomena of *osmosis* and *nucleation*, described in Chapter 24. One cannot in general prevent heat flow when particle flow is allowed, so in this case the two regions have the same temperature when a full equilibrium is attained. However, one can imagine a case where a temperature difference is established in a transient way, such that the relaxation towards uniform temperature is slow compared to the relaxation of other properties. By an argument similar to the one we just gave for pressure, the equilibrium condition is then $\mu_1 = \mu_2$, whether or not the temperature is uniform. The system will reach this condition to high accuracy before the temperatures are equalized. An example is the *fountain effect* in liquid helium, discussed in Section 24.3.1.

17.1.2 The minimum energy principle

The maximum entropy principle dictates what the internal configuration of a system will be when it reaches equilibrium at given values of internal energy U and volume V. Figure 17.3 illustrates this idea by showing an example in which S depends on U and on some internal parameter X (the volume V being fixed). X could be, for example, one of the variables U_1, V_1, or N_1, introduced previously, or it could be one of many internal parameters describing a more complicated system. For given U, there is a set of states having various values of S and X. Of these, the one with largest S is the equilibrium state of the system when the energy is fixed at U.

A related idea is to consider the situation at given S (the volume V still being fixed). The states at various X now have different internal energy, so they are not

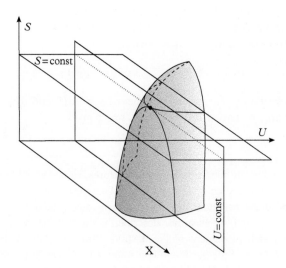

Fig. 17.3 *Maximum entropy and minimum energy. The curved surface shows the entropy of a closed rigid system as a function of its internal energy and an internal parameter X, such as the position of an internal piston. The plane at constant U picks out a set of states; of these, the equilibrium state (for that U) is the one at maximum S. The plane at constant S picks out another set of states; of these, the equilibrium state (for that S) is the one at minimum U.*

available to the system if it is isolated. However, we can imagine the system exploring other states without changing its entropy by doing work on its surroundings. Of all the states thus explored, the one with least internal energy is the equilibrium state when the system has the given S. This is not self-evident, but may be proved as follows.

We suppose the contrary, in order to obtain a contradiction. So, suppose the system is in equilibrium at given S and V, with an internal energy that is larger than the minimum at that S, V. This means that some internal parameter X has a value other than that giving least energy. In this case, withdraw some energy in the form of work, by allowing X to change. The entropy does not change. Next, give this energy back to the system in the form of heat. Thus the entropy increases and the internal energy returns to its starting value, and overall X has changed. But now we have a state at the same U as the starting state, with a different X, and a *larger* entropy. It follows that the starting state cannot have been the maximum entropy state at the given U, so it cannot have been an equilibrium state, so we have a contradiction. It follows that when the starting state is a maximum entropy state for given U, it must be a minimum energy state for given S. This is illustrated by Figure 17.3.

An algebraic proof of this is given in Appendix B. The essential mathematical idea is that the entropy surface cannot have any shape, but must have a *convex* shape as a function of U; this is explained in Section 17.1.3.

We now have two closely related principles, and for clarity it is useful to state them explicitly:

Maximum entropy principle. For an isolated system, the equilibrium value of any unconstrained internal parameter is such as to maximize the entropy for the given value of the total internal energy.

Minimum energy principle. For an isolated system, the equilibrium value of any unconstrained internal parameter is such as to minimize the energy for the given value of the total entropy.

These two principles are similar from a mathematical point of view, but different from a physical point of view, because the internal dynamics can change the entropy but not the internal energy. Therefore an isolated[1] system can spontaneously evolve to the maximum entropy point at given U, but not to the minimum energy point at given S. Nevertheless, when a system finds its equilibrium, it will satisfy the minimum energy principle 'whether it likes it or not'. This is useful to know. For example, it justifies the use of the method of virtual work that we employed in Section 14.3 to find the pressure inside a bubble.

17.1.3 Stability

So far we have gained some understanding of how to treat out-of-equilibrium situations, and what is meant by maximization of entropy. We have also established that differences in temperature, pressure, and chemical potential lead to flow of energy, volume, and particles in the expected directions. We can go

---

[1] In Appendix B a simple experimental apparatus is described which allows a system to locate the minimum energy point spontaneously through interaction with an external damper.

further. Suppose two subsystems or regions initially had a temperature difference $T_A > T_B$, so that some heat flowed from A to B. How would this arrival of heat at B affect the situation there? And what change would be left at A? We should expect that the temperature at B will rise, and at A it will fall. Otherwise, immediately after the heat flow, the temperature difference is exacerbated, and even more heat will flow. This 'run-away' situation is an example of *unstable* behaviour. For a stable equilibrium, therefore, it is not sufficient only that $T_A = T_B$. We require also that

$$\frac{dT}{\mathchar'26\mkern-12mu dQ_V} > 0 \quad \Rightarrow \quad \frac{\mathchar'26\mkern-12mu dQ_V}{dT} > 0.$$

In other words, *the heat capacity at constant volume must be positive*. In terms of entropy, the requirement is

$$\left(\frac{\partial^2 S}{\partial U^2}\right)_{V,N} < 0, \quad \text{[for stable equilibrium]} \tag{17.6}$$

where the symbol S denotes the equilibrium entropy of either region, which is why we have not attached a tilde to it. You can obtain this by writing $(\partial S/\partial U)_{V,N} = 1/T$ and then differentiating with respect to U to get the heat capacity; but it is instructive to look at inequality (17.6) directly. It says that, as we plot S as a function of U, with the other extensive variables fixed, the graph must always bend away from the vertical axis, losing gradient as it goes (Figure 17.4). Keep in mind that we are now discussing the *equilibrium* value of S, not some out-of-equilibrium situation. Divide the system into two equal parts, then the same graph also represents twice the entropy of either part. Pick a point on such a graph and imagine that a fluctuation occurs, so that one part of the system gains ΔU and another loses ΔU. The region on the receiving end has an entropy increase $\Delta S_1/2$, where ΔS_1 is the entropy increase the whole system would have if its energy increased by ΔU (see Figure 17.4). The region losing energy suffers an entropy decrease by the amount $\Delta S_2/2$, where ΔS_2 is as shown on Figure 17.4. The stability condition $d^2S/dU^2 < 0$ guarantees that $\Delta S_1 < \Delta S_2$, so that overall the system *loses* entropy. Therefore subsequent processes will tend to damp out this fluctuation and restore the entropy back to its initial, larger, value. Hence the configuration is stable.

If, on the other hand, the graph had an increasing gradient, then a fluctuation such as the one just considered offers a way for the system to increase its entropy. It will happily accept the increase and go further—the starting configuration cannot have been a stable equilibrium after all.

An argument of this kind can be made concerning all the flows of heat, work, and particles inside an isolated system. Whenever the flow can take place, it is accompanied by a 'traffic control system', regulating the flow so that run-away implosions or explosions are avoided. These 'traffic control systems' are what we

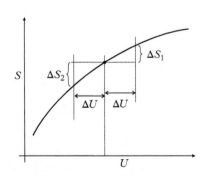

Fig. 17.4 *The functional form of $S(U)$ for an isolated system in equilibrium must be such that the gradient $(\partial S/\partial U)_{V,N}$ decreases monotonically with U (functions with this property are said to be* concave*). Otherwise the state is unstable: see text.*

call heat capacities and compressibilities. Any physical system capable of arriving in a stable thermal equilibrium must be possessed of them, such that[2]

$$C_p \geq C_V \geq 0 \quad (17.7)$$
$$\kappa_T \geq \kappa_S \geq 0. \quad (17.8)$$

More generally, the 'traffic control system' principle is known as *Le Chatelier's principle*. It can be stated:

> If a system is in stable equilibrium, any perturbation produces processes which tend to restore the system to its original equilibrium state.

Exercise 17.7 presents some further information.

17.2 Phase change

Suppose that the entropy of a system is predicted to have a form, such as that shown in Figure 17.5, which breaks the stability condition (17.6) over some finite range of U. What will happen? We have already noted that a system placed at a point in the unstable region will not stay there. We will show that the system will divide itself into two parts, one of which rapidly gains energy at the expense of the other.

In an attempt to maximize its entropy, while conserving energy, the system could adopt the following strategy. 'Look', says the managing director, 'there's a lot of entropy available at higher energy, so I want N_1 particles to go exploring up in energy to some state A, I don't know where yet. You can obtain the energy you need from the other N_2 particles who will have to go to state B of lower energy'. The accounting (i.e. conservation of energy and particles) is correct as long as

$$\left. \begin{array}{l} U = N_1 u_A + N_2 u_B \\ N = N_1 + N_2 \end{array} \right\} \quad (17.9)$$

$$\Rightarrow \quad \frac{N_1}{N} = \frac{U - U_B}{U_A - U_B}, \quad \frac{N_2}{N} = \frac{U_A - U}{U_A - U_B}. \quad (17.10)$$

After the particles have thus rearranged themselves, the total entropy is

$$S = N_1 s_A + N_2 s_B = \frac{(U - U_B) S_A + (U_A - U) S_B}{U_A - U_B}, \quad (17.11)$$

where $S_A = N s_A$ is the entropy that the whole system would have if it were wholly at state A, and $S_B = N s_B$ is the entropy that the whole system would have if it were

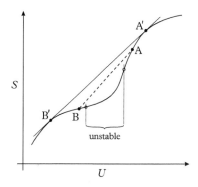

Fig. 17.5 *An entropy function $S(U)$ that exhibits the unstable behaviour $(\partial^2 S / \partial U^2)_{V,N} > 0$ over some finite region of U. The system finds a stable maximum entropy by splitting into two and following the upper tangential line (see text).*

[2] We have already proved $C_p \geq C_V$ and $\kappa_T \geq \kappa_S$ in Chapter 13.

wholly at state B. Equation (17.11) is a linear function of U which interpolates directly from A to B: see the dotted line on Figure 17.5. The system still has the same energy U, but its entropy is now on this line so it is higher than it was. By trying further pairs of states A, B the system can attempt to increase the entropy further, and you can see from the graph that the highest entropy solution is when the line meets the $S(U)$ curve at a tangent, indicated by the points A' and B'.

Thus, what the original wiggling curve of $S(U)$ implies is that the system in question has a region of parameter space where it will prefer to exist as two parts with different properties: one part with a higher entropy per particle, the other with a lower entropy per particle. These two parts are said to be different *phases* of the underlying material.

If the system starts in state B' and energy is supplied in the form of heat, then part of the system will stay unchanged, while part will adopt a quite different state. The total entropy will increase linearly with U for a while, so *the temperature remains constant*. This will continue until an energy $U(A') - U(B')$ has been supplied. This energy is called a *latent heat*. At this point the whole of the system will be in state A'. After that, further injections of heat will raise the temperature. This is precisely what is seen in a first-order *phase change* (e.g. melting, boiling), the subject of Chapter 22.

17.3 Free energy and availability

We shall now apply the maximum entropy principle to the case of a closed system at fixed *temperature* and volume. To do this, imagine the system in question is in thermal contact with a large reservoir; see Figure 17.6. Note that we have in mind any system whatsoever. Heat can flow freely between reservoir and system, and we suppose the complete setup of reservoir plus system is isolated. The total energy available is U_{tot} (a constant), but this has to be shared out so that the system possesses energy U, and the reservoir $U_{\text{tot}} - U$. When some heat ΔQ passes from the reservoir to the system, the reservoir's entropy decreases by precisely $\Delta Q/T_R$ (we assume the processes within the reservoir are reversible) and the system's entropy S must increase by at least this amount, so that the total entropy $\tilde{S}_{\text{tot}} = S_R + S$ increases. Therefore

$$\Delta S \geq \frac{\Delta Q}{T_R}. \tag{17.12}$$

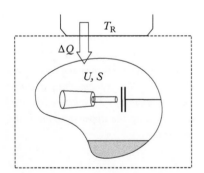

Fig. 17.6 *A generic system receiving heat from a reservoir at fixed temperature T_R. The structures shown inside the system are a visual indication that the argument applies to any system, of any internal construction.*

Inside the system, various processes may be going on if the system is relaxing towards equilibrium, but we assume for the moment the system does no work on other things. Therefore its internal energy increases by

$$\Delta U = \Delta Q \leq T_R \Delta S$$
$$\implies \Delta U - T_R \Delta S \leq 0. \tag{17.13}$$

This simple result tells us something useful about the relaxation to equilibrium of a general system, when it exchanges heat with surroundings at a fixed temperature T_R. Note, we have not assumed anything about the temperature of the system, and indeed it may have different parts at different temperatures, or indeed have no well-defined temperature. But its total internal energy and entropy are both defined by the sum of those quantities over the parts of the system. The result shows that such a system will *not* relax towards a state which maximizes its own entropy, but towards a state which minimizes the quantity $U - T_R S$, because then the combined entropy of system plus surroundings is maximized. This quantity is an example of something called *availability*, which we define more fully in equation (17.16).

Note that the above argument makes no mention of pressure or volume, and in fact it only mentions properties that any system must possess (namely, entropy, energy and, in equilibrium, temperature). We need not be describing a simple compressible (pV) system; the argument applies equally to electric cells, magnets, or even complicated machines, as long as they gain entropy either within themselves or by heat transfer from a given reservoir. The availability, like internal energy and entropy, will depend on the values of whatever internal degrees of freedom the system has, and when the system is out of equilibrium it remains well defined. Note, however, that its definition involves not just the system properties but also the temperature T_R of the surroundings.

Now let's consider a system which can exchange both energy and volume with a large reservoir. For example, consider a system surrounded by air at fixed temperature and pressure. Note, such a system exchanges heat and mechanical work with its surroundings, but it might have an internal construction of any type—it might be an electrical device or a chemically reacting compound or a complicated collection of mechanical parts, for example. The argument only needs to assume that the system has a well-defined volume and is surrounded by a reservoir whose pressure p_R and temperature T_R do not change as the volume varies, and as heat is exchanged. In this case the mechanical work done by the surroundings when the system's volume changes by ΔV is $\Delta W = -p_R \Delta V$, and we have

$$\Delta U = \Delta Q - p_R \Delta V \leq T_R \Delta S - p_R \Delta V$$
$$\implies \Delta U - T_R \Delta S + p_R \Delta V \leq 0. \qquad (17.14)$$

Note, the system's pressure does not appear in this expression, and indeed it may not even be well defined.

Both equations (17.13) and (17.14) can be captured by the expression

$$dA \leq 0, \qquad (17.15)$$

where the *availability* A is defined by

$$A \equiv U + p_R V - T_R S, \qquad (17.16)$$

so that

$$dA = dU + p_R dV - T_R dS. \qquad (17.17)$$

As we have already emphasized, availability is a function of the properties of both system and environment. The physical content of equation (17.15) is that *during any kind of physical process in a given environment, the availability can only ever decrease.*

When a system is relaxing towards equilibrium, some sort of physical process is happening, and we have just seen that the availability is then either staying constant or decreasing. As A decreases, the entropy of system plus surroundings is increasing, and this will continue until the latter is maximized, which happens when the availability is minimized, subject to whatever constraints the system is under. Therefore the condition for thermodynamic equilibrium is $dA = 0$. We will study this condition in conjunction with free energy later. Here we will use it to motivate another way of thinking about availability.

Availability can be physically interpreted as the maximum useful work that can be extracted from a system in given surroundings. Figure 17.7 shows the general idea. To get *maximum* work we know the processes had better be reversible (Section 11.1), so we shall assume this from the outset. Then all the variables refer to equilibrium states, so we can use $dU = TdS - pdV + \sum_i x_i dX_i$ in equation (17.17), giving

$$dA = (T - T_0)dS - (p - p_0)dV + \sum_i x_i dX_i,$$

where X_i and x_i are all the further variables which describe the internal state of the system. However, when the system only exchanges energy with its surroundings through heat exchange and volume change, these internal processes must give no net change in U (by energy conservation), so we have $\sum_i x_i dX_i = 0$ for the process under discussion,[3] hence

$$dA = (T - T_0)dS - (p - p_0)dV.$$

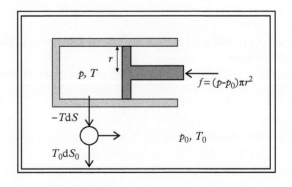

Fig. 17.7 *The concept of available work in given surroundings is illustrated by this example. If the simple compressible system has a pressure greater than that of its surroundings, $p > p_0$, then a movement outwards of the piston through a small distance dx provides the energy $(p - p_0)\pi r^2 dx$ to whichever system is providing the balancing force f on the piston. Also, by using the system to drive a heat engine, one can obtain further work $(-TdS - T_0 dS_0)$, where dS is negative if the heat flows out of the system.*

[3] In fact, we have been assuming this all along when we discussed simple systems such as gases in this book, because, for example, any gas can exchange energy between one region and another inside its own volume, and diatomic gases can move energy between translational and rotational forms, etc., but these processes do not contribute to the energy exchanged with other systems.

If the system expands then it gives out work energy $p\mathrm{d}V$ but only $\mathrm{d}w_1 = (p - p_0)\mathrm{d}V$ is counted as useful work if the surroundings are fixed (think of a steam engine in the presence of a surrounding atmosphere). However, we can exploit the temperature difference $T - T_0$ to get a little more work by using a heat engine, which delivers a further $\mathrm{d}w_2 = -T\mathrm{d}S - T_0\mathrm{d}S_0 = -(T - T_0)\mathrm{d}S$. Hence, the total useful work obtained is $\mathrm{d}w_1 + \mathrm{d}w_2 = -\mathrm{d}A$. As the system expands and cools it continues to provide work until $p = p_0$ and $T = T_0$. Treating A (for given p_0, T_0) as a function of state that can in principle be considered as a function of p and T, one can use the total change, $\Delta A = A(p, T) - A(p_0, T_0)$, to find out how much work can be extracted in total.

This interpretation of A as the 'useful work available' makes it easy to learn the sign of the possible spontaneous changes in A: 'things can only get worse'.

17.3.1 Free energy and equilibrium

Consider now a situation like that shown in Figure 17.6, except that now the system has a well-defined and uniform temperature T at all parts that exchange heat with the surroundings, and this temperature is equal to that of the surroundings. In other respects the system may or may not be in equilibrium. In this case equation (17.13) can be written

$$\Delta U - T\Delta S \leq 0. \tag{17.18}$$

This equation is useful because all the quantities on the left-hand side are properties of the system, not the reservoir. For such a system the Helmholtz function $\tilde{F} = U - TS$ is well defined, both for equilibrium and non-equilibrium states, because we assume the case where T is well defined (and constant), and U and S can always be obtained by summing the internal energies and entropies of the parts of the system. Therefore (17.18) can be written

$$\mathrm{d}\tilde{F} \leq 0, \tag{17.19}$$

where we have introduced a tilde as a reminder that this may be an out-of-equilibrium or an equilibrium property, as in equation (17.1). In equilibrium, $\mathrm{d}\tilde{F} = 0$, which means that the equilibrium state is one in which any small change in the system does not affect $\tilde{F}$ to first order. This result is very useful because it describes the possible processes, and the equilibrium condition, in terms of properties of the system alone, not the reservoir. As we have already emphasized for availability, the result applies to all types of system, not just simple compressible systems.

Now consider a system which can exchange both energy and volume with a large reservoir, such that the system has a temperature T at all parts which exchange heat with the surroundings, with $T = T_R$; and the system also has a uniform pressure p at all surfaces where it can receive mechanical work from the surroundings, with $p = p_R$. For example, consider a system surrounded by air at fixed temperature and pressure, and exchanging heat and mechanical work reversibly; but in other respects it may or may not be in equilibrium or undergoing reversible evolution, and it may have an arbitrary internal construction. An example is a chamber of gas with a movable piston, in thermal contact with a heat and pressure reservoir, and also containing other things such as a rubber balloon or a collection of reacting chemicals. In this case, equation (17.14) becomes

$$d\tilde{G} \leq 0, \quad \text{[allowed processes at fixed } T \text{ and } p\text{]} \tag{17.20}$$

since the Gibbs function $\tilde{G} = U - TS + pV$. Hence we find that to maximize the total entropy of system plus surroundings, it suffices to minimize the Gibbs function of the system.

We have thus shown that F and G are not only energy-like functions, they are also entropy-like functions, which govern the movement of the system towards equilibrium.

Let's summarize what we have learned about the equilibrium conditions:

(1) Isolated system: fixed U, V, N: maximum entropy.

$$d\tilde{S} = 0. \tag{17.21}$$

(2) Rigid system in thermal contact with a reservoir: fixed T, V, N: minimum Helmholtz function.

$$d\tilde{F} = 0. \tag{17.22}$$

(3) Flexible system in thermal and mechanical contact with a reservoir: fixed T, p, N: minimum Gibbs function.

$$d\tilde{G} = 0. \tag{17.23}$$

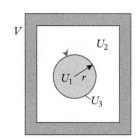

Fig. 17.8 *A composite system in a chamber of fixed volume V. The system comprises a rubber balloon and two fluids: one inside and one outside the balloon. The total internal energy $U = U_1 + U_2 + U_3$, where U_1, U_2 are the internal energies of the fluids and U_3 that of the rubber. The amount of matter is constant. When the chamber is thermally isolated, the equilibrium maximizes $\tilde{S}(r)$ at given U, V, N. When the chamber is in thermal contact with a reservoir at temperature T, the equilibrium minimizes $\tilde{F}(r)$ at given T, V, N.*

In these expressions we have retained the tilde, to signify the generality of the result, but of course at the equilibrium point, these functions take on their equilibrium values, and this introduces a further simplification. Among all the possible state changes encompassed by the expression $d\tilde{S} = 0$, some are changes to unusual non-equilibrium states that are hard to describe, but the expression also says what happens to $\tilde{S}$ under changes to nearby *equilibrium* states—that is, to states which would be in equilibrium if some further constraint were imposed on the system, concerning one of its *internal* variables. For example, consider a spherical balloon floating inside a closed vessel (Figure 17.8). The total entropy of the gas inside the balloon, the rubber of the balloon itself, and the rest of the gas in the vessel can be considered to be a function of the total energy, volume, and amount of matter, and also of the radius r of the balloon:

$$\tilde{S} = \tilde{S}(U, V, N; r). \tag{17.24}$$

When U, V, N are all fixed, $\tilde{S}$ becomes a function of r alone. The condition $d\tilde{S} = 0$ then tells us that the equilibrium state will be at a stationary value of $\tilde{S}$ as a function of r, at the given U, V, N. We can now, if we choose, regard r itself as a system parameter which could in principle be constrained to take one value or another, so that $\tilde{S}(U, V, N; r)$ can be thought of as the entropy of an *equilibrium* state, in the presence of such a further constraint, at each value of U, V, N, r. As long as we only choose to consider states of this kind, then the tilde can be dropped, and this is the standard practice in notation in thermal physics. We have felt it clearer to introduce the tilde, in order to help the reader understand what equations (17.21)–(17.23) are asserting, but we will not always use one, and most books and research literature do not.

All of equations (17.21)–(17.23) are examples of the condition $dA = 0$, as the reader can readily prove. (When U and V are fixed, (17.17) becomes $dA = -T_R dS$, so $dA = 0$ when $dS = 0$; when V but not U is fixed, (17.17) becomes $dA = dU - T_R dS$, which is $dA = dU - TdS = dF$ when T is constant and equal to T_R, so now $dA = 0$ when $dF = 0$; etc.) The versions in terms of S, F, and G, as opposed to the version using availability, are useful because they only use system properties, and they also reveal the general pattern. In each case, the function which reaches a stationary value is the one whose proper variables, in the case of the pV system, are fixed by the environment. On reflection, we could have foreseen this. A system with a property such as volume V fixed must be one that 'chooses its own pressure'. In the process of relaxation towards equilibrium, whatever function X is being extremized must therefore be one that does not have a first-order dependence on p at the extremum. Therefore the formula for dX between equilibrium states cannot have a dp term in it: the formula must be of the form $dX = -pdV + \ldots$, or possibly $dX = pdV + \ldots$, but not $dX = Vdp + \ldots$ or $dX = -Vdp + \ldots$. By arguing in this way from all the constraints imposed on the system by its environment, one can uniquely identify the thermodynamic function that reaches an extreme value when the maximum entropy principle is applied to system and environment together. A little thought then suffices to identify whether the function is maximized or minimized. This is explored in more detail in Appendix B.

Consider again the vessel containing gas and a balloon, shown in Figure 17.8. For the case of a fully isolated system (i.e. fixed U, V, N), we have already explained that the equation $d\tilde{S} = 0$ tells us how to determine what the balloon does: it will expand or contract until the joint of entropy of the air inside it, the rubber, and the air outside it, reaches a maximum, subject to the constraints of fixed U, V, N of the whole system. If, instead, the vessel remains rigid but is placed in good thermal contact with a reservoir at temperature T, then the balloon will expand or contract until the system Helmholtz function is at a minimum, subject to the constraints of fixed T, V, N. Figure 17.9 illustrates this idea (compare it with Figure 9.2). If the reservoir temperature in the second case were the very one which obtains for the equilibrium of the system in the first case, then the same equilibrium state would be found in both cases. Exercise 17.6 gives an example.

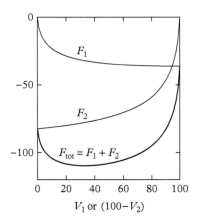

Fig. 17.9 *Free energy example for a pair of systems exchanging volume in isothermal conditions. Systems 1 and 2 here could be parts of a composite system that is in thermal contact with a reservoir at fixed temperature. The equilibrium condition is that F_{tot} is minimized (cf. with Figure 9.2).*

The general idea of an energy that is minimized at equilibrium is very useful, so it is given a name: *free energy*. Thus, for a closed system at externally fixed temperature and volume, the free energy is F; for a closed system at externally fixed temperature and pressure, the free energy is G; for an open system at externally fixed temperature, pressure, and chemical potential, the free energy is Ω. We have already commented that the phrase 'free energy' is sometimes used to mean the Helmholtz function per se. Now we see that, strictly speaking, this is correct terminology only when the conditions are those of fixed T, V, N.

Once we see the pattern, it is natural to consider whether the enthalpy is also a free energy. Does this have a minimum principle associated with it? The answer is yes, in a certain sense. The argument is less physically motivated than for the other free energies, because it is hard to imagine a system which can relax to equilibrium while its entropy remains fixed by its surroundings. However, the minimum enthalpy principle can be understood as a purely mathematical statement about the set of states of given S, p, N; namely: the one that minimizes the enthalpy is the equilibrium state. A similar statement holds for internal energy at fixed S, V, N, as we have already shown in Section 17.1.2. See Appendix B for more information.

17.3.1.1 Free energy as a form of energy content

We have already explained, in Section 13.2.1 (equation 13.18), that the Helmholtz function acts like potential energy in isothermal conditions. The same goes for the Gibbs function in isothermal isobaric conditions. Free energy can be interpreted as the energy one would have to supply to bring the system reversibly into the given environment (assuming the environment was already there). For an isolated system, you need to supply its internal energy U. For a system at fixed p, imagine the environment consists of a large atmosphere of gas at pressure p. In order to introduce the system into such an atmosphere you would have to supply both its internal energy U and also a further pV of work to push the system in against the atmospheric pressure: hence a total of $U + pV = H$.

For fixed T, imagine the environment is a heat bath. You might imagine you would have to provide the complete system energy U in order to introduce the system into the bath, but you don't. Instead, prepare the system at its intended volume but in a very cold state. You need to lend the energy $U_0 < U$ of this cold state, but you will get some of this back! Introduce the system into the bath reversibly by letting the bath warm it up via a heat engine. The bath supplies heat TS, in order that it provides just enough entropy to bring the entropy of the system from zero to its final value, S. All this entropy goes to the system, but not all the energy does. $(U - U_0)$ is provided as heat to the system, and you get back the remaining $TS - (U - U_0)$ as work W. Therefore the total energy you would need to provide in order to introduce the system is $U_0 - W = U - TS = F$.

The argument for the case of fixed T and p, leading to G, follows immediately.

EXERCISES

(17.1) Consider again the two examples illustrated in Figure 3.1: a ball rolling in a bowl, and an atom emitting a photon. Do not discuss the whole universe. Rather, choose some suitable boundary to define the system under discussion, identify the constraints, and hence identify what property of the system dictates the direction of the process.

(17.2) Consider again the apparatus described in Exercise 7.18 of Chapter 7.

   (i) Calculate the initial and final total energy and entropy in the two cases.

   (ii) Suggest a physical process whereby the entropy increases in case A, and explain where the energy lost by the system goes in case B.

   (iii) Now suppose that after the mass is increased to m_2, instead of releasing the piston, it is simply moved to a given height h and fixed there. Consider the total energy $U(S, h)$ and entropy $S(U, h)$. Plot graphs of $S(U_i, h)$ and $U(S_i, h)$ as a function of h, where U_i, S_i are the total energy and entropy after the mass is increased but before the piston is moved (treat 1 mole of gas at STP for the initial conditions, and take $m_2 = 2m_1$). Explain how these graphs relate to the answers to Exercise 7.18 of Chapter 7.

   [Ans. $S = (3/2) \ln(U - m_2 gh) + \ln h + \text{const}$; $U = m_2 gh + [(1/h) \exp(S - \text{const})]^{2/3}$; case A maximizes $S(U_i, h)$; case B minimizes $U(S_i, h)$.]

(17.3) Consider an isolated chamber divided into two parts by a membrane possessing surface tension σ and area A. Let the two parts have equal temperature T, and let them have pressures p_1, p_2 and volumes V_1, V_2. Each part can be treated as a simple compressible system. (i) By considering a fluctuation in which a ripple appears on the membrane, changing its area without affecting V_1 and V_2, show that if the volumes are fixed then in thermal equilibrium the membrane adopts the shape of least area consistent with the fixed volumes. (ii) By considering a fluctuation in which V_1 grows and V_2 shrinks and A also consequently changes, show that the equilibrium condition $dS = 0$ gives $\sigma = (p_1 - p_2) dV_1/dA$ (cf. equation (14.10)). Hence relate the pressure difference at any point on the surface to the local radius of curvature of the surface.

(17.4) After equation (14.15) we observed that the entropy of a surface increases with its area, but in question 17.3 we argued that at maximum entropy, the area is minimized. What is going on?

(17.5) Adapt the argument of equations (17.9)–(17.11) to the case of a system at fixed temperature, and hence confirm the Maxwell construction illustrated in Figure 15.6.

258   Stability and free energy

(17.6) Consider the example system shown in Figure 17.8. Model the two fluids as ideal gases, and suppose the rubber has a constant surface tension σ and heat capacity C.

(i) Show that changes in the total entropy and energy are governed by

$$dS = \frac{CdT}{T} + \frac{N_1 k_B dV_1}{V_1} + \frac{C_1 dU_1}{U_1}$$
$$+ \frac{N_2 k_B dV_2}{V_2} + \frac{C_2 dU_2}{U_2},$$
$$dU = dU_1 + dU_2 + CdT + \sigma dA,$$

where C_1, C_2 are the heat capacities of the two gases, T is the temperature of the rubber, $A = 4\pi r^2$, $V_1 = (4/3)\pi r^3$, and $V_2 = V - V_1$. Hence show that the maximum entropy occurs when

$$(p_1 - p_2)dV_1 = \sigma dA,$$

where p_1, p_2 are the pressures in the fluids.

(ii) Now suppose the chamber is in thermal contact with a reservoir at fixed temperature. Show that, in an isothermal change, the free energy is governed by

$$dF = \sigma dA - p_1 dV_1 - p_2 dV_2,$$

and hence obtain the equilibrium condition again.

(17.7) By starting from the entropy argument given in Section 17.1.3, establish the following stability conditions:

$$\left.\frac{\partial^2 U}{\partial S^2}\right|_{V,N} \geq 0, \qquad \left.\frac{\partial^2 U}{\partial V^2}\right|_{S,N} \geq 0, \qquad (17.25)$$

$$\left.\frac{\partial^2 F}{\partial T^2}\right|_{V,N} \leq 0, \qquad \left.\frac{\partial^2 F}{\partial V^2}\right|_{T,N} \geq 0, \qquad (17.26)$$

$$\left.\frac{\partial^2 H}{\partial S^2}\right|_{p,N} \geq 0, \qquad \left.\frac{\partial^2 H}{\partial p^2}\right|_{S,N} \leq 0, \qquad (17.27)$$

$$\left.\frac{\partial^2 G}{\partial T^2}\right|_{p,N} \leq 0, \qquad \left.\frac{\partial^2 G}{\partial p^2}\right|_{T,N} \leq 0. \qquad (17.28)$$

[See Appendix B for the method.] In summary, *the thermodynamic potentials are convex functions of their extensive variables and concave functions of their intensive variables.*

Reinventing the subject

18

We have now studied most of the basic ideas of thermodynamics, and applied them to numerous physical situations. We took the route of starting out from a definition of temperature in terms of equilibrium, then learning about heat as a form of energy, discovering entropy, and so on. The most important further idea was free energy, and in Chapter 17 we gave an important general insight into the way the maximization of entropy governs equilibrium thermodynamics, and also allows us to explore out-of-equilibrium situations.

In the present chapter we will start over again. We will pretend that we have never heard of temperature, or pressure, or chemical potential. All we know about are extensive things such as energy, volume, and particle number. Note that these can all be defined for systems out of equilibrium just as well as they can for systems in equilibrium. The main thing we shall claim for these quantities is that they are conserved. If one system or part of a system loses some energy, some other system or part must gain energy. If someone's volume falls, someone else expands. If a system loses particles, another system gains them. (We will come to the case where particle number is not conserved after first discussing the case where it is.) In this way we introduce the principle of conservation of energy (the first law of thermodynamics) at the outset.

Next we introduce the second law, in the form of a statement about entropy. We don't claim to know very much about entropy yet. We just assert that it is a property that all systems have, that it is a function of state, and that is obeys the *entropy statement of the second law*:

> There exists an additive function of state known as the *equilibrium entropy S*, which can never decrease in a thermally isolated system.

The word 'additive' here means that the joint entropy of two systems is the sum of their separate entropies. Note that the phrase 'thermally isolated' is not the same as 'completely isolated'.

Consider first of all a simple compressible system which has fixed volume and particle number. It can only exchange energy with its environment. For such a system in equilibrium, the entropy is a function of the only state variable remaining, namely energy: $S = S(U)$. So in this case,

$$dS = \left.\frac{\partial S}{\partial U}\right|_{V,N} dU. \tag{18.1}$$

18.1	Some basic derivations from maximum entropy	262
18.2	Carathéodory formulation of the second law	263
18.3	Negative temperature	265

Thermodynamics: A Complete Undergraduate Course. Andrew M. Steane.
© Andrew M. Steane 2017. Published 2017 by Oxford University Press.

Now suppose we have two such systems next to each other, exchanging energy. Or you could consider two internal parts of a system that have fixed volume and particle number, but exchange energy. By invoking the entropy statement, we know the entropy of the whole is given by the sum of the entropies of the parts, $S = S_1 + S_2$, and we find

$$dS = dS_1 + dS_2 = \frac{\partial S_1}{\partial U_1} dU_1 + \frac{\partial S_2}{\partial U_2} dU_2. \qquad (18.2)$$

Now $dU_1 + dU_2 = 0$, by conservation of energy, so

$$dS = \left(\frac{\partial S_1}{\partial U_1} - \frac{\partial S_2}{\partial U_2} \right) dU_1. \qquad (18.3)$$

Now you and I know some names for the partial derivatives in this expression, but pretend you don't know them. For the time being they are just something to do with entropy. Next we invoke the entropy statement of the second law again. Since S cannot fall, any changes that occur must increase S until it reaches a maximum. Therefore, in thermal equilibrium,

$$\left. \frac{\partial S_1}{\partial U_1} \right|_{V_1, N_1} = \left. \frac{\partial S_2}{\partial U_2} \right|_{V_2, N_2} \qquad (18.4)$$

and also, in the approach to equilibrium the part with the larger value of $\partial S / \partial U$ will gain energy so that $dS > 0$.

By invoking the entropy statement, and the conservation of energy, we have now discovered an interesting property of all systems. Let's give it a name:

$$T \equiv \left[\left. \frac{\partial S}{\partial U} \right|_{V,N} \right]^{-1}. \qquad (18.5)$$

Forget that you already know of this T. Just consider what we have learned from the above. This is that if two closed systems can exchange energy without doing work then they will do so in such a way that the one with higher T will lose energy, the one with lower T will gain it, and the process will reach a dynamic equilibrium when the two systems have the same T. In other words, we have just 'discovered' the property we ordinarily know as temperature.

By similar reasoning, you can now rapidly 'discover' pressure and chemical potential, all now *defined* as the relevant combination of partial derivatives of entropy:

$$p \equiv \left. \frac{\partial S}{\partial V} \right|_{U,N} \left/ \left. \frac{\partial S}{\partial U} \right|_{V,N} \right., \qquad \mu \equiv \left. \frac{\partial S}{\partial N} \right|_{U,V} \left/ \left. \frac{\partial S}{\partial U} \right|_{V,N} \right.. \qquad (18.6)$$

The argument is precisely the one we already gave in Section 17.1 in connection with equation (17.2).

You can now go on to invent the concept of free energy as a way of tracking entropy changes of a system in contact with a large reservoir, and thus having constrained temperature or pressure or both. Constrained chemical potential also arises for a system free to exchange particles with a reservoir. The whole subject of thermodynamics begins to be laid out before you in all its wonderful variety. All from these seemingly innocuous statements about energy conservation and increase of an extensive 'something' called entropy.

Notice that we are now considering the pressure of a fluid as a thoroughly thermodynamic type of quantity. It is not now to be viewed as a result of mechanical forces. It is the result of the urge to increase entropy. To be fair, in an adiabatic expansion, one can regard the pressure as the result of forces doing work just as they do in ordinary (non-thermal) mechanics. In this case, the forces do work, and the system loses precisely that amount of energy. However, in any other type of expansion, this is no longer the case. The change in system energy is not then equal to the work done. The isothermal expansion of an ideal gas is an extreme example of this: here, plenty of work is done, but the system energy does not change at all! This is because the gas draws in from the reservoir precisely the amount of heat energy it needs in order to keep its temperature, and hence its internal energy, constant. So why did it bother to expand at all? The answer is given by entropy: it did so because if the property we call p, obtained from partial derivatives of the entropy of the gas, is greater than that same property evaluated for the environment, then an overall entropy increase is achieved by the expansion. Equally, if we want to make the gas contract, going against its natural inclination to have high entropy, then we shall have to either receive the entropy ourselves, or else, by squeezing the gas adiabatically, cause the entropy to move from position to momentum degrees of freedom. This involves an increased kinetic energy in the gas; we shall have to supply this energy.

Similar considerations apply to an ideal elastic material. Take an elastic band and stretch it. You can feel the force it exerts. This force is not caused by stretching of chemical bonds with their associated potential energy (which ultimately would be electromagnetic field energy). It is caused almost entirely by the molecules' thirst for entropy. Each polymer molecule is like a chain whose links do not stretch, but they can bend easily (like a bicycle chain). The molecules store both kinetic energy and entropy. Their entropy is associated with the number of momentum states associated with their thermal motion, and with the number of shapes consistent with a given distance between their ends. By pulling on the ends of the elastic band, you reduce the shape contribution to the entropy. But if there is no heat exchange, then the total entropy of the molecules cannot fall, so the contribution associated with momentum must rise. Hence they must have more kinetic energy, and the elastic force results because you must supply this energy.

18.1 Some basic derivations from maximum entropy

We will now present derivations of the Clausius and Kelvin statements of the second law, and of Carnot's theorem, from the entropy statement of the second law.

First, suppose there were a process in which heat was transferred from one body to a hotter body, without any net change in the rest of the world. In such a case the hotter body would gain less entropy than the colder body loses, so the entropy of the joint system composed of both bodies would fall. But the two bodies, plus whichever part of the rest of the world was involved in the process, jointly compose a thermally isolated system. Therefore the entropy statement applies, and it tells us that any such process is impossible. This is the Clausius statement of the second law.

Next suppose we extract some heat from a body, and then use that energy to drive a system to perform an equivalent amount of work, for example by reversible compression of a gas, without passing any heat elsewhere. In such a process the entropy of the composite system falls, and that system is thermally isolated, so once again the entropy statement applies and rules that the process is impossible. This is the Kelvin statement.

Finally, suppose that a heat engine operates in a cycle between reservoirs at two fixed temperatures. The engine suffers no net change in a cycle, so its entropy is unchanged. The composite system of both reservoirs and engine can be thermally isolated, and its entropy changes by

$$\Delta S_{\text{tot}} = -\frac{Q_1}{T_1} + \frac{Q_2}{T_2} \geq 0, \tag{18.7}$$

by using the entropy statement and equation (18.5). Therefore,

$$Q_2 \geq (T_2/T_1)Q_1. \tag{18.8}$$

The engine will be as efficient as possible when Q_2 is as small as possible. The result shows that this occurs when the equality holds, which is the condition for no net entropy increase. This is also the condition for the process to be reversible, so we have Carnot's theorem.

In view of the way the entropy method yields the existence and nature of temperature and chemical potential and other such properties, and in view of the rapidity with which it also yields other basic ideas such as Carnot's theorem and the heat ratio equation, the entropy statement of the second law is widely regarded as the best way to view the second law. In this book we have presented the other ideas first because this involves less abstraction. It can seem odd to go about science by merely declaring the existence of some physical property. However, the essential content of the entropy statement is not the existence of some otherwise undefined physical property, but what it asserts about that physical property: that it is additive, and a function of state, and cannot decrease in thermal isolation. The statement does also require some prior understanding of what is meant by a function of state, and what is meant by thermal isolation, so the investment of effort in understanding those concepts is never wasted.

18.2 Carathéodory formulation of the second law

The statement of the second law presented by Constantin Carathéodory is sometimes preferred on the grounds of being more minimal and cleanly defined than other statements.

Carathéodory's approach is motivated by the idea of *adiabatic surfaces*, which we introduced in Section 9.7. The concept is to try to state, as minimally as possible, some requirement on physical systems that will suffice to prove the existence of these surfaces. By this we mean the claim that reversible adiathermal change transports a system state within a sub-space of dimension $d-1$ in the state space, where d is the dimensionality of the state space. Also, we want our basic statement or law to imply some further properties of these surfaces, such as that they do not mutually intersect or fold back. In technical language, we say they *foliate* the space, like pages in a book: each surface lies on top of the one before (but we don't require flat surfaces, so compare it to an old book with warped pages all undulating in synchrony).

The Carathéodory statement or principle (Section 3.3) asserts that, in the absence of heat flow in or out of a system, you can't manage to bring the system to all the states which are there in its state space. And furthermore, no matter where you are in the state space, there is another state nearby—infinitesimally close—that you cannot reach, no matter how you manipulate the system in other ways, i.e. by doing work. Why can't you reach it? Because the principle says so.

In the following we will call this principle CP.

With CP in hand, we employ arguments very much like that we gave in connection with Figure 9.15 in Section 9.7. The difference is that we shall be more careful to state our assumptions, and we shall restrict ourselves to adiathermal processes in the whole of the argument to derive the existence of adiabatic surfaces and hence the entropy function. No heat exchange is mentioned in CP, and we will require none. Once this is done, one can go on to connect entropy to temperature and to heat using standard methods.

Recall the idea of *generalized coordinates*, which are the parameters which can be used together to uniquely specify the system state. We shall take as our set of coordinates the internal energy and the $(d-1)$ parameters whose small changes appear in the expression for reversible work on the system. We will refer to the latter by the name *configuration coordinates*, in order to distinguish them from internal energy. Thus there are $(d-1)$ configuration coordinates and one further coordinate: internal energy. A *configuration* is a set of values for the configuration coordinates; it specifies a line in state space. If we imagine the state space plotted with energy on the vertical axis, then this line is vertical. A *state* is defined by the combination of configuration and internal energy.

The first assumption we will require is that it is possible to move the system between any pair of *configurations* by a reversible adiathermal process. This is not much of an assumption; it is almost obvious. You just thermally isolate the system and then start quasistatically changing its extensive parameters, such as volume, or the product of volume with magnetic field, or length, etc. by pushing or pulling on the system, or putting a current through a field coil, etc. However, in the presence of hysteresis this assumption will break down, so we do need to state it. The second assumption will be that it is always possible to introduce energy into the system adiathermally without changing the configuration. For a fluid, for example, this could be done by stirring. For many systems it could be done by passing an alternating electric current through the system for a while. In such processes no heat passes across the control surface that notionally defines our system. The external agency is busy doing work đW, but the configuration has no net change because this is irreversible work whose expression is not given by $-p\mathrm{d}V$ or $-m\mathrm{d}B$ and their companions, but by the extra friction-related terms that arise for irreversible processes (Eq. 7.9).

Now we get to work with our argument which is based on Figure 18.1.

We start with any pair of states A and C, and define B as a state which can be reached in a reversible adiathermal process from A and has the same configuration (but not necessarily the same internal energy) as C. We wish to prove that this B is unique.

With a view to obtaining a contradiction, let us suppose that B is not unique. That is, depending on the path, one might arrive reversibly and adiathermally from A at another state B' that has the same configuration coordinates as B and C. Let U_B and U'_B be the internal energies of states B and B', and suppose $U_B > U'_B$. (If $U_B < U'_B$ then simply switch the labels B and B'.) We will show that in this case it is possible to pass adiathermally from B to *any* state in its neighbourhood, which violates CP. Hence B must have been unique after all.

To prove this, start from B and pass reversibly to A and thence to B'. Next, change the configuration coordinates to those of any state ϕ near to B'. Since B' and B had the same configuration coordinates, by assumption, this nearby state ϕ has a *configuration* near to B, but we don't yet know its internal energy. Since only an infinitesimal change is required to move from B' to a neighbouring state, the internal energy of ϕ must be less than U_B (since $U_B - U'_B$ is finite). Therefore, one can now provide internal energy and thus move the system from ϕ to a thermodynamic state in the neighbourhood of B. Furthermore, since all d coordinates including U are fully adjustable in this argument, every state near to B can be reached. So we have a contradiction of CP. Therefore, B must have been unique. QED.

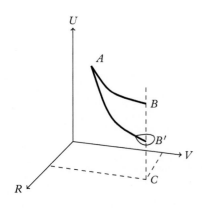

Fig. 18.1 *Argument used to prove the existence of adiabatic surfaces from Carathéodory's principle (see text).*

We now have that the set of states accessible from A by a reversible adiathermal process lie on a surface, with just one state for every configuration. If we now consider another state at the same configuration as A, but at a different internal energy, then we shall have another surface. It is easy to prove that the new surface does not intersect the first one, because if it did then by placing A at the intersection we would have the situation shown in Figure 18.1, which is what we have just ruled out. Therefore we have an orderly foliation of the space. We can label these surfaces by a single number S which increases monotonically with the internal energy at A. Then, for any pair of states B, C, if $S(B) < S(C)$ one can pass from B to C by an adiathermal process by first passing to A, then adding energy by stirring, then passing to C. One cannot then pass from C back to B because if one could then one could arrange to move from C to any state is its neighbourhood, by a natural extension of the argument above, and this is ruled out by CP.

It follows that CP, plus the first law (which is needed in order to claim that U is a function of state) together suffice to allow the entropy function to be derived.

So far we have made no mention of time, nor of spontaneous processes. It remains to add a remark, then: an isolated system can evolve to states that are adiathermally accessible, but not to states that are adiathermally inaccessible. Now that we have defined the entropy, this remark immediately equates to the statement that the entropy of an isolated system can increase but not decrease with time. We can now define things like temperature and heat using standard arguments, and in particular we can arrange the definitions in such a way that $dS = đQ_{\text{rev}}/T$ during a reversible heat exchange. The main point is that one can do all this without appeal to heat engines, and without simply asserting the existence of entropy as is done in the entropy statement of the second law.

18.3 Negative temperature

The definition (18.5) can be straightforwardly generalized, for a closed system of any type, to

$$\frac{1}{T} = \left.\frac{\partial S}{\partial U}\right|_{\text{work}=0}. \qquad (18.9)$$

Here the label on the partial derivative indicates that the path is one in which no work is done on the system. For a compressible system, this indicates constant volume V. For the magnetic system described by equation (14.23), it indicates constant magnetic field B. Note that the definition can be applied satisfactorily to non-equilibrium states as well as equilibrium states, so it is more general than the notion of temperature that can be deduced from the zeroth law.

Usually one has $T > 0$, but the situation $T < 0$ can occur when there is an upper bound on the way the system in question can store internal energy.

Fig. 18.2 *An example $S(U)$ curve for a system exhibiting negative temperature. S is always positive, and U can have either sign (negative values of U are unremarkable; they merely reflect the convention on where the zero of energy is placed). The gradient dS/dU, on the other hand, gives $1/T$, which we ordinarily expect to be positive. In some isolated finite systems there can be metastable equilibrium states having negative temperatures; such states are* hotter *than states at positive temperature (see text). An example case is the ideal paramagnet, where a state at negative temperature can be achieved by abruptly reversing the applied magnetic field.*

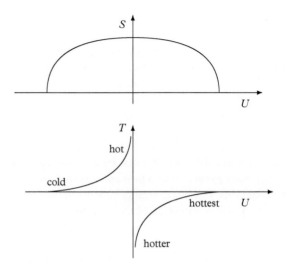

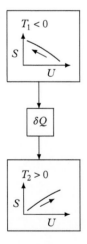

Fig. 18.3 *Heat transfer from a body at negative temperature to a body at positive temperature. Both entropies increase.*

This happens for example for paramagnetic systems. The function $S(U, B)$ can have the form shown in Figure 18.2, in which at high energy there is a region where $(\partial S/\partial U) < 0$. What is happening is that as the internal energy goes up towards its upper bound, there remains a stable internal thermodynamic equilibrium, but the system is becoming more structured not less as U increases. In the magnetic example, the individual magnetic dipoles are lining up more and more towards their state of maximum potential energy. The situation is stable against internal fluctuations because we have not transgressed the stability condition (17.6).

The idea of negative temperatures appears odd because it seems to suggest that the system can be colder than absolute zero. However, this is a false appearance, because in fact if a system at negative temperature is brought into thermal contact with a system at positive temperature, *heat will always flow from the system at $T < 0$ to the system at $T > 0$* (see Figure 18.3). This is because when some heat δQ leaves the system at negative temperature, its entropy increases, and when that heat arrives at the system at positive temperature, its entropy also increases. So this is the direction of the heat flow. One may say, therefore, that a system at negative temperature is 'hotter' than any system at positive temperature, where by 'hot' we mean 'liable to give off heat'.

A system with no upper bound on its internal energy cannot reach a stable negative temperature. This can be proved as follows. If U has no upper bound, then any region of the $S(U)$ curve having $dS/dU < 0$ must lead to another region having $d^2S/dU^2 > 0$ (since S must remain positive at all U), and we have the situation shown in Figure 17.5: a 'wiggle' in the entropy curve. This results in the phenomenon of phase change, and thus the negative temperature is avoided.

Negative temperature can be metastable but it is never a fully stable situation. This is because any system that has a finite upper bound to one form of internal energy also has further degrees of freedom that allow it to store energy in other ways, and in particular there is always kinetic energy which has no upper bound. The relaxation time may be long, but the relaxation to complete equilibrium is guaranteed eventually to extract energy from the negative temperature aspect until it reaches a state where $dS/dU > 0$, i.e. positive temperature.

19 Thermal radiation

19.1 Some general observations about thermal radiation 268
19.2 Basic thermodynamic arguments 275
19.3 Cosmic microwave background radiation 288
Exercises 289

Many physical systems, such as a radio antenna, a laser, a light-emitting diode, a hot metal filament, the human body, a glow-worm, emit electromagnetic radiation. The total emission from a physical object can usefully be separated into two parts: the thermal radiation, and the rest. Thermal radiation is that part of the emitted radiation which is associated with a non-zero temperature of the system, and which vanishes as the system temperature tends to absolute zero. It is caused by random jiggling of the charged particles of which the system is composed. Thermal radiation has many universal properties, that is, properties that do not depend on the details of the system emitting it. They depend only on a small number of macroscopic parameters such as temperature and absorptivity. This is the subject of this chapter.

In order to study thermal radiation, it is useful in the first instance to restrict the discussion to ordinary inert objects which have no radiative emission except the thermal kind. Thus we do not discuss systems such as an r.f. signal generator, a laser, a light-emitting diode, or a glow-worm. This does not mean thermodynamics cannot be applied to such systems, but it has to be applied in a more detailed way. Within the restriction to systems where only thermal emission is involved, we are still able to treat a fairly rich set of systems, including for example the line spectra of atoms in a vapour (under thermal excitation), stars, and ordinary solids and liquids. Thermal radiation is especially important in the physics of the early universe, and in the response of the Earth to heating from the Sun.

We begin with a statement of some facts; we will then show that many of these facts can be linked together or derived by thermodynamic arguments. We will refer throughout the chapter to *electromagnetic* radiation. At extremely high temperatures, above $\sim m_e c^2 / k_B \simeq 6 \times 10^9$ K, thermal radiation can also include electron–positron pairs, but many of the arguments about energy density and flux still apply.

19.1 Some general observations about thermal radiation

When electromagnetic radiation is incident on a surface it may be absorbed, reflected, or transmitted. We define coefficients α, R, τ as the ratio of absorbed, reflected, or transmitted intensity to incident intensity. *Intensity* is here defined as power per unit area, also called *flux*. Then, by conservation of energy,

$$\alpha + R + \tau = 1. \tag{19.1}$$

In ordinary circumstances (such as thermal equilibrium) these coefficients cannot be negative, so they are all in the range 0 to 1. The word 'reflected' here should not be taken necessarily to mean mirror-like reflection (called *specular* reflection): it includes diffuse reflection, also called *elastic scattering*. Sometimes care is needed to distinguish between absorption and reflection. For example, when a laser beam propagates through a piece of coloured glass, then some light is reflected at the air–glass interfaces, and some further light is scattered to the sides throughout the length of the beam (we see this when we can see the beam as a glowing line). The scattered part is partly owing to absorption followed by re-emission, and partly owing to an elastic scattering process where the energy is not temporarily acquired by the glass. Therefore the processes inside the glass should be assigned partly to α and partly to R (both processes remove flux from the incident beam). When, after some absorption, there is a re-emission process (called fluorescence), then it contributes to the radiation (thermal or otherwise) emitted by the glass, but it should not be added as an extra term in (19.1), because its energy has already been included in the accounting for the incident energy via the α term.

An opaque system has $\tau = 0$. A good mirror has R close to 1; a good 'power dump', used for example to absorb a powerful laser beam, has α close to 1. All these coefficients can be functions of the state of the radiation field and of the state of the thermodynamic system in question. For example for a simple compressible system we may have

$$\alpha = \alpha(\omega, \hat{\mathbf{k}}, \mathbf{e}, p, T), \tag{19.2}$$

where ω is the angular frequency of the radiation, $\hat{\mathbf{k}}$ is its direction of propagation, $\mathbf{e}$ is its polarization, and p, T are the pressure and temperature of the system. For more general systems further parameters would be needed to specify the system state.

The amount of absorption exhibited by a given physical entity often depends on the intensity I of the radiation incident on it. For example, α usually tends to zero at high intensity. Although equation (19.2) does not explicitly exhibit a dependence on I, it can still account for such effects via the parameters which describe the internal state of the system. This is because the drop in α (called 'bleaching' or 'saturation') is usually caused by a change in the system state brought about by its interaction with the intense beam. In principle the absorptivity might also depend on the phase of the radiation, but we shall only discuss cases with no such sensitivity.

An object which absorbs all the radiation incident upon it, i.e. having $\alpha = 1$ at all wavelengths and angles, is called a *black body*. A large black hole (that is, a region enclosed by an event horizon in general relativity) is a perfect black body, but in more ordinary circumstances the concept of a black body is an idealization.

A good approximation to a black body at wavelengths in the infrared to ultraviolet part of the electromagnetic spectrum can be constructed as follows. We make a small hole in an opaque wall, behind which is a diffuse black surface, shaped so that multiple reflections would be needed to reflect incoming radiation back towards the hole; see Figure 19.1. In this construction the *hole* is the 'black body'. This hole is not a material body, but as far as the rest of the universe is concerned, radiation incident on the hole is not reflected or transmitted, and that is all we require. A grain of soot is also a good approximation to a black body over a wide range of wavelengths, and so is a star.

A black body at non-zero temperature T emits thermal radiation. Note that this means a black body only looks visibly 'black' at low temperature. The thermal emission means it will glow visibly if the temperature is high enough. A rough indication of the typical wavelength of thermal radiation is given by the formula

$$\hbar\omega \simeq 5k_B T, \qquad \Rightarrow \lambda T \simeq 3 \text{ mm K},$$

where we have used $\lambda = 2\pi c/\omega$ (for a more precise statement see equation (19.40)). This increase in frequency with temperature is called *Wien's displacement law*; it is derived in Section 19.2.3. For example, at room temperature bodies emit in the infrared part of the spectrum, whereas a 'red hot' body glows visibly red, with an emission spectrum that peaks in the infrared but includes a substantial fraction of visible radiation. The Sun appears almost white mainly owing to thermal emission; it has a surface temperature of around 5800 K and the emitted spectrum peaks at around 500 nm. If one examines the constellation Orion with an ordinary pair of binoculars, it is easy to see that the brightest star, Rigel (at the lower right of Orion) is bluish-white and therefore hotter than our Sun, whereas Betelgeuse (at the upper left) is orange-red and therefore colder than our Sun.

Stefan–Boltzmann law

For a black body the power emitted as thermal radiation from a small element of its surface of area A is

$$P = \sigma A T^4, \qquad (19.3)$$

where σ is the *Stefan–Boltzmann constant* defined by

$$\sigma = \frac{\pi^2 k_B^4}{60 \hbar^3 c^2} \simeq 5.6704 \times 10^{-8} \text{ Wm}^{-2}\text{K}^{-4}. \qquad (19.4)$$

The presence of $\hbar = h/2\pi$ in this formula is a signal that quantum theory is required in order to derive it. However, equation (19.3)—the fact that the emitted flux goes as the fourth power of temperature, and that the proportionality constant is universal—can be derived from purely thermodynamic reasoning, as we shall see. It is called the *Stefan–Boltzmann law*.

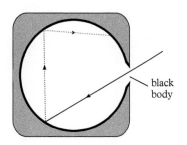

Fig. 19.1 *Making a black body using a hole in a highly absorbing chamber.*

Example 19.1

Treating the Sun as a spherical black body with surface temperature 5800 K, what is the total power output of the Sun?

Solution.
The mean radius of the Sun is 6.96×10^5 km, giving a surface area $A = 4\pi r^2 = 6.09 \times 10^{18}$ m^2. The power output is therefore $\sigma A T^4 = 3.9 \times 10^{26}$ W. (This is in reasonable agreement with the measured value of 3.8×10^{26} W.)

Kirchoff's law

Non-black bodies also emit radiation by virtue of their non-zero temperature. For a flat planar region of surface area A, the power emitted is

$$P = \epsilon \sigma A T^4, \tag{19.5}$$

where ϵ is called the *emissivity* or *emittance* and this equation serves to define ϵ: it is the ratio of flux emitted to that which would be emitted by a black body at the same temperature. It is found (and we shall later provide a thermodynamic argument to prove) that if α is independent of wavelength and direction, then

$$\epsilon = \alpha. \tag{19.6}$$

That is, the emissivity of a given surface is equal to its absorptivity. This is called *Kirchhoff's law*. It is an example of a non-trivial piece of physics—a connection between physical properties whose relationship is not self-evident—that emerges from the laws of thermodynamics. Values of emissivity vary greatly for different materials and surface qualities. Some examples are shown in table 19.1.

The form given in equation (19.6) is valid when both quantities are independent of wavelength, or when the radiation being absorbed has the same distribution as that emitted; and the symbols refer to averages over the distribution. More generally, the relationship is

Table 19.1 *Emissivity of some common materials and surface qualities.*

Surface	Temp. (°C)	ϵ	Surface	Temp. (°C)	ϵ
Brass			Brick		
highly polished	260	0.03	red, rough	40	0.093
dull plate	40–260	0.22	fireclay	980	0.75
oxidized	40–260	0.5	silica	980	0.80–0.85
Steel			Carbon soot	40	0.95
mild, polished	40–260	0.07–0.10	Silver	200	0.01–0.04
sheet, rolled	40	0.66	Water	40	0.96

> **Kirchoff's law**
> $$\epsilon(\omega, \hat{\mathbf{k}}, \mathbf{e}) = \alpha(\omega, \hat{\mathbf{k}}, \mathbf{e}). \tag{19.7}$$

That is, the emissivity of a given surface at any given wavelength, direction $\hat{\mathbf{k}}$, and polarization $\mathbf{e}$, is equal to its absorptivity at that wavelength, direction, and polarization. Both quantities may of course also depend on the state of the system, but we have not mentioned that dependence explicitly in equation (19.7) because it is understood that the equation refers to properties of a system in some given state. The notation is such that ϵ and α are dimensionless and in the range 0 to 1.

In order to discuss the frequency dependence it is necessary to introduce the notion of a *spectral* quantity. This means a quantity *per unit frequency range*. We can imagine using a filter which only allows radiation in the frequency range ω to $\omega + d\omega$ to pass. The power passing through such a filter would then be $P_\omega d\omega$ where P_ω is the power per unit frequency range (SI unit: watt per inverse second, i.e. Ws)). This is often called 'spectral power'. The total power is

$$P = \int_0^\infty P_\omega(\omega) d\omega.$$

In terms of spectral power the Stefan–Boltzmann law is

$$\int_0^\infty P_\omega^{\text{BB}}(\omega) d\omega = A\sigma T^4,$$

where the superscript BB refers to a black body. The more general definition of emissivity is

$$P_\omega(\omega) = \epsilon(\omega) P_\omega^{\text{BB}}(\omega), \tag{19.8}$$

where for the sake of simplicity we have given the case where the direction dependence is not under study, so the equation refers to the total power per unit frequency range emitted in all directions and polarizations.

Note that whereas α can be defined for bodies that are not in thermal equilibrium and do not have a well-defined temperature, the definition of ϵ involves a comparison between the body of interest and a black body 'at the same temperature', therefore it is only well defined for bodies in or very near to a thermal equilibrium state.

A body whose emissivity is independent of wavelength, direction, and polarization is called a *grey body*.

The essential content of Kirchhoff's law is that *good absorbers are good emitters*. You can observe this by placing different materials on a hot-plate and trying to detect the radiant emission (avoiding convection and conduction). Figure 19.2 shows an example from atomic spectroscopy.

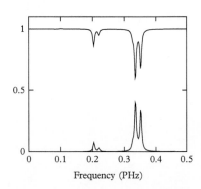

Fig. 19.2 *An example of Kirchoff's law. The upper curve is an absorption spectrum for a caesium vapour at 10 000 K. The lower curve shows the thermal emission from the same vapour. Where there is strong absorption, there is also strong thermal emission—equations (19.7) and (19.8).*

Example 19.2

Find the equilibrium temperature of (a) a spherical grey body orbiting the Sun at one astronomical unit, (b) planet Earth.

Solution.
(a) For simplicity we suppose the body has a uniform surface temperature (a reasonable approximation if it rotates rapidly). Let I be the intensity of sunlight at the body. Then the power absorbed by the spherical grey body is

$$P_{\text{in}} = \alpha \pi r^2 I, \tag{19.9}$$

where r is its radius, and the power emitted is

$$P_{\text{out}} = \epsilon 4\pi r^2 \sigma T^4. \tag{19.10}$$

For steady-state conditions these must be equal, so by equating them and using Kirchoff's law, we find

$$T = \left(\frac{I}{4\sigma}\right)^{1/4} \simeq 278\,\text{K},$$

(using $I = 1.36$ kW m^{-2}). It is interesting to note that the result is independent of the properties of the grey body. Another nice way to express the result is to model the Sun as a black body, so that $I = \sigma T_s^4 R_S^2 / R_{\text{ES}}^2$, where R_S is the mean radius of the Sun and $R_{\text{ES}} \simeq 1.496 \times 10^8$ km is the Earth–Sun distance. Then we find

$$T = \sqrt{\frac{R_S}{2 R_{\text{ES}}}}\, T_S.$$

(b) The calculation from part (a) can also be used to give a rough estimate of the surface temperature of the Earth. For a better estimate, we take into account that the Earth's atmosphere is not a grey body: it has an absorptivity and reflectivity that depend on wavelength. Averaged over the whole Earth, the reflectivity of the ground, sea, and clouds to solar radiation is around $R_{\text{uv}} \simeq 0.31$ (this is called *albedo*). This reflectivity is non-zero in the visible and ultraviolet part of the spectrum, where most of the power in solar radiation lies. To account for this, one should use $\alpha = 1 - R_{\text{uv}}$ in equation (19.9). The thermal emission from Earth's surface and from the atmosphere, on the other hand, is in the infrared part of the spectrum, so ϵ in equation (19.10) should be understood as the emissivity in the infrared, ϵ_{ir}. After taking this into account the result is

$$T \simeq \left(\frac{(1-R_{\text{uv}})}{\epsilon_{\text{ir}}}\right)^{1/4} \sqrt{\frac{R_S}{2 R_{\text{ES}}}}\, T_S \simeq 255\,\text{K}, \tag{19.11}$$

> where we have used $\epsilon_{ir} = 1$, because the absorptivity and hence also emissivity is close to 1 in the infrared part of the spectrum. The result is −18 °C, a reasonable estimate for the temperature of the upper atmosphere, where most of the thermal emission takes place.

Equation (19.11) shows that to keep cool on a hot day, a good strategy is to wear clothes that emit well in the infrared but not in the ultraviolet.

19.1.1 Black body radiation: a first look

Radiation emitted by non-black bodies owing to their microscopic thermal motion is correctly called 'thermal', but beware: when physicists use the phrase 'thermal radiation' in other contexts, they almost always mean that type of thermal radiation which is emitted by black bodies. If one wants to be precise, then the type of thermal radiation emitted by black bodies is called 'black body radiation' or 'cavity radiation'. The second label refers to the fact that radiation inside a large cavity in an opaque body at thermal equilibrium must be black body radiation. We shall prove this in Section 19.2, and also show that black body radiation is *universal*. That is, all black bodies at a given temperature must emit thermal radiation of precisely the same form, i.e. having the same distribution of intensity as a function of wavelength and direction.

Black body radiation is isotropic, unpolarized, and has an intensity distribution conveniently expressed by writing down the *spectral energy density* ρ_ω. This is the energy per unit volume per unit frequency range (SI units J m^{-3} Hz^{-1}, or J s m^{-3}). It is found experimentally, and predicted by quantum theory, that

$$\rho_\omega = \frac{\hbar}{\pi^2 c^3} \frac{\omega^3}{e^{\beta\hbar\omega} - 1}, \qquad (19.12)$$

where $\beta = 1/(k_B T)$. This function is plotted in Figure 19.3.

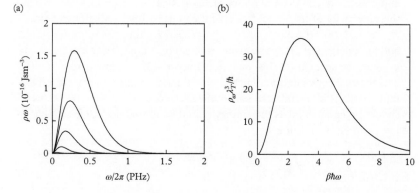

Fig. 19.3 (a) The spectral energy density of black body radiation as a function of frequency in petahertz (1 petahertz = 10^{15} Hz). ρ_ω is shown at five temperatures between 1000 K and 5000 K. (b) $\rho_\omega \lambda_T^3/\hbar$ versus $(\beta\hbar\omega)$ where $\beta = 1/k_B T$ and $\lambda_T = 2\pi\hbar c/k_B T$. Shown this way, a single curve displays the result for all temperatures.

The energy per unit volume is obtained by integrating ρ_ω over all frequencies:

$$u = \frac{U}{V} = \frac{\hbar}{\pi^2 c^3} \int_0^\infty \frac{\omega^3}{e^{\beta \hbar \omega} - 1} d\omega. \tag{19.13}$$

It is instructive to perform this integral by using the change of variable $x = \beta\hbar\omega$, then one finds

$$u = \frac{k_B^4 T^4}{\pi^2 c^3 \hbar^3} \int_0^\infty \frac{x^3}{e^x - 1} dx = \frac{\pi^2 k_B^4 T^4}{15 c^3 \hbar^3} = \frac{4\sigma}{c} T^4, \tag{19.14}$$

in which the T^4 dependence of the Stefan–Boltzmann law is apparent. The integral evaluates to $\pi^4/15$. One finds $u = 0.5\,\text{J/m}^3$ at $T = 5070$ K.

The intensity (= power per unit area) incident on a surface is related to this by

$$I = \frac{1}{4} u c, \tag{19.15}$$

where we include the power incident from all directions on one side of the surface (i.e. half the full solid angle). The factor 1/4 comes from the isotropic nature of the radiation; this is the same factor as arises in the kinetic theory of gases when calculating the flux by effusion through a small hole. Equations (19.3) and (19.4) follow immediately.

19.2 Basic thermodynamic arguments

By quoting equation (19.12) we have stated without proof many of the important properties of black body radiation. We shall now aim to derive as many of those properties as we can using purely thermodynamic reasoning.

Universality of cavity radiation

Consider a large cavity of arbitrary shape whose walls are of arbitrary surface quality, except that they are opaque, in a body in thermal equilibrium at temperature T. Inside the cavity there is thermal radiation. We wish first to establish that this radiation has a universal character, such that its intensive physical properties do not depend on any feature of the cavity except its temperature. For example, there is no dependence on the size or shape or reflectivity of the walls (as long as they are opaque and large enough). The following argument is due to Kirchoff.

Suppose we have two cavities, of arbitrary shape, connected by a small evacuated pipe, and both at the same temperature T; see Figure 19.4. In this situation, there must be no net flux of energy down the pipe, because if there were then to reduce the flux to zero would require a non-zero temperature difference ΔT between the cavities; and if the temperature difference were greater than zero but below ΔT then energy would flow spontaneously from a colder to a hotter body,

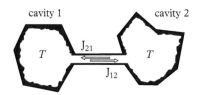

Fig. 19.4 Two cavities at the same temperature, with opaque walls, but with not necessarily anything else in common, such as volume, shape, or the surface quality of the walls. The energy fluxes in the two directions must agree (see text).

which is forbidden by the Clausius statement of the second law. Now, as long as the pipe allows the radiation to pass freely, then the flux in one direction is owing to the radiation in cavity 1, and the flux in the other direction is owing to the radiation in cavity 2. Furthermore, the flux in either case is proportional to the energy density, with the same proportionality constant $c/4$, because we shall show in a moment that cavity radiation is isotropic. It follows that the energy densities of the radiation in the two cavities must be the same: $u_1 = u_2$. Furthermore, we can apply the argument at each wavelength, because it is physically possible to have a filter which only transmits electromagnetic radiation at one wavelength, and such filters could be placed at the two ends of the pipe. The same goes for polarization and direction of propagation.[1] Therefore,

$$\rho_{\omega,1}(\omega) = \rho_{\omega,2}(\omega), \qquad (19.16)$$

where we have only explicitly indicated the frequency because it will emerge that there is no dependence on polarization or direction of travel.

Furthermore, the energy density is spatially uniform, as can be proved by applying the argument with a fixed cavity 1 while allowing the pipe to connect to different parts of cavity 2. Also, the flux is isotropic and unpolarized, as can be proved by pivoting cavity 2 about the end point of the pipe connected to it.

The interesting feature of this argument is that the conclusion does not depend on the shape or quality of the cavities, only that the walls are opaque and the cavity is large. It follows that all large cavities of given temperature have thermal radiation of the same spectral energy density as a function of frequency. In short, *the properties of cavity radiation are universal for large cavities.* The argument is restricted to large cavities because we need to assume that the diameter of the pipe is small enough not to disturb the cavity radiation, but large enough to allow radiation to pass down it freely. This requires that the physical dimensions L of the cavity are large compared to the typical wavelength in the radiation, i.e. $L \gg hc/k_\mathrm{B} T$. In fact, for small cavities at low temperature the distribution is different and depends on details of the cavity shape. For example, the spectral and angular distribution of the thermal radiation inside a centimetre-sized cavity at 1 Kelvin departs significantly from the cavity radiation in a large cavity at the same temperature.

Kirchoff's argument also applies to the flux of other properties such as entropy, and hence shows that the entropy per unit volume $s = S/V$ in cavity radiation is also a function of temperature alone. This fact can alternatively be derived from the energy arguments and thermodynamic reasoning, as we shall show in the argument leading to equation (19.31). The entropy of a radiation field is physically expressed by the fact that the energy is dispersed over a large number of modes of oscillation of the field, oscillating with random relative phases.

---

[1] To select radiation propagating at angle θ to some axis, a pair of lenses around a small pinhole may be used.

Derivation of Kirchoff's law

Consider now a cavity at temperature T with, placed inside it, a grey body of absorptivity α at the same temperature. Let the body be of convex shape. This means we can construct a surface Σ enclosing the body and just outside it, such that all straight lines outwards from the body only cross this surface once. Let the total power from the cavity radiation crossing this surface in the inward direction be P^C, then the total power absorbed by the grey body is

$$P_{\text{in}} = \alpha P^C. \tag{19.17}$$

All thermal radiation emitted by the grey body crosses Σ and is not reabsorbed by it, therefore the total power emitted is

$$P_{\text{out}} = \epsilon P^{BB} \tag{19.18}$$

since this is the definition of emissivity. In thermal equilibrium these powers must agree, so we have

$$\alpha P^C = \epsilon P^{BB}. \tag{19.19}$$

For a black body $\alpha = 1$ and $\epsilon = 1$ (by definition), therefore we obtain $P^{BB} = P^C$. That is, the total power emitted by a black body in thermal equilibrium is equal to the power in the cavity radiation incident on it. Substituting this back into equation (19.19), we have

$$\alpha = \epsilon,$$

which is Kirchoff's law for grey bodies.

If we now consider the case that the surface Σ is a filter that only allows radiation of one wavelength or direction to pass, then the same argument suffices to establish both the general form of Kirchoff's law, equation (19.7), and also

$$P^{BB}_\omega(\omega) = P^C_\omega(\omega),$$

that is, black body radiation (= the thermal radiation emitted by a black body) has the same spectral and angular distribution as cavity radiation. Therefore it is universal too.

A consequence of Kirchoff's law and the universality of black body radiation is that objects inside a furnace are invisible! If a furnace is sufficiently hot that the cavity radiation includes a substantial visible part, then one might expect that by looking through a small window one would be able to see the various objects inside the furnace—pots, coals, metal ingots, or whatever—by the light of this thermal radiation, just as we ordinarily see things in daytime by the light of the Sun. However, if the objects are themselves at the same temperature as the rest of the furnace, then they will all glow in precisely the same way. An opaque ceramic pot, for example, will absorb, reflect, and emit at any given wavelength in the amounts

$$\alpha(\omega)P^{BB}_\omega(\omega), \qquad (1-\alpha(\omega))P^{BB}_\omega(\omega), \qquad \epsilon(\omega)P^{BB}_\omega(\omega).$$

But from Kirchoff's law, $\epsilon(\omega) = \alpha(\omega)$, so the emitted part exactly makes up for the absorbed part, with the result that the total spectral power propagating away from the pot (reflection plus emission) is $P^{BB}_\omega(\omega)$, which is independent of properties of the pot, and therefore the same no matter what pot or other object is considered. All one can expect to see is a uniform glow at the characteristic colour of the thermal radiation (e.g. orange or yellow or white, depending on the temperature). You can see this effect if you look into the white-hot part of a coal fire in a room with the room lights turned off: the individual coals merge into a uniform whiteness.

Another interesting consequence is the following. The colour of ordinary objects typically arises by virtue of the fact that they absorb some wavelengths more than others. For example, a red pot appears red to us because it preferentially absorbs blue light and reflects (or scatters) red light. This implies, from Kirchoff's law, that its emissivity is larger in the blue part of the spectrum. It follows that if such a pot were placed in a cold, dark room, and heated to a 'white hot' temperature, then it would appear bluish rather than red (assuming $\alpha(\omega)$ did not depend strongly on temperature). Similarly, an inert pot which appears blue under normal illumination would glow red when heated to a suitable temperature, in the absence of incident radiation.

Transparent bodies

Similar arguments can be used to treat a surface which is not opaque. Suppose a layer has $R = 0$ and $\tau \neq 0$. If we place this layer, at temperature T, in front of a black surface at the same temperature, then the two together form a black surface since all incident radiation is absorbed either by the layer or by the black surface. Therefore the composite emits black body radiation P^{BB}_ω. But of the radiation emitted by the black surface in this situation, only a proportion τ emerges. It follows that the transparent layer must emit precisely $(1-\tau)P^{BB}_\omega$, i.e. black body radiation but less of it. Now take away the black surface and argue that the properties of the layer are intrinsic to itself; it follows that a transparent object is a poor emitter of thermal radiation.

For example, fused silica at high (or low) temperature is transparent over a wide range of infrared and optical frequencies. Its thermal emission is therefore small. In consequence, when heated in a gas flame it has little means of losing energy and can be heated by a blow-torch to much higher temperatures than is possible for most other materials using the same method. Glass-blowers make much use of this fact. A slight contamination of carbon on the surface of the silica is sufficient to greatly lower the temperature which can be reached.

We have already claimed that the Sun is to reasonable approximation a black body with surface temperature 5800 K. This accounts for the main facts about the quantity and distribution of the energy emitted. However, when the Sun is examined in the radio frequency part of the spectrum, corresponding to wavelengths around 5 m, it appears to have a diameter considerably larger than its visual diameter, and its emission corresponds to a black body spectrum with a very much higher temperature (several million Kelvin). These observations are accounted for by the solar corona, which is a very low density plasma 'atmosphere' surrounding the Sun, at very high temperature. Such a plasma has an absorptivity varying as the square of the wavelength, therefore it can be almost completely transparent to visible light while being opaque to radio waves. Correspondingly, it emits as a black body in the radio region but hardly at all in the visible region (its weak visible emission can be seen during a solar eclipse). The reason for the extraordinarily high temperature has long baffled physicists. Various mechanisms are now proposed that could pump energy into the corona, but the complete picture is not yet clear. The near transparency over a wide range of frequencies helps: like pure fused silica, the corona doesn't readily cool itself by radiant emission.

19.2.1 Equation of state and Stefan–Boltzmann law

We have established from the second law that the energy density u of cavity radiation is independent of the size of the cavity (in the large cavity or high temperature limit). It follows that u can be expressed as a function of a single parameter such as temperature:

$$u = u(T). \tag{19.20}$$

This can be compared and contrasted with the case of an ideal gas which obeys Joule's law. For such a gas, the energy per particle $U_1(T)$ depends on temperature alone, giving $u = (N/V)U_1(T)$. Thus for fixed N, the energy density in an ideal gas does depend on volume as well as temperature, but for fixed number density N/V it does not.

We would now like to derive the T^4 dependence that previously we exhibited in equation (19.14).

Cavity radiation exerts a pressure on the walls of the cavity. This can be calculated either from classical electromagnetism or from a photon model of the radiation. From classical electromagnetism we find that the force per unit area exerted by a plane wave[2] on an absorbing surface is $p = u$, and for isotropic radiation a standard argument involving integration over angles yields

$$p = \frac{1}{3}u. \tag{19.21}$$

The photon model gives the same result, as may be seen by applying kinetic theory: each photon of energy E carries a momentum E/c, and the directions of

[2] For a collimated beam, in time δt a length $c\delta t$ of beam passes any given plane; in this length is carried energy $uAc\delta t$, where A is the cross-sectional area, and hence momentum $uA\delta t$; the rate of bringing up of momentum is therefore uA; if this momentum is absorbed then the force per unit area is u.

travel are isotropic. This results in the total pressure $(1/3)n\langle E\rangle = u/3$, where n is the number density of photons.

This simple and direct relationship between pressure and energy density is an important feature of cavity radiation; it is the same as that for an ideal gas (Eq. 7.28) with adiabatic index $\gamma = 4/3$ (and see Exercise 19.5).

The radiation is a simple compressible system. Equations (19.20) and (19.21) may be combined to yield the energy equation and the equation of state, and hence the complete thermodynamic description. In order to present the argument, first we need to consider the question, whether we ought to regard the radiation as an open or a closed system. During an isothermal expansion, the energy density remains fixed and therefore the total energy in the radiation grows. In this process, nothing passes to or from the cavity walls and thence to the rest of the universe except energy, therefore it is valid to regard the process as one in which the radiation behaves as a closed system, to which the fundamental relation

$$dU = TdS - pdV \qquad (19.22)$$

applies. On the other hand, in the photon model an isothermal increase in U must mean that the number of photons has grown, so if the photons are to be compared with particles in a gas then N is changing so the system is not closed! According to this argument we must use not (19.22), but

$$dU = TdS - pdV + \mu dN. \qquad (19.23)$$

The contradiction is avoided as long as $\mu = 0$ in the photon model, which is indeed what we shall find. Therefore let us adopt (19.22) at the outset, and confirm $\mu = 0$ at the end.

By applying (5.15), we obtain

$$\left.\frac{\partial U}{\partial V}\right|_T = \frac{U}{V} \equiv u.$$

Using this in (19.22), we have

$$u = T\left.\frac{\partial S}{\partial V}\right|_T - p = T\left.\frac{\partial p}{\partial T}\right|_V - p \qquad (19.24)$$

$$\Rightarrow \qquad 4p = T\frac{dp}{dT}, \qquad (19.25)$$

using a Maxwell relation and then (19.21). Hence we have a first-order differential equation for $p(T)$ whose solution is

$$p = aT^4, \qquad \text{where } a \text{ is a constant.} \qquad (19.26)$$

This is the equation of state which we previewed in (6.14). Since the pressure is a function of temperature alone, the system has an infinite isothermal compressibility and an infinite heat capacity at constant pressure.

Using (19.21) we can immediately deduce

$$u = 3aT^4. \tag{19.27}$$

The total energy in a volume V of thermal radiation is therefore $U = uV = 3aVT^4$, and hence we find that the heat capacity at constant volume is

$$C_V = T \left.\frac{\partial S}{\partial T}\right|_V = 12aT^3 V. \tag{19.28}$$

Note that this increases without limit as T increases.

Now consider the entropy has a function of temperature and volume. Using a Maxwell relation and equation (19.26), we have

$$\left.\frac{\partial S}{\partial V}\right|_T = \left.\frac{\partial p}{\partial T}\right|_V = 4aT^3. \tag{19.29}$$

Since S and V are extensive, and T is intensive, we can use equation (5.15), which tells us that $(\partial S/\partial V)_T = S/V$. Hence we have

$$S = 4aVT^3. \tag{19.30}$$

This is usually written in terms of the entropy per unit volume:

$$s = 4aT^3. \tag{19.31}$$

Putting this into the first equation in (19.24), we find

$$u = Ts - p. \tag{19.32}$$

It follows that the Gibbs function of a volume V of cavity radiation is

$$G = U - TS + pV = (u - Ts + p)V = 0.$$

It follows that if the radiation is to be modelled as a gas of particles, then the chemical potential associated with movement of the particles in and out of the system is zero, just as we claimed after equation (19.23).

The Stefan–Boltzmann law is essentially the result of equation (19.27): the energy flux onto any part of the cavity wall is $(1/4)uc$, so we have that the power incident on a small area A is $(1/4)Auc = (3/4)AacT^4$. Since this must equal the power emitted, we deduce that the constant a appearing in equation (19.27) is related to the Stefan–Boltzmann constant by

$$a = \frac{4\sigma}{3c}. \tag{19.33}$$

19.2.2 Comparison with ideal gas

It is instructive to compare the thermodynamic properties of cavity radiation with those of an ideal gas. The main relevant equations are listed in Table 19.2.

The comparison is facilitated by introducing a photon description of the cavity radiation, but this does not mean one should lose sight of the classical electromagnetic wave description altogether—after all, it has led us to the correct equation of state, and energy and entropy equations. Ultimately the radiation should be understood in terms of quantum field theory, which combines wave and particle concepts, but this is true of all electromagnetic phenomena, it is not a special property of cavity radiation. However, there are important aspects of cavity radiation which cannot be correctly predicted by applying thermodynamic reasoning to classical electromagnetism. An example is the distribution of energy in the radiation as a function of frequency; this will be discussed in Section 19.2.3.

The essential idea of quantum field theory is that each mode (such as a plane travelling wave) of the radiation is quantized, such that the energy in the wave is given by

$$(m + 1/2)\,\hbar\omega,$$

where m is an integer and ω is the angular frequency of the wave. Each unit of energy $\hbar\omega$ is called a quantum of energy, and when the wave has m units more than its minimum possible energy, we say there are m 'photons' present. This leads to a physical picture of a classical electromagnetic wave as a stream of photons, and cavity radiation as a gas of photons.

The number of photons in the gas is

$$N = V\int_0^\infty \frac{\rho_\omega}{\hbar\omega}d\omega = \frac{\Gamma(3)\zeta(3)}{\pi^2}\left(\frac{k_B}{\hbar c}\right)^3 VT^3 \simeq 0.24359\left(\frac{k_B}{\hbar c}\right)^3 VT^3, \quad (19.34)$$

Table 19.2 *Comparison between ideal gas and cavity radiation (photon gas). Lower case symbols such as n, u are per unit volume, (i.e. $n = N/V$, $u = U/V$). $b \simeq 2.03 \times 10^7$ m^{-3}K^{-3}. a is the constant in the Sackur–Tetrode equation.*

	Ideal gas	Cavity radiation
Independent variables	U, N, V	U, V
Number of particles	N	$N = bVT^3$
Equation of state	$p = nk_B T$	$p = 0.9\,nk_B T$
Energy	$U = \frac{1}{\gamma-1}Nk_B T,$	$U = 2.7Nk_B T,$
u and p	$p = (\gamma-1)u$	$p = \frac{1}{3}u$
Entropy	$S = Nk_B\left(\ln\left[a\frac{T^{1/(\gamma-1)}}{n}\right]\right)$	$S = 3.6Nk_B$
	$= \frac{U}{T}(\ln T - (\gamma-1)\ln(n/a))$	$= \frac{4}{3}\frac{U}{T}$
Chemical potential	$\mu = (U - TS + pV)/N$	$\mu = 0$
	$= (\gamma u - Ts)/n$	

where we have used the expression (19.12) for ρ_ω and the integral is related to the Riemann zeta function; see Appendix C. For example, in a cavity at 1 kelvin there are 20 photons per cubic centimetre in the thermal radiation. By introducing N and the number density $n = N/V$ we can express the equation of state in the form

$$p = \frac{1}{3}\frac{\Gamma(4)\zeta(4)}{\Gamma(3)\zeta(3)} nk_B T \simeq 0.900393\, nk_B T,$$

which is highly reminiscent of an ideal gas. The energy per photon in the gas is $2.701\, k_B T$. The entropy per photon is $3.601\, k_B$.

The reason why it is possible to ignore thermal radiation when studying the behaviour of objects under ordinary conditions is that the number of photons is small compared to Avogadro's number. For example, a 25 litre bottle at room temperature contains about 10^{13} thermal photons, which is 2×10^{-11} moles of photons. Consequently the contribution of the cavity radiation in the bottle to energy and entropy changes is negligible compared to that of a mole of atoms in the bottle. However, owing to the T^3 factor in (19.34), at very high temperatures the thermal radiation can dominate everything else. For example, the early evolution of the universe can be treated to good approximation by assuming that it was dominated by radiation not matter.

19.2.3 Adiabatic expansion and Wien's laws

An important property of electromagnetic waves in vacuum, and equally of photons, is that they do not interact with one another (except at very high intensities where there is an indirect interaction via electron–positron creation and annihilation). This means that cavity radiation cannot attain thermal equilibrium by internal collisional processes like an ordinary gas, it must be via interaction with the walls of the chamber containing it. For example, if a pulse of laser light is sent into a cavity by briefly opening a window, then the radiation in the cavity is initially not in a thermal equilibrium state; it subsequently relaxes to equilibrium by a process in which the material in the walls redistributes the energy, absorbing more photons of one frequency and emitting more of another. In the limit where the reflectivity of the walls tends to one (i.e. perfect reflection), the relaxation time tends to infinity. This raises the question, whether cavity radiation ceases to be in a thermal equilibrium state if it is in a perfectly reflecting chamber and the chamber is manipulated by slowly compressing or expanding it. In other words, does adiabatic compression of cavity radiation result in a new distribution of energy, different from a black body distribution (and consequently not a thermal equilibrium state), or does the repeated reflection from moving walls cause the distribution to adjust in just the right way to maintain the black body form, but at a new temperature?

This question is readily answered in a quantum analysis of the radiation: the adiabatic process is such that the degree of excitation of each mode (i.e. the number of photons of a given type) is unchanged, while the mode frequency

Fig. 19.5 *Steps of Wien's argument concerning adiabatic compression of cavity radiation.*

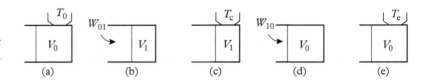

changes, and all mode frequencies change by the same factor if the compression is isotropic. It follows that the distribution of energy maintains a thermal (maximum entropy) form. This proof involves the methods of statistical mechanics which invoke a statistical definition of entropy. We should like to know whether the conclusion can be obtained by more general macroscopic arguments, and thus serve as a constraint on the allowable microscopic theories.

Wilhelm Wien (1864–1928) gave the following macroscopic thermodynamic argument to show that cavity radiation does remain in a thermal state when subjected to adiabatic change. The argument uses the fact that properties such as volume, energy, and entropy are well defined for non-equilibrium as well as equilibrium states. We also make use of a special feature that applies to radiation but not all physical systems, namely that the pressure is uniquely determined by the energy density, even if the system is in a non-thermal state, provided only that energy flux in the radiation remains uniform and isotropic. Another system having this property, and therefore to which the argument applies, is the monatomic (but not the diatomic) ideal gas.[3]

We consider the radiation in a cavity with perfectly reflecting walls, subject to the following sequence of steps (see Figure 19.5).

(a) First make part of the wall non-reflecting and bring the radiation into thermal contact with a reservoir, so that it attains a thermal equilibrium state at temperature T_0, volume V_0.

(b) Reinstate perfect reflectivity of the walls, and compress the chamber reversibly from volume V_0 to V_1. The entropy change of the radiation is $\Delta S = 0$ and its internal energy increases by the work done, $\Delta U = W_{01}$.

(c) Bring the radiation into thermal contact with another body, whose temperature T_c has been chosen such that in the interaction that now takes place, the net flow of heat between the radiation and the body is zero. Then the body does not change state and the radiation has $\Delta U = 0$. If the radiation were not in a thermal state after step (b), then this interaction would result in an irreversible relaxation to equilibrium, so the entropy of the radiation would grow by $\Delta S_c \geq 0$.

(d) Reinstate perfect reflectivity of the walls, and expand the chamber reversibly back to V_0. The radiation undergoes $\Delta S = 0$, $\Delta U = W_{10}$.

(e) Repeat step (c), the other body being at a temperature (typically different from T_c) such as to ensure $\Delta U = 0$ again. If the radiation were not in a thermal equilibrium state after step (d), then its entropy would grow by $\Delta S_e \geq 0$.

---

[3] In the diatomic gas, the pressure is uniquely determined by the translation kinetic energy density, but not by the total energy density.

At the end of this process the radiation is in a thermal equilibrium state with volume and energy,

$$V = V_0, \qquad U = U_0 + W_{01} + W_{10}.$$

Now suppose the cavity walls are diffuse reflectors, that is, the absorption remains zero but the walls are rough, such that the reflected radiation remains uniform and isotropic as the walls move. In this case the pressure at all stages during the compression or expansion is uniquely determined by the energy density. Therefore the pressure at the end of the compression is the same as the pressure at the start of the expansion, and furthermore a step-by-step breakdown of the expansion process (d) must find that it is the reverse of the compression process (b), having the same pressure at each volume, and consequently $W_{10} = -W_{01}$. It follows that $U = U_0$ and therefore the radiation suffers no overall change in its thermodynamic state (since V and U are sufficient to uniquely specify the state). Therefore the net change in its entropy is zero: $\Delta S_c + \Delta S_e = 0$. But both quantities in this sum are non-negative, so we must have $\Delta S_c = \Delta S_e = 0$. Therefore there was no irreversible relaxation in steps (c) and (e), so the radiation must have been in a thermal state already. It follows that its state remains thermal (with slowly changing temperature and volume) throughout an adiabatic compression or expansion, provided the radiation remains isotropic. QED.

19.2.3.1 Wien's laws

We can use the fact that the different waves in the radiation do not interact to obtain some information about the spectral energy density function $\rho_\omega(\omega)$.

We study first the behaviour of a single ray at wavelength λ, during an adiabatic compression. Consider for simplicity a smooth spherical cavity, so that the ray bounces around the wall of the cavity, always meeting it at the same angle θ to the normal, and undergoing a reflection every time interval $\Delta t = 2r\cos\theta/c$ (Figure 19.6). In the case of diffuse reflection the same conclusions will apply on average. If the walls move slowly then the Doppler effect at each reflection gives a change in frequency of the wave by

$$\Delta\omega = \omega \frac{2v\cos\theta}{c} = -\frac{2\omega\cos\theta}{c}\frac{dr}{dt}.$$

Therefore,

$$\frac{\Delta\omega}{\Delta t} = -\frac{\omega}{r}\frac{dr}{dt}.$$

For a slow compression the effect on ω of a step change $\Delta\omega$ each Δt is equivalent to a continuous change at the rate $d\omega/dt = \Delta\omega/\Delta t$, so we have

$$\frac{d\omega}{dt} = -\frac{\omega}{r}\frac{dr}{dt},$$

which has the solution

$$\omega r = \text{const.} \qquad (19.35)$$

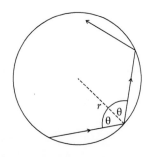

Fig. 19.6 *A light ray reflecting around the interior of a sphere.*

Another way to express this is to say that the number of wavelengths across the diameter of the cavity remains fixed during the compression, just like the behaviour of a standing wave with nodes at the walls.

Using (19.30) we find that during an adiabatic change,

$$T^3 V = \text{const.} \quad \Rightarrow \quad T \propto 1/r,$$

i.e. the temperature falls by the linear expansion factor. Combining this with (19.35), we find

$$\frac{\omega}{T} = \text{const.} \tag{19.36}$$

In this equation, ω refers to the frequency of the particular ray we picked, and T refers to the temperature describing the whole radiation field, which we have already shown remains black.

Now consider an adiabatic compression causing a temperature change (not necessarily small) of the whole system from T_1 to T_2. We examine the effect on a set of spectral components whose centre frequency and width change from $\omega_1, d\omega_1$ to $\omega_2, d\omega_2$ as a result of the compression. By applying the previous result,

$$\frac{\omega_1}{T_1} = \frac{\omega_2}{T_2}, \quad \frac{\omega_1 + d\omega_1}{T_1} = \frac{\omega_2 + d\omega_2}{T_2} \quad \Rightarrow \quad \frac{d\omega_1}{T_1} = \frac{d\omega_2}{T_2}. \tag{19.37}$$

The essential point in the next and final part of Wien's argument is to see that the previous result $u \propto T^4$ for the total energy density can be applied equally to the energy density in any small frequency range, since during an adiabatic expansion each set of components acts independently (and we have already shown the overall state remains thermal). Therefore,

$$\frac{\rho_\omega(\omega_1, T_1) d\omega_1}{T_1^4} = \frac{\rho_\omega(\omega_2, T_2) d\omega_2}{T_2^4}.$$

By combining this with (19.37), one obtains

$$\rho_\omega(\omega_1, T_1) = \frac{T_1^3}{T_2^3} \rho_\omega\left(\frac{T_2}{T_1}\omega_1, T_2\right).$$

Now set T_2 equal to some fixed temperature, which we may as well choose to be 1 in our chosen units, and let $T = T_1$, $\omega = \omega_1$, then we have

---

Wien's distribution law

$$\rho_\omega(\omega, T) = T^3 \rho_\omega\left(\frac{\omega}{T}, 1\right). \tag{19.38}$$

In other words, the spectral distribution of energy density is equal to T^3 multiplied by a function of ω/T alone. This is called *Wien's distribution law*.

It follows that the effect of a change of temperature is to stretch the distribution as a function of ω without changing its form as a function of ω/T. In particular, the position of the maximum must change as

---

Wien's displacement law

$$\frac{\omega_{\max}}{T} = \text{const.} \qquad (19.39)$$

This is called *Wien's displacement law*.

Note that whereas an argument from adiabatic expansion was used to obtain Wien's laws, the result is a property of the distribution function itself, so it applies to any equilibrium state, no matter how the system was brought into the state. It is truly remarkable that we can find out so much about the distribution of energy in cavity radiation from general thermodynamic arguments. Wien's laws are fully in agreement with observations to date.

The measured value of the constant in (19.39) is found to agree accurately with the prediction from equation (19.12), which is

$$\frac{\omega_{\max}}{T} = 2.82144 \frac{k_B}{\hbar} = 3.694 \times 10^{11} \text{ rad s}^{-1}\text{K}^{-1}. \qquad (19.40)$$

So far we have studied the energy density per unit frequency range. One can also take an interest in the energy density per unit wavelength range, given by

$$\rho_\lambda = \rho_\omega \left|\frac{d\omega}{d\lambda}\right| = \frac{2\pi c \rho_\omega}{\lambda^2}.$$

The λ^2 factor skews the distribution a little compared to ρ_ω, with the result that the value of λT at the peak depends on which distribution one picks:

$$\lambda T = \begin{cases} 2.897 \text{ mm K} & \text{at the maximum of } \rho_\lambda \\ 5.099 \text{ mm K} & \text{at the maximum of } \rho_\omega. \end{cases}$$

For this reason a distribution which peaks in the visible by one measure may peak in the infrared by the other measure.

19.2.3.2 The ultraviolet catastrophe

The predictions for cavity radiation were among the early triumphs of thermodynamics, because they showed that the ideas could be applied to a physical system very different from the fluids and steam engines for which the theory was first developed. The attempt to understand the distribution function also revealed a profound limitation in classical physics, and greatly stimulated the development of quantum theory. According to classical physics the energy in a wave motion such as an electromagnetic wave can take on a continuum of values, and consequently when it exchanges energy with a source in conditions of thermal equilibrium, the amount of energy in the wave is expected to scale linearly with temperature (the proof of this can be found in introductory texts on statistical mechanics). Therefore, in Wien's distribution law (19.38) we must have

$$T^3 \rho_\omega \left(\frac{\omega}{T}, 1\right) = Tf(\omega),$$

where $f(\omega)$ is an unknown function. The solution is $\rho_\omega(\omega/T, 1) \propto (\omega/T)^2$ and therefore

$$\rho_\omega(\omega, T) \propto \omega^2 T$$

using (19.38). The total energy in the cavity radiation at a given temperature is then the integral of this with respect to ω, which is infinite. This unphysical conclusion was called the *ultraviolet catastrophe*, because it is the high-frequency part of the proposed distribution function which is erroneous and which leads to the divergence of the integral. The word 'catastrophe' suggests just how severe the problem was: it was a conclusion to which physicists were forced if they accepted the extremely impressive, accurate, and wide-ranging theory of electromagnetic interactions provided by Maxwell's equations and the accompanying Newtonian concepts. Planck famously took the first step towards the correct model by proposing that sources in the walls only exchanged energy with the radiation in quanta of finite size $\hbar\omega$. Einstein subsequently proposed that the radiation was itself quantized, and ultimately quantum field theory had to be developed to understand this idea in full. That was the work of many people, with de Broglie, Schrödinger, Heisenberg, and Dirac all making significant contributions.

The ultraviolet catastrophe is sometimes presented as a failure or limitation of thermodynamic reasoning, but in fact the thermodynamic reasoning was correct. It was the premise of classical microphysics which was erroneous. Indeed, we can never be confident of having discovered the exactly correct microtheory, whereas the thermodynamic reasoning is correct if its few assumptions are correct. One of the assumptions leading to Wien's distribution law that is questionable in extreme conditions is that the radiation in each frequency range acts independently.

19.3 Cosmic microwave background radiation

A highly significant example of cavity radiation is the cosmic microwave background (CMB) radiation. This is one of the key pieces of evidence for the hot big bang model of the early history of the universe. The radiation is present in every direction in the sky that is not directly obscured by a star or other object, and it is found to be highly uniform. It is isotropic in a reference frame relative to which the solar system moves at 370 km s^{-1}. Its spectral distribution in that frame is thermal to very high precision, with a temperature 2.725 K.

The hot big bang model explains these observations as follows.

In its early history the universe was a hot plasma, whose temperature fell as the universe expanded. After some time, the matter content of the universe consisted of a plasma of mainly protons and electrons, with also some nuclei of helium and a few other light elements. These charged particles readily scattered photons and the radiation field was very close to a thermal equilibrium state. It is not yet fully understood how equilibrium was originally attained over the distance scales involved; this is called the *horizon problem*, but given that equilibrium was attained, the consequences can be calculated.

The plasma was everywhere expanding and cooling. When the temperature fell below about 3000 K the plasma underwent a fairly abrupt transition to a neutral gas of (mostly) hydrogen atoms. This temperature is known because the process can be modelled to first approximation by applying the Saha equation (and more accurately by a more complete model). To apply the equation one needs to know either the electron density or the density of protons and hydrogen atoms. The latter can be estimated from observation of the current universe, modified by a factor to account for expansion.

After this transition, owing to the near absence of unbound charged particles, the universe became transparent, so the transition is called 'decoupling' or the 'time of last scattering'. The thermal radiation thereafter propagated freely, that is, subject only to gravitational effects. An accurate description of the latter requires general relativity, however the main conclusion can be stated simply. The effect of cosmological expansion is essentially to change all the observed frequencies in the radiation by the same scale factor, just like an adiabatic expansion. Therefore the radiation has remained thermal and still has a black body spectrum. This has been confirmed to very high accuracy by observations in the present. The temperature is found to be 2.725 K. It follows that the scale factor of the cosmological expansion has increased by a factor roughly 1100 since the time of last scattering.

The CMB photons currently arriving on Earth were emitted some 14 billion years ago by electrons located on the surface of a vast expanding sphere centred on Earth (and a similar statement would apply at any other location in the universe). Such photons, which make even dinosaur bones seem rather young,[4] are in our vicinity all the time.

....

EXERCISES

(19.1) (a) Give the SI units for the following quantities: energy density u (that is, energy per unit volume), and spectral energy density ρ (that is, energy per unit volume per unit frequency range).

(b) Write down the expression relating u and ρ.

(c) If ρ is the energy density per unit frequency range, and η is the energy density per unit wavelength range, write down the expression relating ρ and η for electromagnetic radiation.

[4] Harvey Leff, *Am. J. Phys* **70**, 792 (2002).

(d) Derive the relationship between energy density u and energy flux ϕ for a collimated beam of light. (Energy flux is also called intensity, it is power per unit area.)

(19.2) Bearing in mind that glass is a good absorber of infrared radiation, sketch on the same, labelled graph:

(a) the spectral emissive power of a bathroom mirror at room temperature, as a function of wavelength, in the wavelength range 400 nm to 100 μm.

(b) the spectral energy density of radiation incident on a detector placed close to the mirror, when the mirror (still at room temperature) is used to reflect sunlight onto the detector.

(19.3) A confused student puts forward the following argument. 'Red paint appears red (under white light illumination), and this is to do with the fact that it absorbs other colours such as green and blue. Therefore, according to Kirchoff's law, it should be a good emitter of green and blue, and a relatively poor emitter of red. Therefore it should appear blue-green, and not red after all'. This is a contradiction. Where did the student go wrong?

(19.4) Find the temperature at which the number of photons in a chamber of volume 25 litres is equal to one mole.

(19.5) Show that if the pressure of a simple compressible system is related to its energy density by $p = \Gamma u$, where Γ is a constant, then the adiabatic index is $\gamma = \Gamma + 1$.

(19.6) Show that the Helmholtz function of the cavity radiation in a cavity of volume V is $F(T, V) = -aVT^4$, where a is a constant. Hence obtain the equation of state and the heat capacities.

(19.7) Show that the entropy of such radiation, as a function of its natural variables, is

$$S = \frac{4}{3}\left(\frac{4\sigma}{c}\right)^{1/4} V^{1/4} U^{3/4}. \tag{19.41}$$

(19.8) Show that a simple compressible system whose energy density and pressure each depend on temperature alone, i.e. $u = u(T)$, $p = p(T)$, and such that $du/dT \to 0$ and $dp/dT \to 0$ as $T \to 0$, must have entropy $S = V dp/dT$ and chemical potential $\mu = 0$. [Hint: generalize the argument of equations (19.20)–(19.32).]

(19.9) Show that a simple compressible system whose chemical potential is always zero and whose energy density depends on temperature alone must have an entropy density that depends on temperature alone.

(19.10) Describe one kind of electromagnetic radiation which is *not* black body radiation, even approximately, and identify some features which distinguish it from black body radiation.

Radiative heat transfer

20.1 The greenhouse effect	294
Exercises	297

The discussion in Section 19.1 has already introduced some simple examples of radiative heat transfer. The basic idea is that when two bodies at temperatures T_1, T_2 are able to receive each other's thermal radiation, then energy is transferred from the hotter to the colder (we treat the case where no work is done by radiation pressure or other means). The rate of transfer depends on the temperatures, the emissivities, reflectivities, and the geometry. For simplicity in the first instance we shall treat opaque grey bodies, i.e. emissivity independent of wavelength. The generalization to arbitrary emissivity functions involves introducing integrals in a fairly obvious manner. We shall use both absorptivity and emissivity in the equations, in order to clarify which processes are absorptive, which emissive, and then simplify by applying Kirchhoff's law.

Consider first the case of two infinite opaque parallel plates; Figure 20.1. Let I_+, and I_- be the power per unit area in the radiation between the plates proceeding in the positive and negative x directions, respectively. Note, we don't assume the radiation is collimated: it propagates in all directions, but I_+ and I_- are well defined because we can consider a mathematical plane at any position and then I_+ accounts for all radiation crossing this plane from left to right, and I_- from right to left.

Since the parallel plates are opaque, all incident power that is not absorbed must be reflected, so we may write

$$I_+ = \epsilon_1 \sigma T_1^4 + (1-\alpha_1)I_-$$
$$I_- = \epsilon_2 \sigma T_2^4 + (1-\alpha_2)I_+. \tag{20.1}$$

These equations are easily solved for I_+ and I_-. After applying Kirchhoff's law one then finds that the net transport of energy in the positive direction is

$$I_+ - I_- = \sigma(T_1^4 - T_2^4)\frac{\alpha_1 \alpha_2}{\alpha_1 + \alpha_2 - \alpha_1 \alpha_2}. \tag{20.2}$$

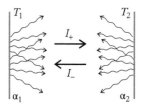

Fig. 20.1 *Radiative heat transfer between two opaque grey plates.*

Example 20.1

A thermos flask has a pair of silvered walls at temperatures $T_1 = 90°C$ and $T_2 = 20°C$, separated by vacuum. Find the rate of heat transport per unit area. What thickness of cork insulation (thermal conductivity $0.04\ \mathrm{Wm^{-1}K^{-1}}$) would be needed to achieve the same result?

> *Solution.*
> From Table 19.1 we have $\epsilon = 0.01 - 0.04$ for silver; we will take $\epsilon = 0.02$. Using Kirchhoff's law, this is also α. Using this in (20.2), we find
>
> $$I_+ - L = \sigma \left((363\,\text{K})^4 - (293\,\text{K})^4\right) \frac{0.02}{2 - 0.02} = 5.7\,\text{W}\,\text{m}^{-2}.$$
>
> In instead one used cork insulation, the thickness required is $L = \kappa \Delta T / \mathcal{J} = (0.04 \times 70/5.7)\,\text{m} = 0.5\,\text{m}$. Clearly, the thermos is a good insulator!

Next consider the case of a body located inside a cavity. In order to consider the energy flux, construct a pair of closed surfaces, Σ_1 and Σ_2, whose shapes are arbitrary except that both are convex and Σ_2 totally encloses Σ_1 (see Figure 20.2). Then any radiation which passes out through Σ_1, travelling in any outward direction, must subsequently pass out through Σ_2. The converse is not true: some of the radiation crossing Σ_2 in an inward direction subsequently emerges back out of Σ_2 without crossing Σ_1. We should like to know what proportion.

Suppose Σ_1 is the surface of a black body that is in equilibrium in the cavity, and Σ_2 is the inside surface of the cavity. Let A_1, A_2 be the areas of these surfaces. Then the total power emitted by the black body is $\sigma A_1 T^4$ and all of this power subsequently arrives at Σ_2 and is absorbed there. The total power emitted from the cavity wall is $\sigma A_2 T^4$, but not all of this arrives at the black body. Since we have thermal equilibrium there is a steady state, so the amount of power arriving at the black body and absorbed by it must be $\sigma A_1 T^4$. Therefore a proportion A_1/A_2 of the radiation from the cavity wall is intercepted by the black body. It is obvious that this same proportion would be intercepted by a grey body of the same shape.

This is an example of a more general idea, as follows. Suppose two bodies are exchanging energy by radiation. Define

$$F_{12} \equiv \text{fraction of energy leaving 1 which reaches 2}$$
$$F_{21} \equiv \text{fraction of energy leaving 2 which reaches 1}.$$

Note that these are functions of geometry only. Now, the total energy leaving body 1 per unit time is $A_1 \mathcal{J}_1$, where $\mathcal{J}_1$ is the emitted flux averaged over the area of body 1, and therefore the total arriving at body 2 is $A_1 \mathcal{J}_1 F_{12}$. Similarly, the total arriving at 1 is $A_2 \mathcal{J}_2 F_{21}$. There will in general be other bodies, such as the surroundings, that can also exchange energy with these two. However, we do not need to consider those other bodies, because in conditions of thermal equilibrium, the principle of detailed balance requires that

$$A_1 \mathcal{J}_1 F_{12} = A_2 \mathcal{J}_2 F_{21}.$$

Suppose now that both bodies are black and there is thermal equilibrium. Then $\mathcal{J}_1 = \mathcal{J}_2$, so we obtain

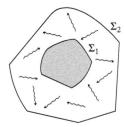

Fig. 20.2 *A pair of convex surfaces, one enclosing the other. We consider the radiative heat transfer between them, first for the case of black surfaces, then for grey surfaces. Thus we obtain the radiative heat transfer between a body and its surroundings when they are not at the same temperature.*

> **Shape factor reciprocity relation**
> $$A_1 F_{12} = A_2 F_{21}. \qquad (20.3)$$

We have thus discovered a useful fact about any geometry. It means that the heat exchange can always be written as $P_{12} = A_1 F_{12}(\mathcal{J}_1 - \mathcal{J}_2)$ (or as $P_{12} = A_2 F_{21}(\mathcal{J}_1 - \mathcal{J}_2)$).

Consider now a grey body at temperature T_1 inside a cavity with grey walls at temperature T_2. Let P_+ be the total power in the outward-going radiation that passes between the enclosed body and the cavity wall, and let P_- be the total power in the inward-coming radiation that passes between the cavity wall and the enclosed body. If the body is at temperature T_1 and the cavity wall at temperature T_2, then these powers must satisfy the equations

$$P_+ = \epsilon_1 A_1 F_{12} \sigma T_1^4 + (1-\alpha_1) P_- = \epsilon_1 A_1 \sigma T_1^4 + (1-\alpha_1) P_-$$
$$P_- = \epsilon_2 A_2 F_{21} \sigma T_2^4 + (1-\alpha_2) P_+ = \epsilon_2 A_1 \sigma T_2^4 + (1-\alpha_2) P_+,$$

where $F_{12} = 1$ and we have used the shape reciprocity relation in the P_- equation. Notice that the only area that appears in the final result is A_1. The right-hand sides of these equations are just like (20.1), with the replacement $\sigma \to A_1 \sigma$, so the solution for the net power transport is

$$P_+ - P_- = \sigma A_1 (T_1^4 - T_2^4) \alpha_{12}, \qquad (20.4)$$

where we have introduced

$$\alpha_{12} \equiv \frac{\alpha_1 \alpha_2}{\alpha_1 + \alpha_2 - \alpha_1 \alpha_2}.$$

For a black body of arbitrary (not necessarily convex) shape, the total radiant emission is $\sigma A_{\text{env}} T^4$, where A_{env} is the surface area of a minimal convex envelope around the body—the shape that would be adopted by an elastic membrane stretched over it; see Figure 20.3. To see this, argue that if the non-convex black body were placed in a cavity then it would intercept the same proportion of the radiation from the cavity walls as would a black body that just filled the convex envelope. This is because for straight-line propagation, any incoming rays that hit the body must hit its convex envelope, and any that do not hit the body must miss its convex envelope. It follows that these two bodies would absorb the same net power from the cavity radiation. Since the absorbed powers in the two cases are the same, so are the emitted powers.

Next, consider a useful further idea: that of a 'radiation shield'. We consider the case of three surfaces Σ_1, Σ_2, Σ_3, each enclosing the one before. For example, Σ_1 could be a body, Σ_3 its environment, and Σ_2 a thin opaque 'radiation shield' interposed between them, for the purposes of reducing the net heat flow. Then by applying (20.4) twice, we have

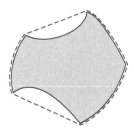

Fig. 20.3 *A non-convex body (shaded) and its minimal convex envelope (dashed).*

$$P_{12} = \sigma A_1(T_1^4 - T_2^4)\alpha_{12},$$
$$P_{23} = \sigma A_2(T_2^4 - T_3^4)\alpha_{23}.$$

Now assume that a steady flow has been reached, such that the net power loss from the outer surface of 2 is equal to the net power arriving at its inner surface, i.e. $P_{12} = P_{23}$. Then one can use the equations to eliminate T_2, and one finds

$$P_{13} = \sigma(T_1^4 - T_3^4)\frac{A_1 A_2 \alpha_{12}\alpha_{23}}{A_1\alpha_{12} + A_2\alpha_{23}}. \tag{20.5}$$

This value is smaller than that which would obtain without the shield. A case that often arises is where the environment is black to good approximation. Then $\alpha_3 = 1$ which leads to $\alpha_{23} = \alpha_2$ and $\alpha_{13} = \alpha_1$. The effect of the shield can be expressed by examining the ratio between the shielded and non-shielded results for P_{13}. This is

$$\text{ratio} = \frac{A_2\alpha_2}{A_1\alpha_1 + A_2(\alpha_1 + \alpha_2 - \alpha_1\alpha_2)} \simeq \frac{\alpha_2}{2\alpha_1}, \tag{20.6}$$

where the approximate result holds for $A_1 = A_2$ and either $\alpha_2 \ll \alpha_1$ or $\alpha_1 \simeq 1$. In particular, if $A_1 = A_2$ and all surfaces are black then the heat transport is halved. Note that the shield is less effective if it has a larger area.

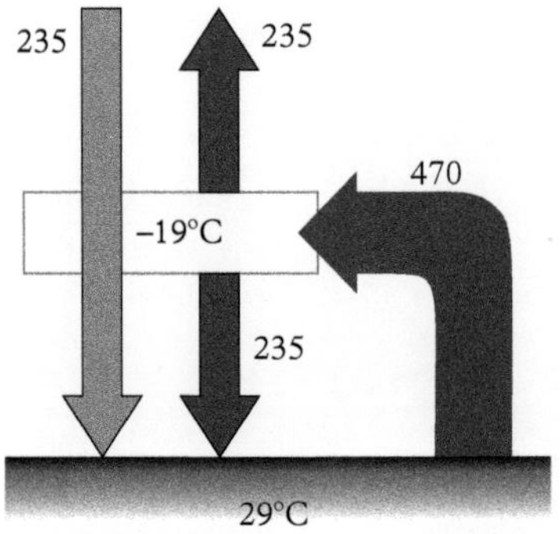

20.1 The greenhouse effect

The *greenhouse effect* is the name given to the warming effect on Earth, or other planets, produced by the wavelength-dependent transmission and emission properties of their atmospheres. This effect is also present, to a small degree, in glass houses, but in the latter the warming effect is achieved primarily by preventing convection. In the case of planet Earth the greenhouse effect is roughly as illustrated in Figure 20.4, which gives a crude model in order to convey the main idea. The incoming solar radiation, after allowing for reflection (albedo) and averaging over the whole surface area of the Earth, provides $I = (1 - R_{\text{uv}})I_0/4 \simeq 235 \text{ Wm}^{-2}$ of power per unit area. This penetrates the stratosphere and arrives at Earth's surface. For steady state, this same power per unit area must be radiated into space. If the stratosphere is opaque in the infrared, then the emission into space comes wholly from the stratosphere, so its temperature is such that it radiates 235 Wm^{-2} from its upper surface into space, and also this same amount from its lower surface towards the ground. It follows that the total radiation power arriving at the ground is twice what it would be if there were no atmosphere, and consequently in steady state the temperature at the ground is elevated by a factor $2^{1/4}$ (sufficient to allow it to emit enough to balance the incoming power). In this model the steady-state temperature of the stratosphere is $-19\,°\text{C}$ and that of Earth's surface is $29\,°\text{C}$. Considering the crudeness of the model, these values are in reasonable

Fig. 20.4 *The greenhouse effect (crude model). The figure shows the essential idea: the atmosphere is transparent to solar radiation (mostly visible and ultraviolet) but almost opaque to infrared radiation, while the Earth's surface is approximately black at all wavelengths. This results in a steady-state energy energy flow pattern as shown. Each arrow is labelled by the power per unit area in units of Wm^{-2}; Earth's surface receives twice the flux it would receive if there were no atmosphere, and therefore in steady state reaches a temperature such that it emits twice what it would otherwise emit.*

agreement with the observed average temperatures. The main idea is that a layer which transmits most of the ultraviolet solar radiation but absorbs well in the infrared can act as an effective 'winter coat', resulting in a substantial temperature difference between the Earth and its upper atmosphere.

For a somewhat more detailed calculation, we treat the stratosphere as a layer with absorptivity $\alpha = \epsilon$ in the infrared, and transmissivity $1 - \alpha$, with the surface of the Earth remaining black; see Figure 20.5. In steady state one has

$$I + \epsilon \sigma T_a^4 = \sigma T_E^4,$$
$$2\epsilon \sigma T_a^4 = \alpha \sigma T_E^4 \quad \Rightarrow \quad T_a = 2^{-1/4} T_E,$$

hence,

$$T_E(\epsilon) = T_E(0)(1 - \epsilon/2)^{-1/4}, \qquad (20.7)$$

where $T_E(0) = 255\,\text{K} = -18\,°\text{C}$ is the value in the absence of an atmosphere. To match the observed average surface temperature of $15\,°\text{C}$, a value of $\epsilon = 0.78$ is required. This is a reasonable estimate of the net effect of the gases, called 'greenhouse gases', which are responsible for the absorption. These are primarily water vapour and carbon dioxide.

More sophisticated models allow for convection, for the temperature profile in the atmosphere, and for the more detailed wavelength dependence of ϵ.

It is estimated that a doubling of the CO_2 content of the atmosphere would raise the temperature, and hence also the water content, with a net result of an increase in ϵ by approximately 0.04. The resulting value, $\epsilon = 0.82$, would raise the surface temperature of the Earth by two or three degrees Celsius. There is ample evidence that a 40% increase in CO_2 levels has taken place during the last hundred years; see, for example, Figure 20.6. If one also estimates the amount of wood, coal, and oil burning taking place throughout the world in that period, one finds that it can account for the increased atmospheric concentration of CO_2 and other greenhouse gases. One thus arrives at the important conclusion that human activity has changed atmospheric conditions by an amount that results in an increase in the average temperature of the Earth by of order one degree Celsius, and continuing such activity will result in further such increases.

This phenomenon is called *global warming*. Although one or a few degrees Celsius might seem like a modest change in the temperature, its effects would be very great, because of a range of consequences, such as increased humidity of the air, flooding as the Arctic ice melts and sea levels rise, extreme weather, social upheaval, and species extinctions throughout the world. The latter is related to both temperature and humidity—humans and other mammals cannot survive in climates moderately more hot and humid than already occur near the Equator, because we rely on the evaporation of sweat to regulate body temperature. Consequently this is an extremely important issue.

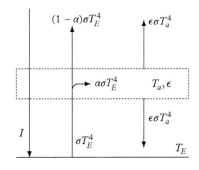

Fig. 20.5 *The greenhouse effect, still idealized but somewhat less crude model (see text).*

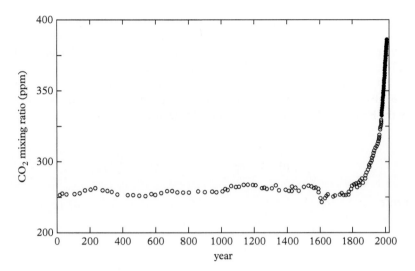

Fig. 20.6 Carbon dioxide concentration in Earth's atmosphere as a function of time, deduced from direct measurements at Cape Grim (filled circles) and air samples from Antarctic ice at Law Dome (open circles). The rise over the period 1900 to present is much greater, both in size and rapidity, than anything that took place before the industrial era. [Data from CSIRO; http://www.csiro.au/greenhouse-gases/]

Because of its importance, one must be clear on the evidence here. The evidence concerning the changing composition of the atmosphere is not in any reasonable doubt. Also, human activity has certainly pumped large amounts of CO_2 and other greenhouse gases into the atmosphere of the Earth. The fact that this must result in climate change is also not in serious doubt, because the calculations based on the physics of radiation are robust, and direct temperature data has also been gathered. It indicates a rise consistent with what is expected from the rise in greenhouse gases. Therefore, overall, the case that human activity is now having a significant impact on global climate is well established, to the extent that it is beyond reasonable doubt. There is no great difficulty in the basic physics here; the difficulty is all in estimating the economic and other costs of climate change. This is no longer just a case study in thermal and atmospheric physics, therefore. It is a question of weighing one person's interests against another's. It calls upon our willingness to recognize and respond to the interests of people and other animals for whom we hold the planet in trust.

Economic planning aims to take into consideration the fact that present human activity produces both benefits and costs to future inhabitants of Earth. This is often calculated by means of measures such as global economic growth (GEG), but measures of that kind are highly questionable in this context, because of the following considerations. First, GEG calculations often put no value at all on considerations whose value is hard to assess, such as the cost of a species going extinct, or the loss of a language or culture. Secondly, GEG often measures mere economic activity, irrespective of what the activity is, so that what might in fact amount to long-term environmental devastation is termed 'growth' because it involves human activity that other humans are willing to pay money for. The fact that there are other humans who eventually pay in other ways, unseen and gradual, may go unnoticed and unaccounted for. So one of the pressing needs of our time is a better way to assess the value and the cost of what people do.

....

EXERCISES

(20.1) World energy consumption, averaged over a year, is currently around 18 TW (18×10^{12} W). The average solar irradiance at Earth's surface is around 20 MJ/m^2 per day. What fraction of the surface area of the Earth would be required to provide all energy needs from solar power, if the conversion efficiency is 10%?
[*Ans.* 0.15%]

(20.2) In global economic calculations, the effects of policy decisions have to be estimated for timescales of many decades. Typically the amount of any given benefit or cost is multiplied by a factor $\alpha(t)$ which gets smaller with time t. What value or functional form would you consider appropriate for such a factor? Try to discover what factors are being used to frame public policy in your country, and in world economic meetings.

(20.3) A thermocouple is used to measure the temperature of air inside a long pipe with black walls. The pipe walls are at 100 °C, and the air is at 20 °C. Treating the thermocouple as black, what temperature will it indicate if it exchanges heat primarily by radiation and by convection in the air? (The convection may be modelled by Newton's law of cooling, with a coefficient $h = 14$ W/m^{2}K.) What will be the temperature reading if the thermocouple is surrounded by a tightly fitting shiny envelope of emissivity 0.05?
[*Ans.* 51 °C, 21 °C]

(20.4) Show that the solid angle subtended by a sphere of radius r at a distance R from its centre is $4\pi r^2/R^2$. Hence, by reasoning directly from the geometry, show that for a spherical body at the centre of a spherical cavity, the proportion of the radiation emitted from the cavity wall that hits the sphere is r^2/R^2, if the emission is isotropic. Derive the same result using the shape factor reciprocity relation.

(20.5) A hot sphere of radius R_0 is mounted in the centre of an evacuated spherical cavity of radius R_1, which is maintained at temperature T_1. Show that, if a thin spherical shield hanging freely is interposed half way between the sphere and the wall of the cavity, the rate of loss of heat by radiation from the sphere is reduced by a factor $R_s^2/(R_s^2 + R_0^2)$, where R_s is the radius of the shield. All the surfaces can be assumed to be black.

(20.6) A small spherical meteor is at a distance from the Sun of 50 Sun radii. Estimate its temperature, given that the surface temperature of the Sun is 6000 K.

(20.7) A large evacuated chamber is maintained at a constant temperature T_0. Two black bodies A and B are situated in the enclosure, not touching, with body A maintained at temperature T_1 ($T_1 > T_0$). When B is at T_0 the rate of loss of heat from A is W_1 and when B is at T_1 the rate of heat loss from A is W_2. Show that A experiences no net loss of heat by radiation when the temperature of B is T_2, given by

$$T_2^4 = \frac{W_1 T_1^4 - W_2 T_0^4}{W_1 - W_2}.$$

(20.8) A long cylinder of radius R_1 and emissivity ϵ is positioned inside an evacuated enclosure whose walls are held at temperature T_0 and are black. The cylinder attains a steady-state temperature T_1 when power W_1 per unit length is dissipated in it. Derive an expression for T_1 in terms of Stefan's constant and the parameters specified.

An intermediate cylinder of external radius $R_2 = 1.2R_1$, whose surfaces are black, is next suspended in the enclosure, surrounding the inner cylinder and centred on the same axis. It is now found that the power dissipation in the inner cylinder needed to maintain it at the same steady-state temperature T_1 is reduced to 75% of its former value. Deduce the value of ϵ.

Chemical reactions

When there are several chemical species present, the fundamental relation for a simple compressible system becomes

$$dU = TdS - pdV + \sum_i \mu_i dN_i, \qquad (21.1)$$

21.1 Basic considerations	299
21.2 Chemical equilibrium and the law of mass action	301
21.3 The reversible electric cell	307
Exercises	309

where the sum is over the different chemical species, often called *components*. By generalizing the arguments of Section 13.1.1, one finds

$$G = \sum_i \mu_i N_i, \qquad (21.2)$$

where $G \equiv U - TS + pV$ is the Gibbs function. Using this, one obtains

Generalized Gibbs–Duhem relation
$$0 = SdT - Vdp + \sum_i N_i d\mu_i. \qquad (21.3)$$

21.1 Basic considerations

There are two main considerations when we want to understand chemical reactions. The first issue is, does the reaction occur at all? The second is the reaction rate.

Consider a chemical reaction such as

$$Ca(OH)_2 + CO_2 \rightarrow CaCO_3 + H_2O.$$

We pose the question, does this reaction occur at standard temperature and pressure?

To answer, first we need to know the physical arrangement of the reactants. If they are kept apart, then of course the reaction will not proceed! We shall assume that the conditions are ordinary: the slaked lime (calcium hydroxide) is in powder form and carbon dioxide gas is allowed to blow over the surfaces of the powder granules.

Thermodynamics: A Complete Undergraduate Course. Andrew M. Steane.
© Andrew M. Steane 2017. Published 2017 by Oxford University Press.

In this situation, the bulk of each of the reactants (slaked lime or carbon dioxide) is in a pure form. That is, most of the molecules are not at the surfaces; they are in the bulk of each granule or in the bulk of the gas. This means that the chemical potential of each reactant is to good approximation equal to the value it would have in a pure sample without anything else present. The same goes for the products, because calcium carbonate does not dissolve in water. Therefore we can find out whether or not the reaction will proceed simply by using a table of values of the chemical potential of each substance in its pure form, at standard temperature and pressure. The following values would be obtained from such a table:

$$\mu_{Ca(OH)_2} = -897 \text{ kJ}/N_A$$
$$\mu_{CO_2} = -394 \text{ kJ}/N_A$$
$$\mu_{CaCO_3} = -1129 \text{ kJ}/N_A$$
$$\mu_{H_2O} = -237 \text{ kJ}/N_A,$$

where N_A is Avogadro's number.[1] In order to decide whether the reaction proceeds from left to right or in the other direction, it suffices to compare the sum of the chemical potentials of the reactants with that of the products:

$$\mu_{Ca(OH)_2} + \mu_{CO_2} = -1291 \text{ kJ}/N_A$$
$$\mu_{CaCO_3} + \mu_{H_2O} = -1366 \text{ kJ}/N_A.$$

Since the products have a lower combined chemical potential, the reaction will spontaneously proceed from left to right. Indeed it describes the well-known process of setting of mortar.

In quoting the values of μ for the various compounds we follow standard practice in the way tables of chemical potentials are compiled. The tables do not usually show the complete absolute value of any chemical potential. Rather, the difference between the μ of the pure compound and the summed μ's of its constituent elements (each in the form of a pure substance) is given. This is sufficient because in standard chemical reactions the amount of each chemical element is conserved (we are not concerned here with radioactive decay). Therefore it is possible to imagine that any given reaction proceeds first by decomposing every reactant into its elements, and then the elements are recombined. We should then have two chemical potential differences to calculate, that between the reactants and the pure elements, and then between the pure elements and the products. It is therefore enough to tabulate these. Of course, in such a table the value reported for a pure element will be zero.

To be precise, then, the value we have quoted for $\mu_{Ca(OH)_2}$ (namely −897 kJ/N_A) is really the value of

$$\mu_{Ca(OH)_2} - \mu_{Ca} - \mu_{O_2} - \mu_{H_2},$$

each μ being that of a pure sample at standard temperature and pressure. The conservation of the total amount of each element guarantees that all the pure element μ values must cancel in a complete calculation for any given chemical reaction, so we don't need to keep track of them explicitly.

---

[1] 1 kJ/N_A = 1.66054×10^{-21} J; a chemical potential of this amount per particle corresponds to 1 kJ per mole.

A convenient notation to keep track of the particle numbers in a general reaction is as follows. Consider for example the reaction

$$O_2 + 2H_2 \rightarrow 2H_2O. \tag{21.4}$$

The equation says that whenever one oxygen molecule and two hydrogen molecules disappear, two water molecules must appear. We can express this in general by bringing everything to the right-hand side and writing

$$\emptyset \rightarrow 2H_2O - O_2 - 2H_2$$

or

$$0 = \sum_i \nu_i N_i, \tag{21.5}$$

where N_i stands for a molecule of substance i, and the coefficients ν_i are negative for the reactants and positive for the products. In our example, assigning index numbers 1,2,3 to oxygen, hydrogen, and water respectively, we have

$$\nu_1 = -1, \ \nu_2 = -2, \ \nu_3 = 2.$$

These ν_i are called 'stoichiometric coefficients'. With this notation in hand, we can write the condition for the reaction to proceed spontaneously as

$$\sum_i \nu_i \mu_i < 0. \tag{21.6}$$

In all the above, we describe the chemical reaction in terms of chemical potential. One could equally refer to the Gibbs function per particle, or the Gibbs function per mole. It is this function that is significant here, not internal energy, for example, nor enthalpy. This is the reason why the molar Gibbs function of many substances in their standard state has been tabulated for easy reference, whereas the internal energy typically has not.

21.1.1 Reaction rate

Many chemical reactions have a rate that is proportional to

$$e^{E_a/k_B T}, \tag{21.7}$$

where E_a is an *activation energy*, typically of the order of 0.5 eV. Thus in general, reaction rates increase rapidly with temperature (Exercise 21.2).

21.2 Chemical equilibrium and the law of mass action

Another interesting situation is that of *chemical equilibrium*. This is where a reaction is free to take place in either direction, and is found to be balanced so that each direction proceeds at the same rate. Clearly from (21.6), the condition for this to be possible is

$$\sum_i \nu_i \mu_i = 0. \tag{21.8}$$

For example if the reaction is[2]

$$3\text{Fe} + 4\text{H}_2\text{O} \rightleftharpoons \text{Fe}_3\text{O}_4 + 4\text{H}_2,$$

then the condition for equilibrium at constant p, T is

$$3\mu_{(\text{Fe})} + 4\mu_{(\text{H}_2\text{O})} = \mu_{(\text{Fe}_3\text{O}_4)} + 4\mu_{(\text{H}_2)}. \tag{21.9}$$

Think of a balance beam and you get the idea.

Equation (21.8) is a generalization of the equilibrium condition (12.22) which we used to obtain the Saha equation, and it is proved by the same argument. If a reaction can proceed freely in either direction at equilibrium in an isolated system, then it must be that the system entropy does not change when the reactants combine to form the products. In this case (21.8) is the statement of energy conservation in an isolated system. More generally, since chemical potential is a function of state, if the intensive variables (pressure, temperature, concentrations etc.) are such that (21.8) holds in an isolated system, then it will also hold in a non-isolated system at the same pressure, temperature, concentrations, etc., and the reaction rates in each direction will still be balanced. Therefore (21.8) remains the condition for chemical equilibrium no matter what are the external constraints. In practice, constant pressure and temperature are commonly employed.

It is rare for such an equilibrium to exist for reactions of the type we discussed in Section 21.1, involving bulk pure substances at standard temperature and pressure. Rather, we are now concerned with cases such as a non-standard temperature, a mixture of gases, or chemicals in solution, where each μ is a function of all the concentrations. Then the chemical reaction in question will adjust the relative concentrations of the various substances until the condition is satisfied. Chemical potentials typically increase with density, so for example if a reaction $A \rightleftharpoons B + C$ is possible, then if initially μ_A is large and μ_B, μ_C are small then particles will move from substance A to substances B and C, causing μ_A to fall and μ_B, μ_C to rise, until a balance is achieved.

In the case of gas-phase reactions, or for chemicals in solution at low concentration, it can happen that the chemical potential of each substance depends only on the density of that substance, to good approximation. In this case the condition (21.8) can be converted into a useful statement about the densities. When μ has a logarithmic dependence on density (as it does for an ideal gas, for example), then the statement (21.8) concerning a *sum* of chemical potentials becomes a statement concerning a *product* of densities. In this form it is called the *law of mass action*.[3]

---

[2] The symbol $\rightleftharpoons$ is here used to indicate a reaction that can take place in either direction, and such that there is no obstruction or need for a catalyst.

[3] The name 'law of mass action' originated from the observation that the reaction rates here depend on the relative amounts of the various substances present in a given container, not just on external conditions such as temperature. Obviously it is density not mass that is the relevant property however.

To derive the law of mass action, first we write the chemical potential for each substance as a function of number density.[4] The number density n_i of substance i can be written

$$n_i = n^\ominus e^{(\mu_i - \mu_i^\ominus)/k_B T}, \tag{21.10}$$

where $n^\ominus$ is some standard density that is the same for all species, and $\mu_i^\ominus(T)$ is the value of the chemical potential of species i at the standard density. The superscript $\ominus$ merely indicates that the quantity in question is to be taken at the standard density; equation (21.10) will enable us to separate out the way the reaction depends on number density from the way it depends on other things. For an ideal gas (21.10) is the same as equation (12.12), and for any substance in solution it is a good approximation at low concentration. We thus obtain

$$\mu_i = \mu_i^\ominus(T) + k_B T \ln(n_i/n^\ominus). \tag{21.11}$$

Using this, (21.9) reads

$$-3\mu_1^\ominus - 4\mu_2^\ominus + \mu_3^\ominus + 4\mu_4^\ominus = k_B T \left(3 \ln \frac{n_1}{n^\ominus} + 4 \ln \frac{n_2}{n^\ominus} - \ln \frac{n_3}{n^\ominus} - 4 \ln \frac{n_4}{n^\ominus} \right),$$

where we have labelled the four substances Fe, H_2O, Fe_3O_2, H_2 with index numbers $1, 2, 3, 4$ respectively, and we have gathered terms related to density on the right. It follows that

$$\frac{3\mu_1^\ominus + 4\mu_2^\ominus - \mu_3^\ominus - 4\mu_4^\ominus}{k_B T} = \ln \left(\frac{n_3 n_4^4}{n_1^3 n_2^4} \right) + (7 - 5) \ln n^\ominus$$

$$\Rightarrow \quad \frac{n_3 n_4^4}{n_1^3 n_2^4} = K(T), \tag{21.12}$$

where

$$K(T) = (n^\ominus)^{-2} e^{-\Delta \mu^\ominus / k_B T}, \tag{21.13}$$

$$\Delta \mu^\ominus = \mu_3^\ominus + 4\mu_4^\ominus - 3\mu_1^\ominus - 4\mu_2^\ominus. \tag{21.14}$$

The form (21.12) is useful because it brings out the fact that K is a constant for the given reaction at a given temperature, and furthermore we can calculate it from the standard chemical potentials. The idea is that if we change one of the concentrations, say n_1 for example, then we have discovered something about how the other concentrations will adjust if equilibrium is maintained. K is called the *equilibrium constant* and (21.12) is the *law of mass action* for this reaction. You can readily check that for a general reaction the law takes the form

$$\prod_i (n_i)^{\nu_i} = K(T), \tag{21.15}$$

where ν_i are the stoichiometric coefficients and

[4] The law of mass action is often presented in terms of pressure rather than number density; we prefer number density as the more physically intuitive idea; the extension to a more precise treatment in any case invokes the concept of activity, which we quote at the end, cf. equation (21.23).

$$K(T) = \left(n^\ominus\right)^{\sum_i \nu_i} e^{-\Delta\mu^\ominus/k_B T}, \tag{21.16}$$

$$\Delta\mu^\ominus = \sum_i \nu_i \mu_i^\ominus. \tag{21.17}$$

The physical dimensions of K vary from one reaction to another.

Example 21.1

A one litre glass vessel is filled with two moles of hydrogen gas and one mole of chlorine gas and maintained at standard pressure. The temperature is such that the equilibrium constant for the reaction

$$H_2 + Cl_2 \rightleftharpoons 2HCl$$

is $K = 10$. Find the amount of hydrogen chloride present when the vessel reaches equilibrium.

Solution.
Since we may assume each gas expands to fill the available volume independently of the others, the number densities are all given by the number of moles divided by the same volume. Therefore we may as well calculate with the numbers of moles directly. Let x, y, z be the number of moles of hydrogen, chlorine, and hydrogen chloride molecules respectively. At chemical equilibrium, the law of mass action gives

$$\frac{z^2}{xy} = K = 10.$$

If z moles of hydrogen chloride are formed, then clearly the number of moles remaining of the other two substances are

$$x = x_i - \frac{z}{2}, \qquad y = y_i - \frac{z}{2},$$

where $x_i = 2$ and $y_i = 1$ are the initial values. Hence,

$$\frac{z^2}{(2 - z/2)(1 - z/2)} = 10.$$

This is a quadratic equation for z, readily solved to give $z = 1.584$ or 8.416. The second solution is unphysical (it corresponds to a negative amount of hydrogen and chlorine remaining), so we conclude that 1.584 moles of hydrogen chloride will form (and there will remain 1.208 moles of hydrogen and 0.208 moles of chlorine). In this example, because the equilibrium constant was moderately high, plenty of the product (HCl) formed, and the least-abundant reactant was mostly used up.

21.2.1 Van 't Hoff equation

By combining equations (21.16) and (21.17) the law of mass action can be written

$$\ln K = \left(\sum v_i\right) \ln n^\ominus - \frac{\sum v_i \mu_i^\ominus}{k_B T}$$

$$= \left(\sum v_i\right) \ln n^\ominus - \frac{\sum v_i G_{m,i}^\ominus}{RT}, \qquad (21.18)$$

where $G_{m,i}^\ominus = N_A \mu_i^\ominus$ is the molar Gibbs function of the i'th substance at the standard density. By differentiating with respect to T, we find

$$\frac{d \ln K}{dT} = -\sum_i \frac{v_i}{R} \frac{d}{dT}\left(\frac{G_{m,i}^\ominus}{T}\right)$$

$$= \sum_i \frac{v_i}{RT^2} H_{m,i}^\ominus, \qquad (21.19)$$

by using the Gibbs–Helmholtz equation (13.20). The quantity

$$\sum_i v_i H_{m,i}^\ominus \equiv \Delta H_m^\ominus \qquad (21.20)$$

is called the *enthalpy of reaction*. This is equal to the heat absorbed when the chemical reaction takes place in a flow process in which v_i moles of each substance are involved (cf. Section 16.4.1). For this reason it is sometimes called the *heat of reaction*, but note the sign: this is the heat that is *absorbed* by the reactants if the reaction is to proceed at constant temperature. A reaction that generates heat has a negative $\Delta H_m^\ominus$ (the products have less enthalpy than the reactants).

Equation (21.19) is often written

Van 't Hoff equation

$$\frac{d \ln K}{d(1/T)} = -\frac{\Delta H_m^\ominus}{R}. \qquad (21.21)$$

This equation has two main uses. First, by studying the reaction constant at a range of temperatures, one can deduce the heat of reaction, for example from the slope of a graph of $\ln(K)$ versus inverse temperature. Secondly, since the heat of reaction is not a strong function of temperature in practice, one can integrate the equation by making the approximation that $\Delta H_m^\ominus$ is independent of temperature, and thus predict the reaction constant at temperatures not previously studied. This enables one to deduce, for example, how much a change in temperature will favour the production of the products. For constant $\Delta H_m^\ominus$, equation (21.21) gives

$$K = K_0 e^{-\Delta H_m^\ominus/RT}, \qquad (21.22)$$

where K_0 is a constant of integration.

A reaction is called *exothermic* if heat is given off when reactants turn into products, i.e. when $\Delta H_m^\ominus < 0$. A reaction is called *endothermic* when $\Delta H_m^\ominus > 0$. According to equation (21.19), for an exothermic reaction, K decreases with temperature. This is an example of Le Chatelier's principle (Section 17.1.3): an exothermic reaction will tend to raise the temperature of the reaction chamber when reactants turn to products, which will in turn cause K to fall and therefore the forward reaction is slowed. Conversely, when products turn to reactants the temperature falls, which will raise K and enhance the forward reaction. Overall, then, a stable equilibrium can exist. The same argument applies to an endothermic reaction with signs reversed.

21.2.2 Chemical terminology

We have presented the law of mass action in terms of densities, because this is a good way to understand it. However, in order to be more precise, rather than relying on the approximation in (21.11), a new quantity called *fugacity* f is introduced, defined as

$$f = e^{\mu/k_B T},$$

so that the molar chemical potential[5] can be written

$$\mu = \mu^\ominus(T) + RT \ln(f/f^\ominus). \tag{21.23}$$

The word *fugacity* has the same Latin root as 'fugitive' and measures the 'tendency to flee'. Chemists also call f *absolute activity*. The ratio $a = f/f^\ominus$ is known as the *activity* of the substance, and the sum

$$\mathcal{A} = -\sum_i \mu_i \nu_i \tag{21.24}$$

is called the *affinity* of the reaction. The product (not necessarily in equilibrium conditions)

$$\mathcal{Q} = \prod_i a_i^{\nu_i} \tag{21.25}$$

is called the *reaction quotient* because it takes the form of a ratio. The law of mass action then takes the form

$$\mathcal{Q} = \exp\left(\frac{\Delta \mathcal{A}}{RT}\right), \tag{21.26}$$

where

$$\Delta \mathcal{A} = \sum_i (\mu_i^\ominus - \mu_i) \nu_i. \tag{21.27}$$

The main lesson of this is that the quantities describing the reaction are exponential in the affinity difference from the standard, and exponential in the inverse temperature. Since molar chemical potential is the same as molar Gibbs function, one may write

$$\Delta \mathcal{A} = \Delta H - T \Delta S. \tag{21.28}$$

---

[5] Throughout this book we use the chemical potential defined through the contribution μdN to dU, where N is the number of particles. Writing $N = mN_A$, where m is the number of moles, this can also be written $N_A \mu dm$ and thus one arrives at the notion of a 'molar chemical potential', $\mu_m \equiv (N_A \mu)$, which is also the molar Gibbs function for that species. In this section we adopt the convention, widely used in chemistry, of calling μ_m simply 'chemical potential' and dropping the subscript m.

In most cases, ΔH and ΔS are approximately independent of temperature, so the entropy contribution dominates ΔA at high temperature. The reaction will then favour whichever side has the higher molar entropy.

21.3 The reversible electric cell

Many electrical devices are powered by batteries or cells which convert chemical energy into electrical energy. In most batteries the chemical reactions are not taking place in equilibrium and are far from reversible. They can still be treated by thermodynamics, but in this section we restrict ourselves to the simpler case of reversible chemical equilibrium. A good example is the *Daniel cell*. This consists of a porous pot containing a solution of copper sulphate in water, the pot being surrounded by a solution of zinc sulphate in water. The porous pot serves to prevent the solutions mixing directly, while allowing ions to move between them. A copper electrode is placed in the copper sulphate, and a zinc electrode is placed in the zinc sulphate, and these electrodes are in turn connected by wires to a load such as a motor, a light bulb, a volt meter, etc., so as to form an electrical circuit. It is observed that an electric potential difference is produced between the electrodes, and current flows in the circuit.

The underlying process is as follows.

In one solution, the copper sulphate dissociates into copper ions Cu^{++} and sulphate ions SO_4^{--}, and in the other the zinc sulphate dissociates into zinc ions Zn^{++} and sulphate ions. The overall chemical reaction taking place is

$$Zn + CuSO_4 \rightleftharpoons ZnSO_4 + Cu$$

and the cell is an arrangement to separate this reaction into two parts. When the cell operates in the normal direction (i.e. without another electric source connected so as to drive the reaction the other way), the copper ions Cu^{++} gain two electrons at the electrode surface and become copper atoms which are deposited on the electrode. Meanwhile zinc atoms on the zinc electrode lose two electrons and pass into the solution as zinc ions.

Thus, at the copper electrode:

$$Cu^{++}_{(aq)} + 2e^- \rightarrow Cu_{(s)},$$

where the subscripts indicate aqueous solution and solid, respectively. When the electrons leave the electrode, if they are not replaced (e.g. if there is an open circuit), then there will quickly build up a potential difference between the electrode and the solution, matching the binding energy offered to the electrons by the copper ions. The open-circuit potential difference between the electrode and the solution is $+0.340$ V. Meanwhile, at the zinc electrode:

$$Zn_{(s)} \rightarrow Zn^{++}_{(aq)} + 2e^-$$

and the open circuit potential difference between the zinc electrode and the solution is −0.7618 V.

The porous nature of the pot (or, more generally, any suitable connection) allows ions to flow from one solution to the other. Note, it is ions not electrons that flow directly between the solutions, because there are no free electrons left over. If any potential difference were to appear between the solutions, then zinc or copper ions would flow in one direction, and sulphate ions in the other, until the difference vanished. Once the two solutions reach an equal electric potential, the e.m.f. between the two electrodes is 0.34 − (−0.7618) = 1.1018 V.

We will now use the Daniel cell to illustrate the way electrical measurements may be used to derive chemical information. To be specific, we will show how to obtain the enthalpy change in the chemical reaction (the heat of reaction) from measurements of the e.m.f. $\mathcal{E}$ of the cell.

In general, both $\mathcal{E}$ and the chemistry will be a function of the concentration of the solutions, as well as other things such as temperature. We will take an interest in the case of saturated solutions. This can be arranged by placing enough copper sulphate and zinc sulphate crystals in the two parts so that not all of the crystals dissolve. In this case $\mathcal{E}$ becomes a function of temperature alone.

The circuit is operated in a reversible way by connecting an opposing electric source, so that only a small current flows. In the limit of small current, the energy lost to ohmic heating in the internal resistance is negligible. If Z is the electric charge passing around the circuit, the conservation of energy then gives

$$dU = TdS + \mathcal{E}dZ - pdV, \tag{21.29}$$

where U is the internal energy of the whole cell. Therefore,

$$dH = TdS + \mathcal{E}dZ + Vdp,$$

where $H = U + pV$. In conditions of constant pressure, this becomes

$$dH = TdS + \mathcal{E}dZ. \tag{21.30}$$

We would like to know the change in enthalpy in conditions of constant temperature. To this end, consider $H = H(T, Z)$, so

$$\begin{aligned} dH &= \left.\frac{\partial H}{\partial T}\right|_Z dT + \left.\frac{\partial H}{\partial Z}\right|_T dZ \\ &= T\left.\frac{\partial S}{\partial T}\right|_Z dT + \left(T\left.\frac{\partial S}{\partial Z}\right|_T + \mathcal{E}\right) dZ \\ &= T\left.\frac{\partial S}{\partial T}\right|_Z dT + \left(-T\left.\frac{\partial \mathcal{E}}{\partial T}\right|_Z + \mathcal{E}\right) dZ, \end{aligned} \tag{21.31}$$

where in the second step we have used a Maxwell relation. If follows that, for an isothermal charging or discharging by ΔZ,

$$\Delta H = \left(\mathcal{E} - T\frac{d\mathcal{E}}{dT}\right)\Delta Z. \tag{21.32}$$

By measuring $\mathcal{E}$ and $d\mathcal{E}/dT$ at a known T one finds ΔH for any given ΔZ. The latter is related to the amount of material involved: one mole of copper or zinc passes into solution when the charge passing around the circuit is $\Delta Z = 2N_A e$ (the factor 2 being the valence of the metal ions in this case). Hence the heat of reaction can be obtained precisely by using electrical measurements, without the need to perform heat measurements (calorimetry).

EXERCISES

(12.1) A reaction $A + B \to C + D$ is exothermic and produces a positive entropy change. Is such a reaction (i) possible at high temperature, (ii) possible only at low temperature, (iii) not possible at any temperature, or (iv) possible at any temperature? [Hint: consider ΔG]

(12.2) By what factor does the reaction rate given by equation (21.7) increase when the temperature changes from 300 K to 310 K, for a reaction whose activation energy is 0.5 eV?

(12.3) Show that the law of mass action applied to the process of ionization of hydrogen reads

$$\frac{n_e n_p}{n_H} = n^{\ominus} e^{-\Delta\mu^{\ominus}/k_B T}, \tag{21.33}$$

where $\Delta\mu^{\ominus} = \mu_p^{\ominus} + \mu_e^{\ominus} - \mu_H^{\ominus}$ and $n^{\ominus}$ is the standard density introduced in equation (21.11). Obtain this combination by using equation (12.8) for a gas at the standard density $n^{\ominus}$, and allowing for the binding energy. Hence obtain the Saha equation.

(12.4) The equilibrium constant for the gaseous reaction

$$H_2 + I_2 \rightleftharpoons 2HI$$

at a temperature of 720 kelvin is $K = 46.8$. A flask initially containing only HI is prepared and maintained at 720 kelvin. What is the concentration of HI in the flask when chemical equilibrium is attained? [Ans. 77% by volume]

(12.5) An important step in the commercial production of sulphuric acid involves synthesis of sulphur trioxide, in the reaction

$$2\,SO_2 + O_2 \rightleftharpoons 2\,SO_3.$$

At 298 kelvin, one finds $\Delta H = -1.9824 \times 10^5$ J and $\Delta S = -189.79$ J/K per mole of oxygen molecules in this reaction. Assuming these values are independent of temperature, find the reaction quotient at $T = 300$ K and at $T = 1200$ K. Hence establish whether the reaction favours production or dissociation of SO_3 at these two temperatures.

(12.6) Prove (21.26), for example by substituting from (21.27) and (21.25).

Phase change

22

Phase changes have already been mentioned in Chapter 15 and Section 17.2. In this and Chapter 24 we consider the familiar types of phase change in more detail, and in Chapter 25 we consider more general types of phase change. In general, a phase change refers to any process in which one or more entropy related properties *change abruptly* as a function of an intensive property such as temperature or pressure. In such processes the physical system is undergoing some sort of reorganization of its internal structure.

22.1 General introduction	311
22.2 Basic properties of first-order phase transitions	317
22.3 Clausius–Clapeyron equation	320
22.4 The type-I superconducting transition	326
Exercises	329

22.1 General introduction

Consider the example of boiling water. Suppose we have a kettle full of water at 20 °C, at a pressure of one atmosphere. We supply heat using an electric heater immersed in the water, and the temperature begins to rise, while the pressure is maintained constant by the surrounding atmosphere. Each injection of heat is accompanied by a temperature increase. However, when the temperature reaches 100 °C, something new happens: the temperature refuses to change, even though we continue to supply heat. The heat capacity of the system, $C_p = đQ_p/dT$, has become infinite! Of course we know what is happening, the heat is being used up in converting liquid water into steam, but clearly this is an interesting thermodynamic phenomenon. Eventually when all the water has been transformed into steam, further inputs of heat will cause the temperature to increase beyond 100 °C.

The liquid→vapour transition is called *boiling* or *evaporation*; the solid→liquid transition is called *melting*; a direct transition between solid and vapour is called *sublimation*. An example of the latter is observed in carbon at one atmosphere of pressure. There are more names for precisely the same transitions but going in the other direction: vapour→liquid is called *condensation*, liquid→solid is called *freezing* or *fusion*. The word 'vapour' is used mostly as just another word for a gas, but some people like to reserve the word 'vapour' for a case where the gas is in equilibrium with its associated liquid or solid, or is at a temperature low enough that it would go through a vapour–liquid phase change if compressed isobarically. We will use 'vapour' and 'gas' interchangeably.

At very high temperature a gas will become ionized and turn into a plasma; this is another phase transition. Phase transitions are also involved in changes of crystalline structure or magnetic domain structure in solids.

Thermodynamics: A Complete Undergraduate Course. Andrew M. Steane.
© Andrew M. Steane 2017. Published 2017 by Oxford University Press.

The question arises, whether the presence of more than one phase implies that a further state variable, in addition to two (e.g. temperature and volume), is needed to uniquely specify the thermodynamic state of the system. The further variable might specify the proportion of material in each phase, for example. However this is not needed. Suppose one takes a 'black box' point of view, in which the system, e.g. water and water vapour, is contained in an opaque cylinder which only interacts with the rest of the world by movement of a piston and by exchanging heat. This is a closed system. In this case the fundamental relation for dU only involves two terms ($TdS-pdV$). This means that U can be expressed as a function of two natural variables, and all the other thermodynamic properties (i.e. those that influence its exchange of energy and entropy with the rest of the universe) can be obtained from this function. Therefore we have a simple system. The state diagram in Figure 15.2 illustrates this point: the equilibrium states fall on a *surface* in state space, not a higher-dimensional region. More generally, for a fixed total quantity of matter inside the cylinder, other properties such as heat capacities depend on no more than two independent variables. The proportion of one phase or the other in the mix is an internal parameter, like the parameter r discussed in Section 17.3.1. Its value in equilibrium is set by the maximum entropy or minimum free energy rule, and it can be tabulated or plotted as a function of state, as it is in Figure 16.7 for example.

22.1.1 Phase diagram

A *phase diagram* shows the regions of parameter space in which a given substance is found to exist in one phase or another in its more stable form. This can be shown as a function of any suitable set of variables. For a simple compressible system the diagram is usually presented in terms of p and T; a typical example is shown in Figure 22.1. The lines separating the regions of the diagram are called *coexistence curves* or *phase boundaries*. On a pV diagram, such as Figure 22.2,

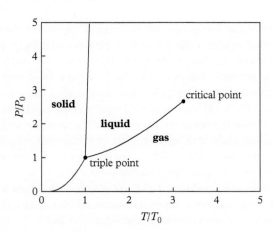

Fig. 22.1 *Phase diagram for a typical substance.*

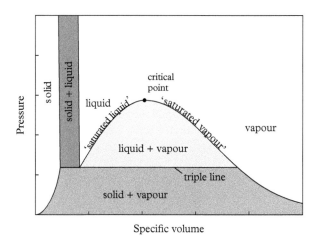

Fig. 22.2 *pV phase diagram for a typical substance. Mixed-phase regions are shaded. The qualitative behaviour is shown, but the relative volumes indicated are not to scale. In practice the volume increase in the gaseous phase can be by several orders of magnitude.*

there are coexistence regions rather than coexistence curves. They are also called 'mixed phase' regions.

The pT and pV indicator diagrams are both projections of the pVT surface (Figure 15.2) onto one plane or another. The third most commonly used projection is the TV diagram; it is qualitatively similar to the pV diagram (this can be deduced by inspecting Figure 15.2).

The *triple point* is the unique[1] combination of temperature and pressure where all three phases (solid, liquid, vapour) can coexist in mutual equilibrium. This can happen for a range of volumes, so it appears as a 'triple line' on a pV phase diagram.

The *critical point* was mentioned previously in Chapter 15; it is the combination of temperature and pressure at which the liquid and vapour phases become indistinguishable and the phase transition vanishes. More precisely, the critical point is at a horizontal point of inflexion on the critical isotherm (i.e. $(\partial p/\partial V)_T = 0$ and $(\partial^2 p/\partial V^2)_T = 0$). The existence of a critical point makes the distinction between liquid and vapour ill defined at temperatures above the critical temperature. Also, it makes it possible to carry a fluid from a state which is unquestionably gaseous to one which is unquestionably liquid, without encountering any phase transition. One simply carries the system around the critical point by following a path such as that shown in Figure 22.3.

The boundary of the coexistence region on a pV diagram may be divided into two parts, either side of the critical point, and these have also earned further names. A pure vapour on the boundary has a temperature such that any small reduction in temperature would cause liquid to start to form as the system passes into the mixed phase region. Such a vapour is said to be 'saturated'. Similarly, a pure liquid such that any small increase in temperature would cause vapour to start to form is said to be a 'saturated liquid'.

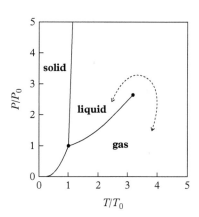

Fig. 22.3 *Carrying a system between liquid and gaseous states without a phase transition.*

[1] Unique, that is, as long as surface effects are negligible, which is the case when the surfaces are flat; see Chapter 24.

One might wonder whether a critical point may also exist on the liquid–solid coexistence curve. Experimental searches suggest that there is no such second critical point. The reason is that the difference between a liquid and a solid is of a different kind to that between a gas and a liquid. If there is a critical point, then it is possible to move between one type of state and the other smoothly, as we have just discussed. In the case of different types of fluid, this is possible because the difference between the gaseous and liquid forms is essentially one of density and the role of interactions. In the case of a solid (we have in mind ordinary crystalline solids, not a glass) there is a further ingredient, namely the regular arrangement of the atoms or molecules in a crystalline structure. It is hard to see how the system could pass between the unstructured and structured forms smoothly—to be precise, without a step change in entropy.

This general idea is discussed by invoking the concept of *symmetry*. Fluids (whether liquid or gaseous) in equilibrium have a very high degree of symmetry: any location is as likely to hold a particle as any other. This is not so for solids, where locations on the crystal lattice hold a particle, other locations do not. Another way of stating the same thing is to take a time average over a time long enough for particles to move a few mean free paths (this is a very short time in practice), and then one finds that fluids are uniform with respect to all translations in space and all rotations, but solids are not. A crystalline solid therefore has much *less* symmetry: it is symmetric under only a discrete set of translations and rotations. The transition from gas to liquid can take place gradually (by working around the critical point) because it amounts to purely a change of density, with gradually growing inter-particle interaction strengths. The transition from liquid to solid cannot take place gradually as a function of equilibrium parameters such as pressure, volume, and temperature because it involves a change in the degree of symmetry—the technical term is *symmetry breaking*. A symmetry is a discrete property: it is either there or it is not.

By invoking the concept of symmetry, the above argument gives a good insight into the physical ideas, but the case is not so strong as to amount to a proof. One can imagine that crystalline structure might appear in small regions which grow smoothly in size as the conditions change, for example. This is indeed what happens as a function of time during a phase change, but it remains open to question whether or not the *equilibrium* state can ever take a form intermediate between amorphous and crystalline.

22.1.2 Some interesting phase diagrams

Figure 22.1 shows the main features of the phase diagram for a typical pure substance. The solid–liquid coexistence line is typically steep and with a positive slope. The substance typically solidifies at low temperatures. Water and helium provide interesting counter-examples.

The phase diagram of water[2] is shown in Figure 22.4. The solid–liquid coexistence curve has a negative slope, which is uncommon, and is related to the fact

---

[2] The name 'water' may legitimately be used to refer to all the various phases of H_2O, but since it is applied only to the liquid form in everyday speech, we will adopt the term 'ice' for the solid phase.

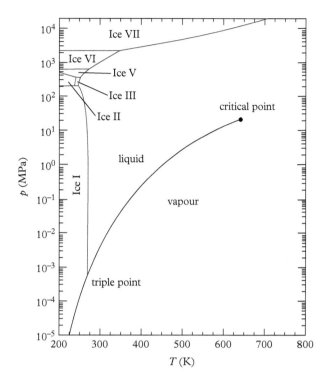

Fig. 22.4 *Phase diagram for H_2O.*

that ice is less dense than liquid water (cf. Eq. 22.4). Note that there are many forms of ice, which differ in their crystalline structure. These are called different *allotropes*. The existence of many different allotropes for a given solid is common. For example, graphite is the most stable form of carbon at ordinary temperature and pressure, whereas diamond is the most stable form at high pressure. The reason that diamonds do not spontaneously turn into graphite at room temperature is simply that the process is very slow.

Helium is interesting because it is the only chemical element which can exist in liquid form at absolute zero temperature. There are two stable isotopes of helium, and both are liquid as $T \to 0$ if the pressure is not too high. Helium-4 (having two neutrons and two protons in the nucleus) is the more common isotope. Its phase diagram is shown in Figure 22.5. Note that the liquid–solid coexistence line meets the pressure axis horizontally; this is predicted by the Clausius–Clapeyron equation (22.4) discussed in Section 22.3.

Liquid helium itself undergoes a further phase transition, called the lambda transition, below a temperature of 2.172 K. This is an example of a *continuous phase transition*, to be discussed in Chapter 25. Below the transition temperature the liquid is called liquid helium II, which can be modelled approximately as a mixed-phase liquid, or a liquid formed of two interpenetrating parts, one of which has no viscosity.

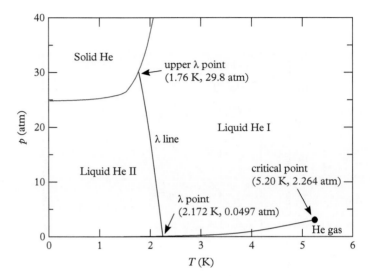

Fig. 22.5 *Phase diagram of ^{4}He. (Data adapted from J.C. Davis group, Cornell; http://nptel.ac.in/courses/112103016/module2/lec3/3.html).*

22.1.2.1 Evaporation

An everyday observation is the tendency of a pool of water to evaporate on a hot day. The temperature of the pool does not need to reach the boiling point for this to happen. The situation is one in which more than one substance (or 'chemical component') is present: water and the other components of the air. Suppose the temperature is 20 °C and the pressure 1 atm. The vapour pressure of water (i.e. the pressure it would exhibit if it were the only substance present and the liquid and vapour forms were both present) at this temperature is well below 1 atm, but it is not zero. Therefore there is a *partial pressure* of water vapour in the air. If the air is warmed, or moved away by the wind and replaced by dry air, then more water will evaporate until the partial pressure is restored. If the water is heated, then the equilibrium vapour pressure rises. When the water temperature reaches 100 °C, the equilibrium vapour pressure reaches one atmosphere.

The difference between boiling and evaporation is that in the latter the molecules in the bulk of the liquid are surrounded by liquid at a pressure well above the vapour pressure, so they show no interest in changing phase. Evaporation only occurs at the surface. In the case of a liquid at the boiling point, by contrast, throughout the bulk of the liquid the temperature is high enough, and the pressure low enough, to allow conversion between phases.

In both cases (evaporation or boiling) a phase transition is occurring, involving a latent heat. This provides the cooling method exploited by mammals: when the surrounding air is dry, evaporation of sweat occurs, the latent heat being provided by the body of the animal, which consequently cools a little. There is a reduction in the entropy of the animal, but the process takes place spontaneously because this is more than made up by the increase in the entropy of the water that has evaporated.

22.2 Basic properties of first-order phase transitions

The familiar phase transitions (solid–liquid–vapour) involve a step change in the volume and entropy of the system as a function of temperature at given pressure; see Figure 22.6. Phase transitions of this type are said to be of *first order*. The term refers to the fact that S and V are the first derivatives of the Gibbs function G. The abrupt change of S implies that the transition involves a *latent heat*,

$$L = T\Delta S. \tag{22.1}$$

This is an extensive quantity. The latent heat per unit mass is the amount of heat that would have to be supplied to cause unit mass of material to change phase at constant pressure. In a first-order phase transition, other thermodynamic functions change by:

$$\Delta U = L - p\Delta V, \quad \Delta F = -p\Delta V, \quad \Delta H = L, \quad \Delta G = 0,$$

as the reader should verify. Here U, F, H, G refer to the complete system. We are here treating a simple compressible system, and it can be regarded as simply a 'black box' or the 'contents of a cylinder'. We do not need to know what proportion of the contents may be of one phase or another when defining quantities such as p, V, T and U.

The latent heat involved in the liquid–vapour transition can be estimated by using the fact that the volume change is large, and the liquid is like a dense vapour (it has no crystalline structure) so the entropy change is dominated by the contribution from mere volume. If we ignore intermolecular forces to get a rough idea, then we can treat the liquid form as simply a gas at high density, so we expect from equation (13.39),

$$\Delta S \simeq N k_B \ln \frac{V_{\text{gas}}}{V_{\text{liquid}}}.$$

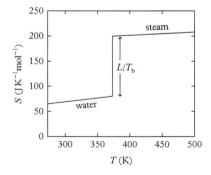
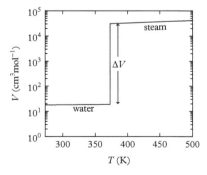

Fig. 22.6 *Entropy and volume changes in a first-order phase transition. A mole of water is used as an example. A mole of water has a mass of 18.015 grams.*

Table 22.1 *The boiling point at 1 atmosphere, latent heat, and L/RT_b for some common substances, illustrating Trouton's rule.*

	Ne	Ar	Xe	CH$_4$	C$_6$H$_6$	N$_2$	He	H$_2$O
T_b (K)	27.1	87.3	165	112	354	77.34	4.22	373.2
L (kJ/mol)	1.77	6.52	12.6	8.18	30.7	5.58	0.084	40.7
L/RT_b	7.85	8.98	9.21	8.80	10.5	8.67	2.39	13.1

Since the formula involves a logarithm, even a rough value of the volume increase will suffice. In passing from liquid to vapour at ordinary pressures the volume increase is typically by a factor of around 1000, giving $\Delta S \simeq 7Nk_B$, so $L \simeq 7RT_b$ per mole. This relationship is called *Trouton's rule*. It is usually quoted with a slightly larger prefactor, which roughly accounts for the effects of intermolecular forces that hold the liquid particles together:

$$L \simeq 9RT_b. \tag{22.2}$$

Table 22.1 shows that this is a reasonable estimate for many ordinary gases; exceptional cases include helium and water.

Since for a first-order phase transition S goes through a step change at constant pressure, its derivative C_p (the heat capacity at constant pressure) becomes infinite, as we have already noted. One may ask whether it is possible to have a truly infinite quantity, or whether some approximation is involved. The issue is whether or not there is a tiny temperature change during the phase transition. A glance at Figure 17.5 shows that there is not: the entropy increases strictly linearly with U during the transition, so T is constant. However, the very concept of temperature is only precisely defined in the thermodynamic limit, i.e. the limit of very large systems. Therefore although the mean temperature of the system does not change during a phase transition, the actual temperature is subject to fluctuations, as is the pressure, and this blurs the degree to which we can talk about a step change in S as a function of T.

Indeed, in phase change phenomena the issue of fluctuations becomes of central importance, because fluctuations which are small away from a phase change can become large near a phase change, and extend to long distances through the system. This makes concepts such as pressure, density, and temperature of limited usefulness. For example, when water approaches its boiling point, there is not a smooth change in the fluid, but turbulent motion and the generation of bubbles of vapour in the body of the liquid. This is associated with the fact that the isotherms are horizontal on a pV diagram; in other words $(\partial p/\partial V)_T = 0$ so $\kappa_T = \infty$. Although the total volume of the system may be constrained by fixed walls, so that V is well defined, within such a chamber the division of material between liquid and vapour phases fluctuates greatly: cf. equations (27.15)

Table 22.2 *First-order phase transitions: properties and examples.*

First-order phase transitions

General properties
1. there is a discontinuity in various properties
2. there are metastable phases (superheating and supercooling)
3. in almost all cases, there is a discontinuity in S and therefore a latent heat

Examples	Latent heat	Discontinuities also in:
Liquid–vapour	yes	density, viscosity, etc.
Solid–liquid	yes	crystalline order, rigidity, etc.
White tin–grey tin	yes	lattice type, etc.
Type I superconductor in finite magnetic field (normal metal)	yes	conductivity, magnetic moment, etc.
Ferromagnetic reversal, as a function of B	no	magnetization direction

and (27.24). To treat the behaviour we should really divide the system up into small cells, and ultimately to capture the physics precisely, these cells may need to be so small that the language of thermodynamics—the language of heat, work, temperature, and so on—is no longer well defined. Nevertheless, thermodynamic language still offers the right global perspective, and it can describe many features, such as the lines on the phase diagram, very precisely.

In addition to first-order phase transitions, there are also further phase transitions in which the entropy does not have a discontinuity, but one of its derivatives may, or else everything is continuous but there is a divergence in properties such as heat capacity at the transition. A rich variety of possibilities exists. All non-first-order transitions are grouped together under the name *continuous phase transitions*; they are the subject of Chapter 25. The first-order phase transitions are distinguished by the properties listed in Table 22.2, where some examples are also given.

Item 2 in the table—the existence of metastable phases—refers to the fact that for temperatures and pressures near a first-order phase transition, there is a region where a mixture of phases is fully stable but the system may remain for a long time in just one phase, in a metastable condition. This significantly affects the way the transition takes place; it will be discussed in Chapter 24.

Continuous phase transitions, by contrast, involve two phases that are identical at the transition point, therefore there is no abrupt entropy change, so there is never a latent heat and there are no metastable phases.

22.3 Clausius–Clapeyron equation

The Clausius–Clapeyron equation is an equation describing the coexistence curve for a first-order phase transition. It has been experimentally tested to a high degree of accuracy for a wide range of conditions and substances. It provides one of the most thorough and precise tests of the second law.

We shall present three derivations. The first has the advantage of conceptual simplicity, the second demonstrates the application of a Maxwell relation, and the third has the advantage of bringing a deeper insight into the underlying physics. We shall treat the case of a pV system; the result readily generalizes to other types of system.

Derivation 1 Suppose a reversible heat engine is operated between two adjacent isotherms in the phase-change region (see Figure 22.7). The heat ratio formula is

$$\frac{Q_1}{Q_2} = \frac{T + dT}{T}$$

and the work performed in a cycle is the area of the shaded region, which is

$$đW = dp \Delta V,$$

where ΔV is the volume change when a given amount of material changes from one phase to the other. By conservation of energy, $đW = Q_1 - Q_2$, from which

$$dp \frac{\Delta V}{Q_2} = \frac{Q_1}{Q_2} - 1 = \frac{dT}{T}$$

$$\Rightarrow \quad \frac{dp}{dT} = \frac{Q_2}{T \Delta V}. \tag{22.3}$$

The heat Q_2 (or Q_1) is, by definition, the latent heat for the given amount of material. Therefore we have[3]

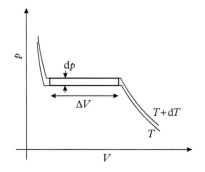

Fig. 22.7 *Simple derivation of the Clausius–Clapeyron equation.*

> **The Clausius–Clapeyron equation**
> $$\frac{dp}{dT} = \frac{L}{T \Delta V}. \tag{22.4}$$

This relates the slope of the coexistence curve to the latent heat and the volume change.

Derivation 2 Consider the Maxwell relation $(\partial p/\partial T)_V = (\partial S/\partial V)_T$. During a liquid–vapour phase change in a cylinder of constant volume, a small increase in temperature will cause more liquid to evaporate until the new vapour pressure is attained, so in this case $(\partial p/\partial T)_V = dp_{vap}/dT$, where p_{vap} is the vapour pressure. If, on the other hand, we expand the chamber at constant temperature, the pressure will stay constant, and liquid will evaporate to fill the new volume. The heat required to achieve this reversibly is, by definition, the latent heat, so we have

---

[3] In some treatments this is called the Clapeyron equation and (22.11) is called the Clausius–Clapeyron equation.

$T dS = L dm$, where L is the specific latent heat and dm is the mass that has evaporated. The volume change is $dV = (\Delta V) dm$, where ΔV is the volume change per unit mass, given by $\Delta V = \rho_v^{-1} - \rho_l^{-1}$, where $\rho_{v,l}$ are the densities. Hence we find that in this process, $(\partial S/\partial V)_T$ may be identified as $L/T\Delta V$. Applying the Maxwell relation, we then obtain the Clausius–Clapeyron equation (22.4).

Derivation 3 The third method of derivation is to consider the chemical potentials of the two phases, μ_1 and μ_2. Or, in a treatment that avoids explicitly introducing chemical potential, it is sufficient to discuss the Gibbs function, as we shall show shortly. Recalling the discussion of Section 17.1, we claim that if two distinct phases of the same substance can coexist in thermal equilibrium, then they must have the same chemical potential:

$$\mu_1 = \mu_2. \tag{22.5}$$

This equation has to be understood correctly: it relates the chemical potential of phase 1 to the chemical potential of phase 2 when the system parameters are somewhere on the coexistence line. Now consider two neighbouring points on the coexistence line, at p, T and $(p+dp), (T+dT)$ (see Figure 22.8). Applying (22.5) to each point individually, we have

$$\left. \begin{array}{l} \mu_1(p, T) = \mu_2(p, T) \\ \text{and} \quad \mu_1(p + dp, T + dT) = \mu_2(p + dp, T + dT) \end{array} \right\} \Rightarrow d\mu_1 = d\mu_2,$$

where $d\mu_i$ refers to the change in μ_i when the pressure and temperature change by dp, dT (note, this is *not* a change *across* the coexistence line; it is a change *along* the coexistence line). Now use the Gibbs–Duhem relation (13.5):

$$s_1 dT - v_1 dp = s_2 dT - v_2 dp$$
$$\Rightarrow \quad \frac{dp}{dT} = \frac{s_1 - s_2}{v_1 - v_2} = \frac{L/T}{\Delta V}, \tag{22.6}$$

which is the Clausius–Clapeyron equation.

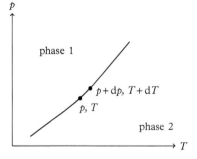

Fig. 22.8 *Neighbouring points on a coexistence line.*

Example 22.1

The pressure in the cabin of an airliner is 0.9 atmospheres. At what temperature will water boil at this pressure?

Solution.
The boiling point is 100 °C at 1 atmosphere, and for the modest pressure change $\delta p = 0.1$ atm we can approximate the coexistence curve as linear, so the temperature change is given by $\delta T = \delta p T \Delta V / L$. The molar latent heat of water is 40.7 kJ/mol and the volume change can be taken to be the molar volume of the vapour (since that of the liquid is much smaller), $\Delta V \simeq 0.0224$ m^3. This gives $\delta T = 2.05$ K. Therefore the water will boil at 98 °C.

Since equation (22.5) is fundamental to the fuller discussion of first-order phase transitions, although we have proved it already in Chapter 17 we shall now present a second proof.

Consider the total Gibbs function of a closed pV system. This obeys

$$dG = -SdT + Vdp.$$

The system could, for example, consist of a substance in one or more phases, in a closed chamber maintained at constant pressure and temperature. Now suppose we change the volume of the chamber while maintaining the pressure and temperature constant. This is possible if the volume change is accommodated by some of the substance changing from the less dense phase to the more dense phase, or vice versa. However, since $dp = 0$ and $dT = 0$, clearly $dG = 0$.

Assuming that surface effects are negligible, we can write $G = N_1 g_1 + N_2 g_2$, where g_1, g_2 are the Gibbs functions per particle of the two phases and N_1, N_2 the number of particles in each phase. During the volume change, $dN_1 \neq 0$ and $dN_2 \neq 0$, but since the system is closed the total particle number is conserved, so

$$dN_1 = -dN_2.$$

Using $G = N_1 g_1 + N_2 g_2$, we have

$$\begin{aligned} dG &= dN_1 g_1 + N_1 dg_1 + dN_2 g_2 + N_2 dg_2 \\ &= dN_1(g_1 - g_2) + N_1 dg_1 + N_2 dg_2. \end{aligned} \quad (22.7)$$

Now,

$$dg_i = -s_i dT + v_i dp, \quad (22.8)$$

where s_i, v_i are the entropy and volume per particle in phase i. This is because adding particles to any given phase without changing p and T will change U_i, S_i, V_i, and N_i together and therefore leave g_i unchanged, so $g_i = g_i(T, p)$. For further discussion of this point, refer to Section 13.1.1.

It follows that during the volume change at constant pressure and temperature, not only has $dG = 0$ but also $dg_1 = dg_2 = 0$. Substituting these facts into (22.7), we obtain

$$g_1 = g_2. \quad (22.9)$$

This is equation (22.5) since the Gibbs function per particle is the chemical potential (by the Euler relation (13.4)). The Clausius–Clapeyron equation now follows, as before.

The derivation of (22.9) involved discussion of a change taking place at constant pressure and temperature. One may ask what would result if the constraints were different, such as constant volume and temperature. A similar analysis via the Helmholtz function leads to the conclusion that $f_1 + pv_1 = f_2 + pv_2$, i.e. $g_1 = g_2$, the same result. However, there is no need to carry out such an analysis, since the chemical potential is a function of the intensive variables alone (Gibbs–Duhem relation). Therefore, if $\mu_1 = \mu_2$ for two phases at some particular pressure and temperature, then this remains true no matter what external constraints the system may be under.

22.3.1 Vapour–liquid and liquid–solid coexistence lines

The liquid–solid phase boundary is steep because ΔV is small. An important prediction of the Clausius–Clapeyron equation is that the slope of the coexistence line is positive if the substance grows in volume when heat is added. Conversely, a substance which becomes more dense on melting will have a coexistence line of negative slope. An example of the latter case is seen in the melting of ice. Since ice is less dense than water near $0\,^\circ\text{C}$, as the pressure is increased above 1 atmosphere, ice melts at a lower temperature. You can test this by pressing a drawing pin (thumb tack) into an ice cube inside a freezer (for a fair test avoid heat flow into the pin).

For the liquid–vapour coexistence line, for conditions not too near the critical point we can assume the specific volume of the liquid is small compared to that of the vapour, i.e. $\Delta V = V_v - V_l \simeq V_v$. If we approximate the vapour as an ideal gas, then the Clausius–Clapeyron equation reads

$$\frac{dp}{dT} \simeq \frac{Lp}{RT^2}, \qquad (22.10)$$

where L is the molar latent heat. In a crude approximation, we ignore the temperature dependence of the latter, and then we can integrate the equation, obtaining

$$p = p_0 e^{-L/RT}. \qquad (22.11)$$

The vapour pressure is thus predicted to be approximately an exponential function of inverse temperature. This function is plotted in Figure 22.9. It is remarkably accurate even near the triple point and the critical point. In the latter case this is because the effect of a falling L near the critical point is offset by the higher-order terms in the equation of state (Exercise 22.7).

In fact the latent heat is a function of temperature—see Figure 22.10. We can relate this to the difference in the heat capacities of the two phases, and also learn how it modifies the coexistence curve. The calculation is easiest if we start from the entropy change $S_v - S_l = L/T$ and differentiate *along the phase boundary*. That is, we consider the change in $S_v - S_l$ under the constraint that the system stays on the phase boundary. This constraint removes one degree of freedom from the system, so that everything depends on a single variable and we can deal in total derivatives:

$$\frac{d}{dT} = \left.\frac{\partial}{\partial T}\right|_p + \frac{dp}{dT}\left.\frac{\partial}{\partial p}\right|_T. \qquad (22.12)$$

Fig. 22.9 (a) The function appearing in equation (22.11). The value of p_0 was taken such that $p = p_0$ at $T = L/9R$. This means that if L is given by Trouton's rule then p is in atmospheres. (b) The measured vapour pressure of water as a function of inverse temperature (full line). The line ends at the triple point and the critical point. The dashed line shows the prediction of equation (22.11), with L and p_0 chosen to match observations at 1 atm.

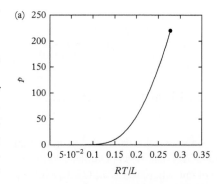

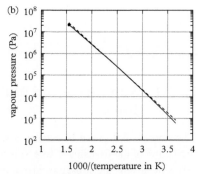

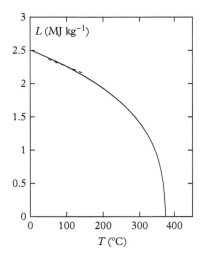

Fig. 22.10 Specific latent heat of water as a function of temperature. L falls to zero at the critical point. The dashed line shows an example of equation (22.15).

(The derivation of this expression is presented in Exercise 22.9.) Then,

$$\frac{d}{dT}\left(\frac{L}{T}\right) = \left.\frac{\partial}{\partial T}\right|_p (S_v - S_l) + \frac{dp}{dT}\left[\left.\frac{\partial S_v}{\partial p}\right|_T - \left.\frac{\partial S_l}{\partial p}\right|_T\right]$$

$$= \frac{C_{pv} - C_{pl}}{T} - \frac{dp}{dT}\left[\left.\frac{\partial V_v}{\partial T}\right|_p - \left.\frac{\partial V_l}{\partial T}\right|_p\right],$$

using a Maxwell relation. This is exact. Now assume $V_v \gg V_l$ and model the vapour as an ideal gas; then we have

$$\frac{1}{T}\frac{dL}{dT} - \frac{L}{T^2} \simeq \frac{C_{pv} - C_{pl}}{T} - \frac{Lp}{RT^2}\frac{R}{p} \qquad (22.13)$$

$$\Rightarrow \qquad \frac{dL}{dT} \simeq C_{pv} - C_{pl}, \qquad (22.14)$$

i.e. when a high-density phase changes to an ideal gas, the gradient of the latent heat as a function of temperature is equal to the difference in heat capacities of the two phases. Figure 22.10 shows an example. The heat capacity of the liquid is generally greater than that of the vapour, so this says that the latent heat falls with temperature, which makes sense since then the system is moving towards the critical point.

If we now assume that $C_{pv} - C_{pl}$ is constant, then we have[4]

$$L = L_0 + (C_{pv} - C_{pl})T. \qquad (22.15)$$

Substituting this into the Clausius–Clapeyron equation and using the ideal gas approximation again, the equation is easily integrated to give

$$p = p_0 \exp\left(-\frac{L_0}{RT} + \frac{\Delta C}{R}\ln T\right), \qquad (22.16)$$

where $\Delta C \equiv C_{pv} - C_{pl}$ and we have absorbed the constant of integration into p_0. This is a more accurate expression than (22.11) if the ideal gas approximation holds well for the vapour, but this is not the case near the critical point—see Exercise 22.7.

---

[4] In this equation, L_0 is simply a constant of integration; it can be obtained from $L_0 = L_m - (C_{pv} - C_{pl})T_b$, where L_m is the measured latent heat at some temperature T_m.

22.3.1.1 Liquid–solid coexistence line

A rough approximation to the liquid–solid coexistence line is obtained by neglecting the temperature dependence of both L and ΔV. Then we have

$$p = p_0 + \frac{L}{\Delta V} \ln \frac{T}{T_0}.$$

22.3.2 Gibbs phase rule

When a system is composed of several different types of material, such as different chemicals or isotopes, we say it has several *components*. In this situation the state space grows in dimension. Assuming the components simply mix without chemical reaction, then the number of degrees of freedom available to the intensive properties of a closed, simple compressible system in equilibrium is

$$f = C - P + 2, \qquad (22.17)$$

where C is the number of components and P is the number of phases present and in equilibrium with one another. This is called the *Gibbs phase rule*. It is a simple idea but can be helpful in making sense of the complicated phase diagrams that arise for multi-component systems.

Examples. For $C = 1$, a single-component system, then if only one phase is present, $P = 1$, there are two degrees of freedom, such as for example p and T, and there are possible equilibrium states throughout the p–T plane. If two phases are present, $P = 2$ gives $f = 1$: the state must be somewhere on the relevant coexistence curve, i.e. on a line rather than a plane. If three phases are present simultaneously then there is no remaining freedom and the system must be at the triple point. This means the intensive variables such as pressure and temperature and the densities of each phase are at their triple point values. The volume of the complete system is still variable, depending on how much of each phase is present; this is not addressed by the Gibbs phase rule.

For $C = 2$, such as a mixture of water and ethanol, the further degree of freedom could be the mole fraction x of one of the components. Allowing for all possible values of the mole fraction, the 'triple point' (when three phases are all present together) becomes a 'triple curve', that is, there is a continuous curve of possible values of pressure and temperature as a function of x. For this reason, care has to be taken to ensure purity of the composition when triple point cells are used to define temperature. Equally, for $C = 2$ and two phases present, one finds the melting or the boiling point at a given pressure is now a function of x; each coexistence line has become a coexistence surface. An example is the change in the freezing point of water when salt is added; we will study this in Chapter 24.

The proof of the Gibbs phase rule is accomplished essentially by generalizing the above examples. If there are C components, each will have a chemical potential, so together with temperature and pressure there are $C + 2$ intensive variables.

These are not all independent, however, as the Gibbs–Duhem relation shows. When more than one phase is present, no more chemical potentials are introduced (because in phase equilibrium the chemical potentials of different phases agree), but now there are further constraints, because there is a separate Gibbs–Duhem relation (21.3) for each phase. To see this, note that any given phase has a well-defined set of extensive variables giving the entropy, volume, and particle numbers in that phase, and one can apply thermodynamic reasoning to that part of the system alone. Hence there are P such Gibbs–Duhem relations, giving P ways in which changes of the $C + 2$ intensive variables are related to one another. Therefore the remaining number of independent intensive variables is given by (22.17).

22.3.3 Behaviour of the chemical potential

If we plot the chemical potential as a function of T at fixed pressure, near a first-order phase transition, the result is as shown in Figure 22.11. The chemical potential of either phase is well defined up to and somewhat beyond the transition temperature (into the metastable region; see Chapter 24), therefore we can examine what happens as the lines cross. Since

$$\left.\frac{\partial \mu}{\partial T}\right|_p = -s,$$

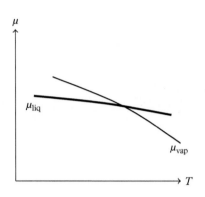

Fig. 22.11 *Chemical potential as a function of temperature. The gradients give minus the entropy per particle.*

the gradients of both lines are negative. The lines cross at the transition point, and the most stable phase is that with the lowest μ. As we move past the crossing point in the direction of increasing temperature, the line with the most negative slope must be the lower of the two, so we can deduce that the more stable phase at high temperature is that with the most entropy per particle.

Similar reasoning can be applied to the variation of μ with pressure at fixed temperature (Figure 22.12). We have

$$\left.\frac{\partial \mu}{\partial p}\right|_T = v = \frac{1}{n},$$

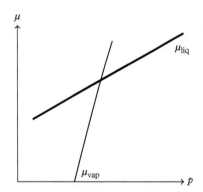

Fig. 22.12 *Chemical potential as a function of pressure (cf. Figure 24.4). Since the gradients give the inverse number density, you can tell that an abrupt density change happens, and you can tell which line is which.*

and again the most stable phase is that with the lowest μ. This time it is the line with least gradient that is lowest above the transition point. Therefore, for given temperature, the more stable phase at high pressure is that with the highest number density.

For further information on Figures 22.11 and 22.12, see Section 24.2.

22.4 The type-I superconducting transition

The phenomenon of *superconductivity* occurs in many metals and other electrically conducting materials at low temperature. It is so named because one of its most striking properties is the vanishing of electrical resistance below a certain temperature called the transition temperature. However, the phenomenon is most conveniently understood as a magnetic phenomenon in the first instance, and the chief magnetic property is the complete expulsion of magnetic field from within the superconducting material (the *Meissner effect*). This is sometimes referred to as 'perfect diamagnetism', but strictly speaking this is not brought about by any magnetization distributed uniformly through the material, but by a cooperative effect that results in a large surface current. In this situation, the surface current is precisely what it needs to be to ensure that $B = 0$ inside the superconductor. It follows that the total magnetic dipole moment of a sample of volume V is

$$m_s = -VH = -VB_{0I}/\mu_0, \qquad (22.18)$$

where H is the field provided by the current in the solenoid. This is the magnetic equation of state in the superconducting condition.

Since B_{0I} is the most convenient magnetic field to take as the applied field in this case, we begin the analysis with the function of state whose fundamental relation is

$$dU_{mI} = TdS + B_{0I}dm \qquad (22.19)$$

(cf. Eq. 14.80). The function U_{mI} is sometimes called internal energy, sometimes magnetic enthalpy. We wish to consider the situation for a system at fixed T and B_{0I}, so we consider the free energy, $G = U_{mI} - ST - B_{0I}m$, whose change is governed by

$$dG = -SdT - m\,dB_{0I}. \qquad (22.20)$$

At fixed T and B_{0I}, this free energy is minimized in equilibrium.

We can use the equation of state (22.18) to obtain the free energy in the superconducting regime, by integrating (22.20):

$$G_s(T, B_{0I}) = G_s(T, 0) - \int_0^{B_{0I}} m_s(T, B)dB = G_s(T, 0) + \frac{V}{2\mu_0}B_{0I}^2. \qquad (22.21)$$

This shows a strong dependence on magnetic field, whereas the free energy G_n of a normally conducting state does not have much dependence on magnetic field. It follows that, at some field $B_c(T)$, G_s will exceed G_n. Consequently, the superconducting state will not be the equilibrium state when B_{0I} exceeds a value $B_c(T)$, called the critical field. The application of a magnetic field is said to 'destroy' the superconductivity, but this is a somewhat misleading phrase. The situation is that of a phase transition: at $B_{0I} > B_c$ the superconducting phase is still possible in principle, but it is no longer the equilibrium phase because there is another condition—normal conduction—that has a lower free energy. (Similarly, in the case of vaporization, a reduction in pressure does not so much 'destroy' the liquid phase as make another phase more liable to be adopted.)

The phase transition takes place when the two phases—one superconducting, one normal—have the same G. The changes in G for the two phases are therefore equal between two neighbouring points on the coexistence line:

$$-S_s \mathrm{d}T - m_s \mathrm{d}B_c = -S_n \mathrm{d}T - m_n \mathrm{d}B_c, \qquad (22.22)$$

where we choose to write B_c for the value of B_{0I} on the coexistence line (this is like writing p_{vap} for vapour pressure). Hence, one finds

$$\frac{\mathrm{d}B_c}{\mathrm{d}T} = \frac{S_s - S_n}{m_n - m_s}. \qquad (22.23)$$

This is a Clausius–Clapeyron equation appropriate to the type of system under study.

Now m_s is given by (22.18) and m_n can be ignored in comparison with m_s, so we have

$$\frac{\mathrm{d}B_c}{\mathrm{d}T} = \frac{\mu_0(S_s - S_n)}{VB_c} = -\frac{\mu_0 L(T)}{TVB_c}. \qquad (22.24)$$

Hence we have a relationship between the gradient of the coexistence line, the field, and the latent heat $L = (S_n - S_s)T$. The latter is positive, so the slope is negative.

In the above, (22.23) is exact (in the thermodynamic limit) and (22.24) is exact when $m_n = 0$ and highly accurate when $m_n \ll m_s$. Note, there is no approximation involved in the employment of thermodynamic arguments here, except that we ignore the influence of fluctuations.

By differentiating (22.23) one finds a useful relationship between the difference in heat capacities and the coexistence curve:

$$\begin{aligned} C_s - C_n &= T \left.\frac{\partial}{\partial T}\right|_B \left((m_n - m_s)\frac{\mathrm{d}B_c}{\mathrm{d}T}\right) \\ &= \frac{TV}{\mu_0}\left[\left(\frac{\mathrm{d}B_c}{\mathrm{d}T}\right)^2 + B_c\frac{\mathrm{d}^2 B_c}{\mathrm{d}T^2}\right], \end{aligned} \qquad (22.25)$$

using (22.18) and $m_n = 0$. The temperature where $B_c = 0$ is called the critical temperature, T_c. Equation (22.25) predicts that the difference in heat capacities at the critical temperature is $\Delta C = (T_c V/\mu_0)(\mathrm{d}B_c/\mathrm{d}T)^2$.

One cannot predict $B_c(T)$ itself from general thermodynamic arguments: one needs either measurements or a theory of superconductivity. However since all entropies vanish as $T \to 0$, we know from (22.23) that B_c must be independent of T in that limit. A reasonable fit for many metals is a formula which has a maximum at $T = 0$, vanishes at $T = T_c$, and is quadratic in between:

$$B_c(T) = B_{c0}\left(1 - (T/T_c)^2\right), \qquad (22.26)$$

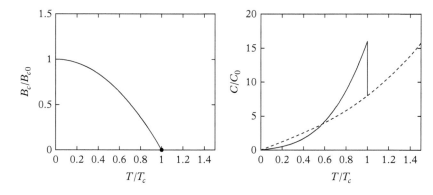

Fig. 22.13 (a) Coexistence line (i.e. the dependence of the critical field on temperature) for a type I superconductor; equation (22.24). (b) Heat capacities C_n (dashed) and C_s of normal and superconducting phases, in units of $C_0 \equiv (1/2)B_{c0}^2 V/\mu_0 T_c$. The difference between the lines is given by equation (22.27).

(see Figure 22.13). Thus one finds that the transition happens at lower temperature as the field is raised from zero, and superconductivity does not occur at all if the applied field is above a value B_{c0}. Using (22.26) in (22.25), the heat capacity difference for states on the coexistence curve is predicted to be

$$C_s - C_n = \frac{2VB_{c0}^2 T}{\mu_0 T_c^2}\left(3\frac{T^2}{T_c^2} - 1\right). \qquad (22.27)$$

This is well obeyed in practice, which confirms the quadratic formula for B_c.

The point where the coexistence line terminates at $(B_c, T) = (0, T_c)$ is a critical point in the thermodynamic sense: here the transition becomes continuous, and it is well modelled as a second-order phase transition.

....

EXERCISES

(22.1) The gradient of the melting line of water on a p–T diagram close to $0\,°C$ is -1.4×10^7 Pa/K. At $0\,°C$, the specific volume of water is 1.00×10^{-3} m^{3}kg^{-1} and of ice is 1.09×10^{-3} m^{3}kg^{-1}. Using this information, deduce the latent heat of fusion of ice.

(22.2) Calculate the amount of energy required to convert 50 g of ice at $-10\,°C$ to steam at $150\,°C$. (The latent heats are 334 kJ/kg and 2257 kJ/kg for melting and vaporization, respectively. The heat capacities may be taken roughly constant at 2108 J K^{-1}kg^{-1}, 4190 J K^{-1}kg^{-1}, 2074 J K^{-1}kg^{-1} for ice, water, steam).

(22.3) Using the Clausius–Clapeyron equation, estimate the temperature at which water boils at the top of Mount Everest (altitude 8854 m). (The air pressure is about 0.5 atm at a height of 18 km.)

(22.4) A pool of liquid in equilibrium with its vapour is converted totally into vapour in conditions of fixed temperature and pressure. What happens to the internal energy, the enthalpy, the Helmholtz function, and the Gibbs free energy?

(22.5) By how much can you lower the melting point of ice by pushing with your thumb on a drawing pin or thumb tack? (A thumb tack is a small metal pin with a large flat head; estimate the force you can apply and make a reasonable estimate of the area in contact with the ice at the sharp end.)

(22.6) Find the depth of a glacier such that the bottom will melt in conditions where the atmospheric pressure is 1 atm and the temperature at the bottom of the glacier is $-5\,°\mathrm{C}$.

(22.7) Consider the Clausius–Clapeyron equation applied to the liquid–vapour transition. Show that the second virial coefficient in the virial expansion for the gas affects dp/dT in the opposite sense to the temperature dependence of the latent heat, so that the two effects tend to compensate one another. (The result is that (22.11) can sometimes be more accurate than (22.16)!)

(22.8) It is quite common to see two of the coexistence lines at a triple point having almost the same gradient (i.e. the angle between them approaches 180°). What does this tell us about the properties of the three phases?

(22.9) Derivation of (22.12). Consider an arbitrary function of state f expressed as a function of p and T (for a closed system): $df = \left.\frac{\partial f}{\partial T}\right|_p dT + \left.\frac{\partial f}{\partial p}\right|_T dp$. Divide this expression by dT under the constraint that both T and p change but dp/dT is fixed. Hence derive equation (22.12).

(22.10) Compare the Clausius–Clapeyron equation with the van 't Hoff equation.

The third law

23

(Planck–Simon statement)
The contribution to the entropy of a system by each aspect of the system which is in internal thermodynamic equilibrium tends to zero as the temperature tends to absolute zero.

23.1 Response functions	332
23.2 Unattainability theorem	333
23.3 Phase change	334
23.4 Absolute entropy and chemical potential	335

The third law of thermodynamics, though it is somewhat less important than the first and second laws, has earned the status of 'organizing principle' because it is very simple and natural (as well as being useful and consistent with observations), but it cannot be derived from the other laws. Historically, for a while it was unclear whether or not the second law was sufficient to impose a condition on entropy at absolute zero temperature, but all arguments to make such a connection have failed and it is generally agreed that the third law is an independent assertion. From a modern point of view, the third law is equivalent to the assertion that the energetic ground state of a quantum system is always non-degenerate. This also is a very natural assertion, but hard to prove,[1] and consequently it has a status similar to that of the third law of thermodynamics.

A system at absolute zero temperature can have a range of values of its other parameters or generalized coordinates, such as volume, magnetic field, internal energy, and so on. The third law asserts that as those other parameters are varied, by compressing the system, for example, or by adjusting the applied fields or chemical concentrations, the entropy remains constant at zero: the thermodynamic state of any aspect that is in thermal equilibrium moves on an adiabatic surface.

In the statement of the third law given above, an *aspect* of a system is a part of the system, or a set of processes, that only interacts weakly with the rest of the system, so that it comes to equilibrium independently of the other aspects. It is useful to separate out one aspect from another, such as, for example, the magnetic from the vibrational, because in some cases such as a glass or a ferromagnet or a nuclear spin system, one aspect might be very slow to reach equilibrium. The third law asserts that the aspects that reach equilibrium go to zero entropy at zero temperature, even if there remain other aspects that are out of equilibrium and may or may not have zero entropy. Also, the equilibrium spoken of in the third law may be metastable. At absolute zero, it becomes stable against thermal fluctuations, because there are none, but there remains quantum uncertainty which in some respects is like a form of fluctuation.

[1] It is natural to conjecture that the ground state of any quantum system is non-degenerate, because if it were not, then any interaction, no matter how small, would be sufficient to lift the degeneracy. However, there remains the possibility that two or more states are strictly degenerate if all interactions coupling them are ruled out by symmetry or other considerations.

The fact that one aspect can be considered independently of another, and the fact that metastable equilibrium is included, are both central to the overall significance and usefulness of the third law. The third law is useful because it has implications for the limits of physical behaviour in conditions close to but not exactly at zero temperature. In the rest of this chapter we present examples.

23.1 Response functions

The first important consequence of the third law is that all heat capacities must tend to zero as the temperature tends to zero. Any heat capacity is, by definition, a quantity of the form

$$C_x = T \left. \frac{\partial S}{\partial T} \right|_x, \qquad (23.1)$$

therefore the change in entropy on cooling from T to absolute zero at constant x is

$$\Delta S = \int_T^0 \frac{C_x}{T} dT. \qquad (23.2)$$

At a finite temperature, S is finite, and at $T = 0$ the third law tells us that $S = 0$. Therefore the change ΔS must be finite. But the integral is only finite if $C_x \to 0$ at least as fast as T. Therefore all the heat capacities vanish at least this fast.

The fact that $S \to 0$ as $T \to 0$ does not in itself imply anything about $\partial S/\partial T$, but it does imply that partial derivatives at constant T must vanish:

$$\lim_{T \to 0} \left. \frac{\partial S}{\partial x} \right|_T = 0. \qquad (23.3)$$

This property allows us to find the limiting behaviour of various response functions. For example, by a Maxwell relation,

$$\left. \frac{\partial S}{\partial p} \right|_T = - \left. \frac{\partial V}{\partial T} \right|_p = -V\alpha_p, \qquad (23.4)$$

so we can deduce that $\alpha_p \to 0$. In other words, at low temperature, a material will neither expand nor contract when its temperature is raised at constant pressure.

The corresponding Maxwell relation for surface tension is

$$\left. \frac{\partial S}{\partial A} \right|_T = - \left. \frac{\partial \sigma}{\partial T} \right|_A, \qquad (23.5)$$

so we deduce that σ becomes constant for temperatures near zero. This has been observed in liquid helium.

For paramagnetic materials, the third law predicts that the susceptibility becomes independent of temperature (see Eq. 14.27i). It follows that Curie's law

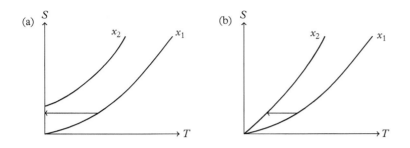

Fig. 23.1 *(a) Impossible and (b) possible behaviour of the entropy as a function of temperature for two values of some state parameter x. In the first case the value of S depends on x at T = 0, which is ruled out by the third law. If it were not ruled out, then the adiabatic change in the value of x indicated by the arrow would allow cooling to absolute zero. In (b) an adiabatic change in x can produce cooling, but not to T = 0.*

($\chi = a/T$) must break down at some temperature, and indeed we have already studied the fact that a phase transition occurs at low temperatures. The ideal paramagnet, like the ideal gas, cannot capture the behaviour all the way to absolute zero temperature.

23.2 Unattainability theorem

Figure 23.1 shows two scenarios, one impossible (ruled out by the third law) and one possible. Both figures plot curves of constant x on an ST indicator diagram, where x is some state parameter (other than S, T) of a thermodynamic system. The curves have positive gradient because heat capacities are always positive in stable equilibrium. The gradients also don't tend to infinity, so the heat capacity tends to zero as it should in both diagrams. However, in the first diagram the entropy at $T = 0$ for one value of x is not the same as the entropy at $T = 0$ for the other value of x, and this is ruled out by the third law.

Now suppose we wish to cool a system. We would adjust one of its parameters, x, in whatever direction produces a lower temperature. If the system temperature is lower than that of the surroundings, then heat flow will tend to warm the system, so should be prevented if cooling is our aim. We conclude that, for a given adjustment in the parameter x, the most cooling will be achieved for an adiabatic process, and therefore a horizontal line on the diagrams in Figure 23.1. In the first case (the impossible one) this could result in cooling to absolute zero. In the second case (the possible one), it cannot.

Now argue, from the third law, that all lines of constant x on such a diagram must meet at $T = 0$, and, from stability, must have positive gradient. Then we can deduce that no adiabatic cooling process between one value of x and another will attain absolute zero temperature, after starting from a finite temperature. The unattainability theorem (Section 3.3) follows. Figure 23.2 illustrates the kind of sequence of steps envisioned in the standard statement of the unattainability theorem.

The unattainability theorem follows from the third law, but is not equivalent to it, because it cannot be used to show that all systems and aspects reach the same entropy at absolute zero. It only requires that $S(T, x_1) = S(T, x_2)$ when $T = 0$.

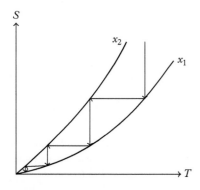

Fig. 23.2 *Cooling by a sequence of isothermal and adiabatic steps.*

A further impossibility can also be conveniently illustrated by an entropy–temperature diagram. On such a diagram, a Carnot cycle appears as a rectangle. The size of the rectangle is limited by the values of the parameter x that are accessible in any given experiment. If one tries to place such a rectangle between the x curves, with one corner at absolute zero temperature, then the size of the rectangle must shrink to zero. Consequently such a Carnot cycle exchanges no heat, and does no work: it has become a non-process, just a single point in the state space. More generally, for a system with further parameters, one may observe that an isothermal process at $T = 0$ has to be an adiabatic process. Therefore during this part of the cycle, there is no way to guarantee that the value of whatever parameters the engine is working on will change. Consequently the engine cannot trace a cycle enclosing a non-zero area ST-space, therefore it extracts no heat and delivers no work overall. It follows that an attempt to get a perfectly efficient conversion of heat to work by this method fails.

23.3 Phase change

The Clausius–Clapeyron equation,

$$\frac{dp}{dT} = \frac{\Delta S}{\Delta V} \tag{23.6}$$

relates the gradient of the phase coexistence line to the entropy and volume changes in a first-order phase transition. The third law requires that $\Delta S \to 0$ as $T \to 0$, so the coexistence line becomes horizontal on a pT diagram if the p axis is vertical. This is borne out by observations on ^{4}He, where the solid–liquid coexistence line extends to absolute zero (Figure 22.5). (Don't be misled by sloppy phase diagrams which appear to show a coexistence line hitting the origin, or somewhere else on the p axis, with a non-zero gradient!)

The corresponding statement for the superconducting phase transition in a magnetic field is that the critical field becomes independent of temperature in the limit $T \to 0$.

The third law also implies that any phase change taking place at absolute zero temperature (while some other parameter is varied) cannot involve a change of entropy.

The metastable phases associated with first-order phase transitions offer a useful test of the third law, because the law asserts that both the fully stable and the metastable phase will tend to zero entropy at absolute zero temperature. Solids such as tin, sulphur, and phosphine have phase transitions (called allotropic transformations) corresponding to reordering of their crystalline state, in which the transition takes place sufficiently slowly that, by rapid cooling, the high temperature phase can be retained in its supercooled state all the way to absolute zero. Alternatively, by cooling more slowly, one may bring the material into its most stable phase and then attain close to absolute zero in that case. The test now proceeds as follows. By warming the system in either case by controlled amounts of heat, and measuring the temperature, one can calculate the entropy increase from absolute zero, in each case, until the substance is brought to its high temperature phase in stable conditions. In the final conditions, both experimental systems are in the same thermodynamic state, so if the calculated entropy increases agree, then one may deduce that the entropies in the different starting states at absolute zero temperature must also have agreed. This has been experimentally confirmed to high precision.

23.4 Absolute entropy and chemical potential

It is sometimes asserted that the zero of entropy is an arbitrary choice, and all the results of classical thermodynamics would be unchanged if we said the entropy tended to some other value at absolute zero, as long as we used the same value for all systems. This is not true, because according to the Gibbs–Duhem equation, we have

$$\frac{S}{V} = \left.\frac{\partial p}{\partial T}\right|_{\mu}, \quad \text{and} \quad \frac{S}{N} = -\left.\frac{\partial \mu}{\partial T}\right|_{p}. \quad (23.7)$$

Therefore, if we have a system of known volume, and we measure the change in pressure and temperature associated with a small change at constant chemical potential, then we can determine the absolute entropy of the system. An example is given by the fountain effect in liquid helium; equation (24.22). Similarly, if we can measure the change in chemical potential and temperature during a small change at constant pressure, then we can determine the absolute entropy per particle, and hence the total entropy.

For thermal radiation, a measurement of the Stefan–Boltzmann constant suffices, since then one can use (19.30) and (19.33) which together give $S = 16\sigma VT^3/3c$. For an ideal gas, the value of the constant a in equation (13.42) can be deduced from chemical equilibrium measurements, or by the study of a plasma using the Saha equation.

24 Phase change, nucleation, and solutes

24.1 Treatment of surface effects   336
24.2 Metastable phases   338
24.3 Colligative properties   347
24.4 Chapter summary   353
Exercises   353

In this chapter we consider further aspects of first-order phase transitions. We begin with a general treatment of surface effects, because this is needed to understand the way surface tension modifies the liquid–vapour phase transition. Similar effects happen in other types of phase transition, for example magnetic transitions, where domain walls contribute to the energy and entropy. We then consider what happens when one substance is dissolved in another, such as salt in water. This can significantly change the melting and boiling point, and also gives rise to related phenomena such as osmotic pressure.

24.1 Treatment of surface effects

Any finite system will have a surface, and a system composed of more than one spatial part will in general have surfaces within it, separating one part from another. With the exception of the mention of surface tension in Chapter 14, we have so far ignored surface effects since for large systems they are negligible compared to the bulk. However we need to understand the influence of surfaces in order to understand some important aspects of phase change.

We begin with general thermodynamic considerations, because some of the proofs in Chapter 17 were only valid in the absence of surfaces and we shall need to know what results we can trust.

We consider the situation presented in Figure 17.2: a system, not necessarily in equilibrium, divided into two parts. We now allow that the boundary between the two parts may itself have a means of storing energy and entropy. For example, for a liquid–gas interface, the energy of the surface is owing to the fact that to make the surface larger without changing the volume it encloses (e.g. by deforming it), one would have to break attractive chemical bonds. The entropy of the surface is owing to the fact that it can support a large number of small ripples. The analysis is essentially the same as that of a set of three systems, but no volume or particles are assigned to the surface. Its properties are related by

$$dU_s = TdS_s + \sigma dA, \tag{24.1}$$

where σ is the surface tension.

The total entropy is

$$S = S_1 + S_2 + S_s \tag{24.2}$$

and now instead of equation (17.2), we have

$$dS = \left[\frac{1}{T_1} - \frac{\alpha}{T_2} - \frac{1-\alpha}{T_s}\right]dU_1 + \left[\frac{p_1}{T_1} - \frac{p_2}{T_2} - \frac{\sigma}{T_s}\frac{dA}{dV_1}\right]dV_1 - \left[\frac{\mu_1}{T_1} - \frac{\mu_2}{T_2}\right]dN_1, \tag{24.3}$$

where α and $1 - \alpha$ give the proportions of the energy arriving in 1 that came from 2 and s. We argue, as before, that the movements dU_1, dV_1, and dN_1 are independent, and furthermore the energy might be divided in any way, so in some cases α might take the value 1, so we deduce $T_1 = T_2$ in equilibrium. It follows that $T_s = T_1 = T_2$ in equilibrium. Thus in equilibrium the temperature is uniform through the system. So is the chemical potential, as one can see immediately, but the pressure may not be.

The area A is not a completely independent parameter. By writing dA/dV_1—a total derivative—in equation (24.3) we have implied that dA is determined by dV_1. This is valid if we take it for granted that the surface is free to adjust its shape at given enclosed volume. We see from (24.1) that since it contributes negatively to the entropy, the surface will, at equilibrium, be configured to have the minimal area consistent with the volume it has to enclose. Once it adopts this shape its area can be expressed as a function of V_1 alone, so the expression (24.3) is valid. By considering the dV_1 term one then finds that at equilibrium,

$$p_1 = p_2 + \sigma \frac{dA}{dV_1}. \tag{24.4}$$

We have already discussed this in Section 14.3 and Exercise 17.3. For a spherical drop, $dA/dV = 2/r$.

In the above we have considered surface tension in a pV system. Similar considerations can arise in other types of system, such as a magnetic system with interior domains separated by domain walls. These walls also carry an associated energy and entropy.

Now let's consider a system composed of a drop of liquid surrounded by vapour; Figure 24.1. The presence of the drop *inside* the system does not affect the way the system can interact with its surroundings, so the fundamental relation for the complete system is

$$dU = TdS - pdV + \mu dN, \tag{24.5}$$

where the variables T, p, μ must be those which apply where the system meets its surroundings, i.e. properties of the *vapour*, if we assume the vapour completely surrounds the liquid drop.

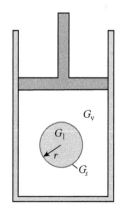

Fig. 24.1 *A spherical liquid drop inside a vapour.*

We can usefully divide the energy and entropy into contributions from the bulk vapour and liquid, and contributions from the surface:

$$U = U_v + U_l + U_s$$
$$S = S_v + S_l + S_s$$
$$N = N_v + N_l$$
$$V = V_v + V_l.$$

No volume or particles are assigned to the surface.

With these replacements the fundamental relation takes the form

$$dU = T_v(dS_v + dS_l + dS_s) - p_v(dV_v + dV_l) + \mu_v(dN_v + dN_l),$$

where we have made explicit that T, p, μ in (24.5) are those of the vapour. However we have already established that at equilibrium $T_v = T_l = T_s$ and $\mu_v = \mu_l$, so we have

$$dU = (T_v dS_v - p_v dV_v + \mu_v dN_v) + (T_l dS_l - p_l dV_l + \mu_l dN_l) + T_s dS_s + (p_l - p_v)dV_l,$$

which after using (24.4), gives

$$dU = dU_v + dU_l + (T_s dS_s + \sigma dA) \tag{24.6}$$

as expected.

24.2 Metastable phases

An important property of first-order phase transitions is that it is possible, and quite common, to obtain a substance in the 'wrong' phase. For example, a vapour below the boiling point (called a *supersaturated vapour*) or a liquid above the boiling point (called a *superheated liquid*). This phenomenon is not restricted to the liquid–vapour transition but applies to all first-order transitions. The 'correct' phase (i.e. the one with the lowest chemical potential) remains the only thermodynamically stable phase, but the situation is metastable and can persist for long periods.

The possibility of a persistent phase other than the most stable one is implied by the behaviour of the entropy, as presented in Figure 17.5. Since this behaviour is at the heart of phase transitions, it is reproduced in Figure 24.2. The condition for stability against *small* fluctuations is

$$\left(\frac{\partial^2 S}{\partial U^2}\right)_{V,N} < 0$$

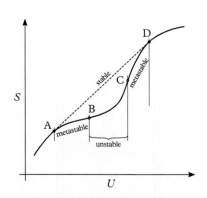

Fig. 24.2 *Entropy as a function of internal energy for a closed system of fixed volume near a first-order phase transition.*

(this is equation (17.6)). This condition is satisfied in the regions AB and CD of Figure 24.2 (as well as below A and above D). If we start at $U < U_A$ and heat the system, then it can pass smoothly into the AB region. If the phase at $U < U_A$

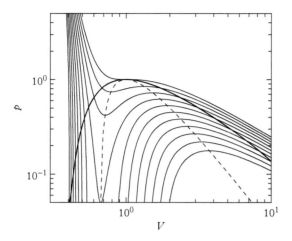

Fig. 24.3 Binodal (thick) and spinodal (dashed) curves for a liquid/gas phase transition, as modelled by the van der Waals equation (the other lines are isotherms). The supercooled or superheated phase follows these isotherms in the metastable region, which is between the spinodal and binodal curves.

is a liquid, for example, then since no abrupt change has taken place it remains a liquid. A large fluctuation is needed in order to allow the system to 'discover' that another configuration is available, namely the two-phase configuration, and that this has higher entropy. In the absence of a large fluctuation, or before one occurs, we can have a superheated liquid. If the internal energy is increased beyond U_B, on the other hand, the entropy is unstable even to small fluctuations. Then fluctuations build up very rapidly and the system passes to the two-phase configuration.

The metastable and unstable regions for a typical liquid–vapour phase change are conveniently shown on a pV diagram such as that shown in Figure 24.3. The coexistence curve is also called a *binodal curve*. This is the locus of points for which $\mu_{\text{liq}} = \mu_{\text{vap}}$; it separates the region where only one phase is stable from the region where two phases can coexist. The binodal line can be found, for example, by the Maxwell construction. Within the coexistence region there is another line called the *spinodal curve*. This marks the boundary of the region where the system can exist fully in one phase, in a metastable condition—that is, a condition that is stable to small fluctuations but not to large fluctuations. Thus, between binodal and spinodal curves the state can be either single phase and metastable or mixed phase and stable, but inside the spinodal curve the state can only be mixed phase (in this region, if a single phase were present on its own, it would be unstable). For the van der Waals theory, the spinodal curve is the locus of the stationary points on the isotherms on a pV diagram. The binodal and spinodal curves meet at the critical point.

It is owing to this metastable behaviour that diagrams such as Figures 22.11 and 22.12 can be drawn. If the less stable phase were not metastable then one could not plot the value of its chemical potential beyond the transition point, because the concept would be undefined. Indeed, for continuous transitions that is what happens: there is no metastable behaviour. For first-order transitions, on the

340 *Phase change, nucleation, and solutes*

other hand, we can track the chemical potential of all phases either by calculation or by measuring the properties of supercooled or superheated substances. In the case of the van der Waals gas, we can obtain the chemical potential from the Gibbs function. For the monatomic case we have

$$G = F + pV = RT\left[\ln\left(\frac{N_A \lambda_T^3}{V-b}\right) - 1\right] - \frac{a}{V} + pV, \tag{24.7}$$

using equation (15.13). We have displayed this as a function of all three variables, for convenience, but of course the van der Waals equation of state gives p in terms of the other two. In order to plot G as a function of pressure at given T, it is convenient to use this equation for G alongside the equation of state, so that one has an implicit equation with V serving the role of parameter. Choosing $T = 0.86 T_c$ for illustrative purposes, and letting V vary from $0.52 V_c$ to $5 V_c$, we obtain the curve shown in Figure 24.4. In the region A–t on the diagram, the system is

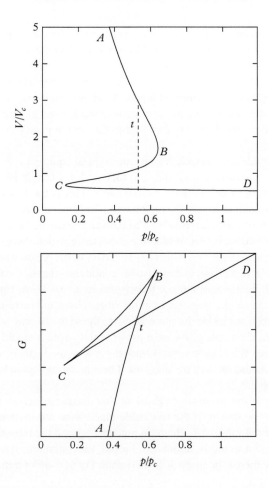

Fig. 24.4 *Behaviour of the Gibbs free energy in the phase transition region of the van der Waals gas. The isotherm at $T = 0.86\, T_c$ is shown.*

wholly vapour. Then, if the change is slow enough, the whole phase transition occurs at t, a single point in the Gp diagram, until the vapour is fully condensed and then from t–D it is liquid. If, instead, one adjusts the system parameters more rapidly, then one may pass into the regions Ct and tB (each approached from the appropriate side). The region tB is a supercooled vapour. The region Ct is a superheated liquid. The region BC has a negative compressibility so is very unstable and does not represent an accessible thermodynamic state. The regions tB and Ct are metastable.

Although the van der Waals model is not quantitatively precise in the phase transition region, it gives a correct qualitative picture and the Gibbs function of real gases and liquids has this behaviour. As the temperature approaches the critical point, the 'triangle' CBt gets smaller and the gradients at t become more and more alike, becoming equal at the critical point, where the triangle vanishes.

It is also possible to have metastable situations in which two phases are in mutual equilibrium at a pressure and temperature where the third phase is the most stable. For example, the continuation of the liquid–vapour coexistence line past the triple point (line tA on Figure 24.5) indicates a set of states where water and water vapour can both be supercooled together, at a pressure and temperature where ice is the most stable phase. This is commonly observed in meteorology: the air in a cloud can be cooled below the temperature at which ice crystals begin to form, and meanwhile there can also be water droplets in the same cloud. The presence of these droplets raises the vapour pressure in the cloud, as we shall see in Section 24.2.1, with the consequence that ice crystals then grow more rapidly.

In principle one can also have corresponding situations along the continuations of the other coexistence lines, i.e. water and ice at low pressure where the vapour is stable (line tB on Figure 24.5), or ice and vapour in the liquid region of the phase diagram (line tC). All these metastable situations can penetrate a considerable but finite distance past the triple point. They terminate when the system state reaches points corresponding to the case B or C in Figure 24.2.

By considering these metastable coexistence lines, one can easily prove that the angles between the equilibrium lines around the triple point are never greater than 180°. For, consider the line tA. Along this line, $\mu_l = \mu_v$ and $\mu_s < \mu_l, \mu_v$. If instead this line penetrated into the vapour region, then one would have $\mu_l = \mu_v$ in a region where $\mu_v < \mu_s, \mu_l$, which is a contradiction. Hence the situation cannot arise, and therefore the angles between coexistence lines around the triple point must all be less than 180°.

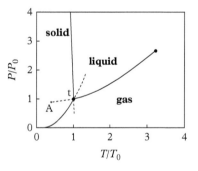

Fig. 24.5 *Phase diagram showing continuation of the coexistence curves into the metastable region.*

24.2.1 Nucleation

The metastability can be enhanced by surface effects, as we now show.

Let us calculate the free energy of a system consisting of a single spherical drop surrounded by vapour, taking into account the effect of surface tension. The system is illustrated in Figure 24.1. It interacts with the rest of the world through heat exchange and mechanical work, so the fundamental relation is

$$dU = TdS - pdV,$$

where $V = V_v + V_l$ is the total volume, and $S = S_v + S_l + S_s$ is the total entropy.

The spherical drop has surface area $4\pi r^2$. We will treat the case where the heat capacity of the surface is negligible. In this case the internal energy of the system (Eq. 24.6) can be written

$$U = U_v + U_l + 4\pi r^2 \sigma, \tag{24.8}$$

where U_v and U_l are the contributions from the bulk, and $\sigma = \sigma(T)$ is the surface tension.

In conditions where the environment is one of fixed temperature and pressure, the availability (the free energy) is the Gibbs function,

$$G = U - TS + pV, \tag{24.9}$$

where T and p are the external temperature and pressure, which are here the same as the temperature and pressure of the *vapour*. The temperature of the drop of liquid is also equal to T, and the pressure inside the drop is raised owing to surface tension: $p_l = p + 2\sigma/r$ (equation (24.4)).

From (24.8) and (24.9), we have

$$G = U_v + U_l + 4\pi r^2 \sigma - T(S_v + S_l) + p(V_v + V_l)$$
$$= G_v(T, p) + G_l(T, p_l) + (p - p_l)V_l + 4\pi r^2 \sigma, \tag{24.10}$$

where G_v and G_l are the Gibbs functions of the bulk vapour and liquid, and we have neglected TS_s since we have already assumed that the heat capacity associated with the surface is negligible. Notice that G_l, the bulk Gibbs function of the liquid, is here to be evaluated *at the temperature and pressure of the liquid*. If the liquid can be treated as incompressible, then

$$G_l(T, p_l) = G_l(T, p) + (p_l - p)V_l, \tag{24.11}$$

since then a change in pressure at constant T will not affect U_l, S_l, or V_l. Using this in (24.10), we have

$$G = G_v(T, p) + G_l(T, p) + 4\pi r^2 \sigma,$$

which is useful because now everything is evaluated at the pressure of the vapour.

Let N be the total number of particles in the system. The number of particles in the drop is $(4/3)\pi r^3 n_l$ so, using $G_v = N_v \mu_v$ and $G_l = N_l \mu_l$, we have

$$G = \left(N - \frac{4}{3}\pi r^3 n_l\right)\mu_v + \frac{4}{3}\pi r^3 n_l \mu_l + 4\pi r^2 \sigma, \tag{24.12}$$

where all quantities are evaluated at T, p, the temperature and pressure of the vapour. Hence we find that although the pressure difference does not affect U_l,

S_l, V_l, it does affect G_l and therefore also the chemical potential of the liquid. Our expression is valid both for equilibrium and out of equilibrium conditions (i.e. when the drop and vapour are in well-defined thermodynamic states but out of equilibrium with one another), and therefore it can be used to understand how equilibrium is established. In equilibrium we shall find $\mu_l(T,p_l) = \mu_v(T,p)$, but we have *not* assumed this in the above.

Equation (24.12) can be written

$$G = N\mu_v - \frac{4}{3}\pi r^3 \Delta g + 4\pi r^2 \sigma, \qquad (24.13)$$

where

$$\Delta g \equiv n_l(\mu_v - \mu_l). \qquad (24.14)$$

This is our main result. Equation (24.13) can be 'read' as the observation that, to form a droplet, the system must pay the free energy cost $4\pi r^2 \sigma$ required to form its surface, in order to gain the free energy pay-off $(4/3)\pi r^3 \Delta g$ offered by its volume. One could, possibly, write down equation (24.13) based on this insight, except that we needed the full analysis in order to learn that both chemical potentials in (24.14) are evaluated at the pressure of the vapour.

Equation (24.13) is plotted as a function of r at given T, p in Figure 24.6, for a case where $\Delta g > 0$. Positive Δg means the chemical potential of the bulk liquid is lower than that of the vapour, so if there were no surface we would expect the stable phase to be liquid. The thermal equilibrium condition (i.e. maximum global entropy) is that G should be minimized, so we find there are two equilibrium configurations: a metastable case (i.e. a local minimum) at $r = 0$, i.e. no liquid at all, and the true stable equilibrium where $r \to \infty$, i.e. the drop grows until the whole system is liquid.

Figure 24.7 helps to get the complete picture. The figure shows the chemical potential for vapour and liquid, at a given T, as a function of the pressure *in the relevant substance*. p_s is the saturation vapour pressure, which is the pressure when $\mu_v(p_s) = \mu_l(p_s)$ at the given temperature (we shall suppress the mention of T, which is assumed to be fixed and the same for vapour and liquid throughout the following). p is the pressure imposed by the external constraints, which means it is the pressure of the vapour. The pressure inside the liquid droplet is above p, so $\mu_l(p_l)$ lies somewhere on the liquid chemical potential line to the right of p_v. It is further to the right for smaller droplets, and tending to p_v for very large droplets. So we see that $\mu_l(p_l) > \mu_v(p_v)$ for small droplets; no wonder, then, that they tend to evaporate. On the other hand, $\mu_l(p_l) < \mu_v(p_v)$ for large droplets, so they grow. There is a size of droplet for which $\mu_l(p_l) = \mu_v(p_v)$: this occurs at the maximum of the G curve shown in Figure 24.6, and we shall consider it in a moment.

What we learn from this is that if the system starts out as completely vapour and then is cooled below the transition temperature of the bulk substance, then the transition to liquid form cannot happen by the steady growth of small drops.

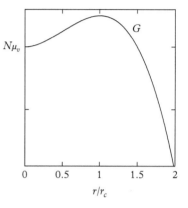

Fig. 24.6 *The variation with drop radius of the Gibbs function of a system consisting of a vapour containing a spherical liquid drop, for fixed pressure and temperature of the vapour, in the supercooled regime. The system is in mechanical equilibrium (the enhanced pressure in the drop is just that needed to counteract the effect of surface tension) but not in phase equilibrium, except that there is a point of unstable phase equilibrium at the maximum of G. The situation at $r = 0$ is metastable.*

Fig. 24.7 Chemical potentials for a supercooled vapour. The full lines show μ_v and μ_l, the chemical potentials of bulk vapour and liquid, as a function of pressure. When the pressure in the vapour is p_v, that in a liquid drop is $p_l = p_v + 2\sigma/r$. The figure shows an example for a small drop, resulting in $\mu_l(p_l) > \mu_v(p_v)$. There exists a drop radius for which $\mu_l(p_l) = \mu_v(p_v)$. This occurs when $p_l - p_v = \Delta g$ (see text).

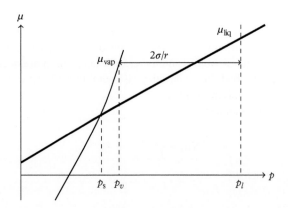

The system has to get over the 'hump' or local maximum of G, which occurs at a *critical radius* given by

$$r_c = \frac{2\sigma}{\Delta g}. \tag{24.15}$$

(Exercise 24.3). If we ignore the motion of the drops (as we have done throughout the above), then for a drop smaller than r_c, the free energy released by forming its volume is not enough to form its surface; should such a drop form it will immediately evaporate again. Once a drop somehow or other gets to a radius above r_c, it will then continue to grow.

In practice the drops will themselves be in thermal motion, which introduces another consideration. The motion of small drops of a given size can be accounted for by regarding them as forming another chemical component in the system, which behaves like an ideal gas of particles of mass $N_r m$, where m is the molecular mass and $N_r = (4/3)\pi r^3 n_l$ is the number of molecules in a drop of radius r. The free energy cost of increasing the size of a drop is like the energy involved in a chemical reaction or an ionization process: it can be provided by the rest of the system if the total entropy thereby goes up. One can find the equilibrium concentrations of drops of different N_r by adapting the generalized Saha equation (12.29). For an estimate, one argues simply that the centre-of-mass motion of droplets of size N_r makes a contribution to the free energy, given by the number of such droplets multiplied by μ_{CM}, given by

$$\mu_{\text{CM}} = k_B T \ln\left[n_r \left(2\pi\hbar^2/N_r m k_B T\right)^{3/2}\right],$$

where n_r is the number density of these droplets in the vapour, and we used equation (12.8). In equilibrium, $\mu_{\text{CM}} = -\Delta G(r)$, where $\Delta G(r) = 4\pi r^2 \sigma - (4/3)\pi r^3 \Delta g$ is the free energy cost of a single non-moving drop of size r. Therefore the equilibrium number density of droplets of radius r is

$$n_r = ((4/3)\pi r^3 n_l m k_B T/2\pi\hbar^2)^{3/2} e^{-\Delta G(r)/k_B T}. \tag{24.16}$$

This argument does not apply to the larger drops (near the critical size, for example) since these are not in equilibrium; they are in a dynamic process of growing or shrinking.

The process of forming drops is called *nucleation*. If there are no impurities, a supersaturated vapour will eventually condense through the mechanism of thermal fluctuations which can supply the required free energy to one part of the system at the expense of another. However this can be very slow; an estimate for water vapour at room temperature and a pressure of twice the saturated vapour pressure is that it would take 10^{40} years to condense without external aid. To bring about condensation, either a higher pressure or the presence of a catalyst is needed. In a pure vapour the pressure becomes sufficient once it is above a few times the saturation pressure.

In practice, condensation almost always involves a catalyst. Nucleation can be catalysed by the presence of minute grains of dust that can adsorb liquid onto their surface, or other types of particle such as charged ions which can attract molecules around them (especially polar molecules). These offer a way for the liquid drops to attain the critical radius. Such particles are called *condensation nuclei* and in practice this mechanism usually is the one by which vapours condense, because it is rare for a vapour to be so pure that it does not contain significant numbers of condensation nuclei of one sort or another.

Similar considerations arise in superheated liquids, in which only bubbles of radius above a finite value will grow (Figure 24.8). This contributes to the violent bubbling that typically is observed in a boiling liquid. The boiling can be rendered more calm by adding small granules, such as glass or ceramic, to provide nucleation centres.

The *cloud chamber* (Wilson) and the *bubble chamber* (Donald Glaser) have been used with great success in particle physics to track the motion of charged particles. The cloud chamber uses a supersaturated vapour, the bubble chamber a superheated transparent liquid such as liquid hydrogen. Each charged particle moving through the chamber creates a string of ions which nucleate droplets (in the cloud chamber) or bubbles (in the bubble chamber); these strings are photographed and thus the paths of the particles are deduced. The 'vapour trail' left by an aircraft in the sky similarly consists mostly of water droplets. This is a 'cloud chamber' writ large: the jet engines ionize the air and thus nucleate the condensation.

24.2.1.1 Vapour pressure outside a droplet

If a liquid drop has the critical radius r_c, then G is at a stationary point so the liquid in the drop is in phase equilibrium with the vapour around it, albeit an unstable equilibrium. The value of r_c can be obtained either by differentiating (24.13) to find the stationary points, or by asserting $\mu_l(p_l) = \mu_v(p)$ in equation (24.11).

It is interesting to compare the vapour pressure in this case with the vapour pressure which would be observed at the same temperature for a flat surface, which is the *saturated vapour pressure* that we have already mentioned in

Fig. 24.8 *A scientist contemplating this glass of beer sees first a tasty and cheering drink, and second a solution of ethanol in water with carbon dioxide which was dissolved at high pressure. The chemical potential of the carbon dioxide is considerably larger in the beer than it is in the surrounding atmosphere, but the CO_2 can only get out by forming little bubbles. The bubbles produced by thermal fluctuation in the liquid immediately vanish, but there are some nucleation centres on the walls and floor of the glass, and there bubbles can form that are large enough to survive. Gravity meanwhile has resulted in a pressure gradient in the liquid, which provides an upthrust on the bubbles, causing them to rise. They grow as they rise, partly because of further evaporation, and partly because the pressure in the liquid is smaller at smaller depth. [Image: shutterstock/9821852/Andrey vaskressenskiy]*

connection with Figure 24.7. By inspecting that figure, or by writing down some algebra, one can deduce that

$$\mu_v(p) - \mu_l(p) = \int_{p_s}^{p} \left. \frac{\partial \mu_v}{\partial p} \right|_T - \left. \frac{\partial \mu_l}{\partial v} \right|_T dp,$$

since by definition the two agree at p_s. After using (13.6), this is

$$\mu_v(p) - \mu_l(p) = \int_{p_s}^{p} \left(\frac{1}{n_v} - \frac{1}{n_l} \right) dp. \tag{24.17}$$

We shall now make the usual approximations of ignoring $1/n_l$ compared to $1/n_v$, and modelling the vapour as ideal, then the integral is easy to perform and we find

$$\Delta g = n_l(\mu_v - \mu_l) = n_l k_B T \ln \frac{p}{p_s}.$$

Combining this with (24.15) gives

$$p = p_s \exp \left(\frac{2\sigma}{r_c n_l k_B T} \right) = p_s e^{2\sigma V_m / r_c RT}, \tag{24.18}$$

where V_m is the molar volume of the liquid phase. This result is known as Kelvin's formula. Physically, the surface tension squeezes the liquid and this results in a raised vapour pressure.

The physical implication of Kelvin's formula is that in the presence of small droplets, the average vapour pressure, taken as a spatial average over the whole vapour, will be elevated above the saturated vapour pressure. Droplets near the critical radius will linger, so that Kelvin's formula gives a good estimate of the pressure that can be found during a process of evaporation or condensation.

For water at room temperature, $2\sigma V_l/RT \simeq 1$ nm, so that a 100 nm radius drop has a vapour pressure increased by only 1%. Therefore for ordinary sized drops the effect is small. However, the formula predicts that the vapour pressure outside a sufficiently small drop is large. This amounts to a shift of the coexistence line on the p, T diagram towards higher pressure.

24.3 Colligative properties

If you take a mixture of salt and ice, initially at room temperature, and leave it, then the ice will start to melt, and, at the same time, the salt will start to dissolve in the water. The temperature of the mixture will gradually fall and may reach ten or twenty degrees below zero Celsius, even in a room at standard temperature and pressure! This was for many years the standard method to produce low temperatures for the manufacture and distribution of ice cream. The behaviour is caused by a lowering of the freezing point of water when salt is dissolved in it. The freezing point can be lowered by up to 21 °C (a minimum is reached at 23% salt by mass).

More generally, when one substance is dissolved in another in liquid solution, some properties of the solution have the interesting feature that they depend on the presence of dissolved particles and their number, but not their identity. An example is the change in the melting and boiling point: when a solute is added to a solvent to a given concentration, the melting point falls by the same amount no matter what the solute is. The boiling point also increases by the same amount no matter what the solute is. Properties of this type are called *colligative*.

The underlying cause of colligative properties is that when a solute substance B is dissolved in a solvent substance A, the chemical potential of A is reduced. Why is it reduced? It is because of the entropy of mixing. In the presence of some B particles, at any given total pressure in the system, the A particles don't have to exert all the pressure, so they can spread out a little, exploring a greater volume, which offers them a bit more entropy. The reduction in chemical potential takes place in all three phases, but mostly in the liquid phase, because the relative concentration of solute is much greater there. This means that at any given pressure and temperature the A particles have less tendency to leave the liquid. Consequently at any given pressure there is both an increase in the boiling point and a lowering of the freezing point—see Figure 24.9. The phase diagram changes as shown in Figure 24.10.

A related effect is *osmosis*, which is the tendency for a solvent to move towards a region of higher solute concentration, diluting it. In the following sections we treat this first, and then we discuss the effect of dissolved particles on phase changes.

24.3.1 Osmotic pressure

If a pure solvent is separated from a solution by a membrane permeable to the solvent, then molecules will move from the solvent to the solution, diluting it,

Fig. 24.9 *Effect on the chemical potential of adding solute to a liquid solvent. The full lines show $\mu(T)$ of the pure solvent at some given pressure. When a solute is added, the solvent's μ is lowered, as shown by the dashed line. This happens in all three phases, but typically the relative concentration of the solute in the liquid is much greater than that in the solid or vapour, so the effect is greatest in the liquid phase. The result is to lower the freezing point and raise the boiling point at any given pressure, as shown.*

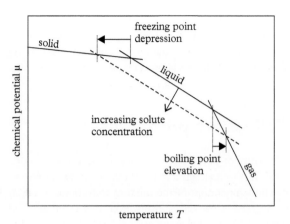

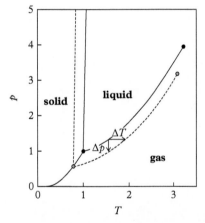

Fig. 24.10 *Phase diagram for a solution. The full lines show the coexistence curves of the pure solvent, the dashed lines show the coexistence curves at some non-zero concentration of solute. The solid and gas phases are not affected much, so the sublimation curve is almost unchanged, while the other curves move so as to favour the liquid phase. ΔT and Δp illustrate the change in boiling point and vapour pressure.*

unless something is done to prevent this. This is because the chemical potential of the solvent molecules A is lower in the solution (note that if it were higher then we would have a runaway process, which is not possible for stable states). The movement can be reduced to zero by increasing the pressure in the solution above that in the solvent, thus elevating the chemical potential in the solution. The excess pressure which must be applied is called *osmotic pressure* and the equilibrium thus attained is called *osmotic equilibrium*.

For equilibrium, the excess pressure Δp must satisfy

$$\mu_A(T, p, 1) = \mu_A(T, p + \Delta p, x_A), \tag{24.19}$$

where x_A is the concentration of the solvent, defined as a mole fraction. That is, the proportions of particles of the two types are x_A and x_B, with $x_A + x_B = 1$.

For many solvent–solute combinations, the dependence of μ on concentration is logarithmic to good approximation. One finds

$$\mu_A(T, p, x_A) = \mu_A^0(T, p) + k_B T \ln x_A. \tag{24.20}$$

A solution for which this is accurate at all concentrations is called an *ideal solution*, and for all solutions it is accurate at low concentration (i.e. when $x_B \ll 1$ so $x_A \simeq 1$). This equation can be derived for a mixture of ideal gases from equation (12.12). Although we are not primarily concerned with gases here, this connection adds some insight so we shall present it. Equation (12.12) gives

$$\mu_A = \mu_A^0(T) + k_B T \ln(p_A/p_0),$$

where the partial pressure of gas A in the mix is $p_A = x_A p$, where p is the total pressure. Therefore,

$$\mu_A = \mu_A^0(T) + k_B T \ln(x_A p/p_0) = \mu_A^0(T) + k_B T \left(\ln(p/p_0) + \ln x_A\right),$$

which gives (24.20).

In the following we will define $x \equiv x_B$ and then $x_A = 1 - x$.
By using (24.20) on the right-hand side of (24.19), we obtain

$$\mu_A(T, p, 1) = \mu_A(T, p + \Delta p, 1) + k_B T \ln(1 - x).$$

Multiplying this by N and making the last term the subject of the equation gives

$$-Nk_B T \ln(1 - x) = G_A(T, p + \Delta p, 1) - G_A(T, p, 1) = \int_p^{p+\Delta p} V_A(T, p') \mathrm{d}p',$$

where the last step uses $(\partial G/\partial p)_T = V$. If the liquid is incompressible, the right-hand side is just $V_A \Delta p$, and if $x \ll 1$ then $\ln(1 - x) \simeq -x$, giving

$$\Delta p\, V_A = Nk_B T x \quad \text{or} \quad \Delta p V_{A,\text{molar}} = RTx.$$

This is reminiscent of the ideal gas equation of state, but now we could be discussing either a gas or a liquid. Since liquids are typically much denser than gases, osmotic pressure can be very large even at moderate concentrations. For example, at 1 % concentration of a solute in water at room temperature, the equilibrium osmotic pressure difference is about 13 atmospheres! This phenomenon is very important in cell biology. Cells removed from their normal environment and placed in unsalty water will often burst, owing to osmotic pressure.

A phenomenon similar to osmosis is the *fountain effect* in liquid helium. When helium is cooled below the lambda point it exists in a phase which can be modelled as a homogenous mixture of two types of liquid: an ordinary fluid which is viscous, and a superfluid which flows without viscosity. If a tube is packed with inert powder then the superfluid can flow through unimpeded, but the ordinary fluid cannot, so this acts as a semipermeable membrane and is called a *superleak*. If such a tube is dipped into a pool of liquid helium, and the top of the tube is warmed slightly, then superfluid will rush through and spray out of the end in an elegant fountain. The reason is that at any given temperature and pressure there is an equilibrium concentration of the mixture, in which there is a higher concentration of normal fluid when the temperature is higher at any given pressure. The superfluid moves from the colder to the hotter region to 'dilute' the region with a higher concentration of normal fluid (the superfluid has zero entropy so this does not break the second law). If the warmer end of the tube is not open then a pressure difference will build up until the flow stops. This can be calculated as follows.

Let $\mu(T, p)$ be the chemical potential of helium. Suppose initially the whole apparatus is at T, p and then the region on one side of the superleak has its temperature raised by ΔT. Superfluid will flow until the pressure in the hotter region attains a value such that the chemical potentials on either side of the superleak agree:

$$\mu(T, p) = \mu(T + \Delta T, p + \Delta p)$$

$$\implies 0 = \frac{\partial \mu}{\partial T} \Delta T + \frac{\partial \mu}{\partial p} \Delta p. \tag{24.21}$$

Therefore,
$$\Delta p = (S/V)\Delta T, \tag{24.22}$$

where S is the entropy of a volume V of liquid helium at the given temperature and pressure. This could also be derived directly from the Gibbs–Duhem relation. The effect is substantial: at 1.5 K, a 1 mK temperature difference suffices to give the pressure of a 2 cm head of liquid helium.

24.3.2 Influence of dissolved particles on phase transitions

We now turn to the effects illustrated in Figures 24.9 and 24.10. By examining the figures you can see that the presence of the solute reduces the vapour pressure. We present first a simple calculation which applies to the liquid–vapour transition and gives a good insight, and then we generalize to all phase changes.

24.3.2.1 Simple calculation

We quote *Raoult's law* which states that the vapour pressure of an ideal solution is the sum

$$p = p_A x_A + p_B x_B,$$

where p_A, p_B is the vapour pressure of the substance A, B respectively, and x_A, x_B are the mole fractions. In other words, the total pressure is the weighted average of the vapour pressures that each substance would produce on its own. Quoting this law really amounts to solving the whole problem at the outset, so we will need the more thorough treatment below. For the moment let's simply note that if A is the solvent, B the solute, then the essential idea is that if $p_B < p_A$, the presence of B makes $p < p_A$.

Using $x_A = 1 - x_B$, we have $p = p_A + x_B(p_B - p_A)$ and comparing this to the case $x_B = 0$ we find that introduction of the solute lowers the vapour pressure by the amount $\Delta p = x_B(p_A - p_B)$. We shall be interested in cases where $x_B \ll 1$ and $p_B \ll p_A$ (think of salt dissolved in water, for example: the vapour pressure of sodium chloride is very much smaller than that of water in standard conditions). In this case,

$$\Delta p \simeq x_B p_A \simeq x_B p.$$

To find the associated temperature effect (elevation of the boiling point), we use the known slope of the coexistence curve, given by the Clausius–Clapeyron equation (cf. Figure 24.10):

$$\Delta T = \frac{dT}{dp}\Delta p = \frac{T\Delta V}{L}x_B p \tag{24.23}$$

Table 24.1 *Data related to some colligative properties of water.*

	Temperature	Latent heat	Constant
Boiling	372.8 K	2260 kJ/kg	$K_b = 0.51$ K kg/mol
Freezing	273.2 K	333 kJ/kg	$K_f = 1.86$ K kg/mol

(the change in the slope associated with x_B will only affect the result to second order in x_B). Approximating the vapour as an ideal gas and the liquid as of negligible volume, we have $p\Delta V = Nk_B T$, so

$$\Delta T = \frac{Nk_B T^2}{L} x_B. \tag{24.24}$$

This is our final result: it gives the elevation in the boiling point for an ideal gas for a given mole fraction of solute. It is remarkable that the effect does not depend on the type of solvent or solute. In the presence of solute the solvent 'likes' being a liquid because it has more entropy in the phase containing a greater proportion of solute.

More generally, the result is expressed:

$$\Delta T_b = K_b m_B, \tag{24.25}$$

where K_b is the *ebullioscopic constant*, a property of the solvent, and m_B is the *molality* of the solution, defined as the number of moles of solute (after taking dissociation into account) divided by the mass of the solvent:

$$m_B = \frac{x_B}{M_A(1-x_B)}, \tag{24.26}$$

where M_A is the molar mass of the solvent (units kg/mol). Molality has units mol/kg; K_b has units K kg/mol. By substituting (24.25) and (24.26) into (24.24) with $x_B \ll 1$, we have

$$K_b = \frac{M_A N k_B T^2}{L} = \frac{RT^2}{L_{kg}}, \tag{24.27}$$

where L is the latent heat of N particles of solvent and L_{kg} is the specific latent heat. Table 24.1 gives example values for water, which show good agreement with this simple prediction.

24.3.2.2 General calculation

Now we present a more thorough calculation which can be applied to any phase change.

The starting point is equation (24.20) for the chemical potential of the solvent. Raoult's law, which was assumed in the simple derivation, can be derived from (24.20) by considering that at phase equilibrium each component (A or B) is in equilibrium with its own contribution to the vapour.

In conditions of phase equilibrium we have

$$\mu_A^{\text{vap}} = \mu_A^{\text{liq}}$$

and therefore, using (24.20),

$$\frac{\mu_A^{\text{vap}} - \mu_A^0}{k_B T} = \ln x_A \quad \Rightarrow \quad \frac{G_A^{\text{vap}}(T) - G_A^0(T)}{Nk_B T} = \ln x_A, \quad (24.28)$$

where N is the number of solvent particles. This equation says that at the new boiling point T, the Gibbs function of the vapour differs from the reference value of the liquid by $Nk_B T \ln x_A$.

To obtain the boiling point elevation it is useful to move from Gibbs function to enthalpy. To this end, consider the Gibbs–Helmholtz equation (13.20). Applying this to both terms on the left-hand side of equation (24.28), we have

$$-\frac{H_A^{\text{vap}}(T) - H_A^0(T)}{Nk_B T^2} = \left.\frac{\partial}{\partial T}\right|_p \ln x_A,$$

so

$$\int_{T^*}^{T} \frac{\Delta H}{Nk_B T^2} dT = -\ln x_A, \quad (24.29)$$

where T^* is the boiling point of pure solvent, and

$$\Delta H(T) \equiv H_A^{\text{vap}}(T) - H_A^0(T).$$

Equation (24.29) is exact for an ideal solution. At $T = T^*$ the quantity ΔH is easy to interpret: for this case the reference enthalpy $H_A^0(T^*)$ is the enthalpy of the liquid, so $\Delta H(T^*)$ is the latent heat L associated with the phase change.

We now introduce an approximation: we treat $\Delta H(T)$ as constant in the region T^* to T. Under this approximation, the integral in (24.29) can be evaluated, and gives

$$\frac{\Delta H}{Nk_B}\left(\frac{1}{T^*} - \frac{1}{T}\right) = -\ln x_A.$$

Now $x_A = 1 - x_B$ so $\ln x_A = \ln(1 - x_B) \simeq -x_B$ for low concentrations of solute, hence

$$x_B \simeq \frac{\Delta H(T - T^*)}{Nk_B TT^*} \simeq \frac{\Delta H \Delta T}{Nk_B T^2},$$

so the elevation of the boiling point is as given by equation (24.24).

To find the change in the freezing point, replace H_A^{vap} in the above by H_A^{solid}. One finds $H_A^{\text{solid}}(T^*) - H_A^0(T^*)$ is now the negative of the latent heat, so the freezing point is depressed not elevated. The coefficient K_f in $\Delta T = -K_f m_B$ is called the *cryoscopic constant*.

Measurements of the change in boiling or freezing points of a suitable solution are used to determine the relative molecular mass of compounds. This makes direct use of the colligative nature of the phenomena.

24.4 Chapter summary

Surfaces affect the entropy and energy; their impact can be assessed by the standard thermodynamic approach: maximizing the entropy in the case of an isolated system, minimizing the relevant free energy in the case of a general system.

In equilibrium, the temperature and chemical potential are uniform throughout the system even in the presence of surface effects, but the pressure is not.

Surface tension raises the pressure and therefore the chemical potential of a liquid drop. Therefore, for phase equilibrium the chemical potential of the vapour must be higher, consequently the vapour pressure is raised. Condensation only takes place once sufficiently large liquid drops are formed; boiling only takes place once sufficiently large bubbles are formed.

Adding solute (e.g., salt) lowers the chemical potential of a solvent (e.g., water), consequently lowering the melting point and raising the boiling point. The effect has a physical rather than chemical origin, in that at low concentration it depends only on the number not the type of solute particles. This also results in osmotic pressure across a semipermeable membrane between a pure solvent and a solution.

....

EXERCISES

(24.1) *Equality of chemical potentials in the presence of surface effects.* We proved this using entropy in Section 24.1. Obtain the same result from the Gibbs function, as follows. Consider a system consisting of a drop of liquid in the presence of its vapour. Ascribe to the surface of the drop internal energy U_s and entropy S_s, but no volume or particles. Define $G \equiv U + p_v V - TS$, where p_v is the pressure of the vapour. Consider a change in which p_v, T are constant but material moves from one phase to the other. By considering dU, dS, and dV show that

$$dG = (u_l + p_v v_l - Ts_l) dN_l$$
$$+ (u_v + p_v v_v - Ts_v) dN_v + \sigma dA,$$

if the liquid is assumed to be incompressible. Relate $v_l dN_l$ to dA and hence show that $dG = dN_l g_l + dN_v g_v$, where g_l is evaluated at the pressure of the liquid and g_v at the pressure of the vapour. Hence show that $g_v = g_l$ in phase equilibrium.

(24.2) Using

$$\mu_l(T, p_l) = \mu_l(T, p) + \int_p^{p_l} \left.\frac{\partial \mu}{\partial p}\right|_T dp,$$

or otherwise, prove that for a drop at the critical radius given by (24.15), $\mu_l(T, p_l) = \mu_v(T, p)$, so in phase equilibrium (whether a stable or an unstable equilibrium), the chemical potential is uniform throughout the system, in agreement with question 24.1.

(24.3) Derive equation (24.15) from equation (24.13).

(24.4) Find the osmotic pressure for healthy red blood cells removed from the body and placed in fresh water. (It may help you to know that normal saline, which is 9g NaCl dissolved in water to a total volume of one litre, is a close approximation to the osmolarity of NaCl in blood). [*Ans.* 7.5×10^5 Pa $\simeq 7.4$ atm]

Continuous phase transitions

25

25.1 Order parameter		357
25.2 Critical exponents		359
25.3 Landau mean field theory		361
25.4 Binary mixtures		370
Exercises		373

Continuous phase transitions contain a wealth of interesting physics, and come in a rich variety of forms. They have in common that they do not involve a latent heat, and do not exhibit metastable phases. The presence of the transition is signalled by the behaviour of the heat capacity, and by other properties such as superconductivity, magnetization, and superfluidity.

Ehrenfest proposed a classification scheme based on the order of the derivative of G which exhibits a discontinuity:

n^{th} order phase transition:
$\partial^m G/\partial T^m$ continuous for $m < n$; $\partial^n G/\partial T^n$ discontinuous.

According to this scheme a 'second-order' phase transition is one in which $S = (\partial G/\partial T)_p$ is continuous, but $(\partial S/\partial T)_p = \partial^2 G/\partial T^2$ is not. Therefore the response functions (susceptibilities) are discontinuous:

$$C_p = -T \left.\frac{\partial^2 G}{\partial T^2}\right|_p,$$

$$\kappa_T = -\frac{1}{V} \left.\frac{\partial^2 G}{\partial p^2}\right|_T,$$

$$\beta = \frac{1}{V} \frac{\partial^2 G}{\partial p \partial T}.$$

If C_p is continuous but its gradient is not, then the transition is third order, and so on. Ehrenfest's classification scheme seems to suggest that the entropy and its gradient do not show rapid change until the transition point is reached, and then the gradient (or some other derivative) changes abruptly. Many continuous phase changes are not like that, however. An example is the lambda point in liquid helium, where the peak in the heat capacity extends on either side of the critical point, with a width of several tenths of a kelvin. But the concept of a second-order transition is useful nonetheless. It applies quite well to the superconducting transition in a type I superconductor at zero applied field, and it gives a good initial guide to what is going on in other continuous phase transitions.

Example phase transitions that fall approximately into Ehrenfest's scheme are listed in Table 25.1.

Thermodynamics: A Complete Undergraduate Course. Andrew M. Steane.
© Andrew M. Steane 2017. Published 2017 by Oxford University Press.

Table 25.1 *An illustrative list of phase transitions arranged by order. All second-or higher-order transitions are continuous transitions, but not all continuous transitions fall fully into the categorization by order, except in that they are not first order.*

First order
solid–liquid, solid–vapour, liquid–vapour phase change
superconducting transition in a magnetic field
some allotropic transitions in solids (e.g. iron)

Second order
evaporation through the critical point
superconducting transition in zero field
superfluid transition in liquid helium (approximately)
order–disorder transition in β-brass (approximately)

Third order
Bose–Einstein condensation

Transitions not falling into the above categorization scheme
most non-first-order transitions, including:
lambda point in liquid helium and in β-brass
Curie point of many ferromagnets

A further difficulty is that a transition can appear to be continuous while in fact being first order, if the latent heat is small. So all transitions claimed to be second or higher order might subsequently be revealed to be first order. We handle this by the usual thermodynamic method: make an idealization and study that theoretically, then compare with experimental realities. The converse case also occurs: a transition which appears discontinuous at one level of experimental precision might prove to be continuous at a finer level of precision.

There are not many continuous phase transitions that show a *finite discontinuity* in the heat capacity. More common is a *divergence* in both the heat capacity and in various other response functions. For instance, near the critical point, the response of a magnetic system is of the form

$$\left.\frac{\partial M}{\partial H}\right|_T \propto \frac{1}{(T-T_c)^\gamma}, \quad (25.1)$$

where the index γ is called a *critical exponent*.

In a modern treatment we distinguish primarily between the first-order transitions and the rest, which are called *continuous phase transitions*,[1] and we focus on entropy. The definition is as follows.

Discontinuous phase transitions (also called first-order phase transitions):

1. S is discontinuous in almost all cases and hence there is a latent heat.
2. There are one or more other properties showing a discontinuity.

---

[1] Unfortunately the term 'second order' is widely used as synonymous with 'continuous' but that is inappropriate, because then one ends up calling a third-order transition (according to Ehrenfest's terminology) 'second order'.

3. There are metastable phases.
4. Heat capacities are finite for $T \neq T_0$, where T_0 is the transition temperature (and they may or may not diverge at $T = T_0$).

Continuous phase transitions:

1. S is continuous; there is no latent heat.
2. There is a critical point T_c.
3. Response functions such as C_p, κ_T, χ_T show singular behaviour (either divergent, or discontinuous, or with a discontinuity in a derivative).

25.1 Order parameter

When viewed as a function of the intensive variables, a first-order phase transition is an 'all or nothing' transition: depending on the temperature and pressure (or the appropriate parameter in another type of system) the substance is either completely of one character (e.g. crystalline), or completely of another (e.g. amorphous). A continuous phase transition is a 'some or nothing' transition. That is, as the system is cooled, at the transition temperature a change of character *starts* to take place. For example, if a gaseous substance is prepared at $T > T_c$, $V = V_c$ where T_c, V_c are the critical parameters, and then cooled at constant volume, it will undergo a second-order phase transition when it reaches the critical temperature. At lower temperatures the system is not completely in the liquid form; rather, the proportion of liquid grows smoothly as the temperature is lowered. Below the transition we may define a parameter equal to the difference in densities:

$$\tilde{\rho} = \rho_l - \rho_v.$$

Below the critical point, this parameter is non-zero. At the critical point, which is here also the transition point, $\tilde{\rho}$ goes to zero. This is referred to as an 'order parameter': it quantifies the extent to which the system has moved towards the more structured (lower entropy) phase.

More generally, the order parameter associated with a continuous phase transition may be taken as any quantity that vanishes above the critical temperature but is non-zero below it, and which can be used to describe the free energy, as we shall clarify in Section 25.3.

The *Weiss model* of spontaneous magnetization treats a material that is paramagnetic at high temperature but at low temperature undergoes a transition to a state where it acquires a non-zero magnetization even at zero applied field. The order parameter is here the average magnetization of the sample. Its behaviour as a function of temperature is shown in Figure 25.1.

In the *superfluid transition* in liquid ^{4}He, it is a mistake to think that there is a switch, at the transition temperature, between a viscous liquid and a completely

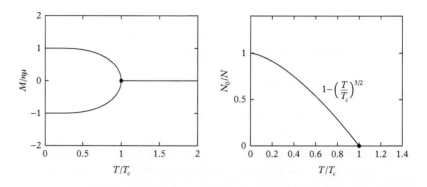

Fig. 25.1 *Two examples of an order parameter. (a) The ferromagnetic transition. The order parameter is the magnetization. It is equal to zero for $T > T_c$ and there are two solutions for $T < T_c$. (b) Bose–Einstein condensation. In a simple model, the order parameter is the population of the ground state.*

superfluid liquid. Rather, below the transition temperature the experimental observations are well modelled by a *two-fluid model* in which the liquid consists of two interpenetrating fluids, one having a non-negligible viscosity, the other having zero viscosity. Above the transition, there is strictly no superfluid component. At the transition, the fraction f of the zero viscosity component starts to increase from zero. f is continuous but df/dT is not. As the temperature falls further below T_c, f grows, reaching 1 at absolute zero. The fraction f can be taken as the order parameter.

Superconductivity can also be modelled using a two-fluid model. In the presence of an applied magnetic field, the transition to the superconducting state is first order, as we discussed in Section 22.4. The transition is continuous in the absence of an applied field.

Bose–Einstein condensation is a transition expected for a gas of bosons at low temperature, if the interactions between the particles are weak enough. In this transition the order parameter is, in a first approximation, the fraction of particles occupying the ground state of motion. In a more careful analysis the order parameter is a rather subtle quantity, related to long-range order in the off-diagonal terms of a density matrix.

All these phase transitions are characterized by the presence of significant fluctuations in the system when it is near the transition temperature. In the case of magnetism, domains in which the atomic dipoles are aligned appear on many distance scales. In a fluid, density fluctuations become larger and correlated over many distance scales. These fluctuations affect the refractive index and hence the visual appearance of anything observed through the fluid. If the fluid is transparent then near the critical point it appears milky, owing to the random variations in refractive index on many length scales. This phenomenon is known as *critical opalescence*. It provides both a useful laboratory tool for studying the fluctuations, and also a beautiful visual demonstration of some of the physics underlying continuous phase transitions. It is easiest to observe in a binary liquid mixture. If you know a friendly chemist, ask for a demonstration!

25.2 Critical exponents

Critical points exhibit many remarkable and instructive properties. Phenomena in quite different experimental systems show themselves to possess detailed equivalences when looked at from a certain point of view. The behaviour can be understood by appealing to the concept of *scale invariance*, which is a reference to the fluctuations appearing on all length scales that we alluded to in Section 25.1. Owing to this scale invariance, the thermodynamic properties exhibit zeros or divergences, around the critical point, that follow simple mathematical functions such as power laws with a given exponent. The exponents are called *critical exponents*; we have already mentioned an example in equation (25.1). What is remarkable is that the values of the critical exponents are the same for a large class of otherwise very different physical systems, and they depend only on general properties such as dimensionality and scaling laws, not on details of the interactions. It is the fluctuations on all length scales that 'drive' the physics. Systems sharing the same values of the critical exponents are said to belong to the same *universality class*.

A full presentation of the theory of critical phenomena is beyond the scope of this book. Here we will simply illustrate the idea by extracting critical exponents predicted by the van der Waals model of a gas in this section, and by showing how they can be found more generally, if approximately, in Section 25.3.

Begin by writing the van der Waals equation of state in critical units:

$$p = \frac{8T}{3V-1} - \frac{3}{V^2} \tag{25.2}$$

and recall that, in these units, the critical point is at $p = T = V = 1$. We will find critical exponents $\alpha, \beta, \gamma, \delta$ which are defined to be those associated with the constant-volume heat capacity, the density difference between liquid and vapour, the isothermal compressibility, and the pressure, respectively. Note, the symbol γ here does *not* refer to the adiabatic index.

We begin with the heat capacity. This is simple because, taking the monatomic case,

$$C_V = \frac{3}{2}k_B, \tag{25.3}$$

which does not depend on temperature at all. This can be written

$$C_V \propto (T - T_c)^\alpha, \tag{25.4}$$

with $\alpha = 0$. So we have that the van der Waals theory predicts that the first critical exponent takes the value $\alpha = 0$.

Next we examine the isothermal compressibility. Using the equation of state, we find

$$\left.\frac{\partial p}{\partial V}\right|_T = \frac{-24T}{(3V-1)^2} + \frac{6}{V^3} = 6(1-T), \tag{25.5}$$

where in the second step we have evaluated the result at the critical volume. Hence we obtain, at the critical volume,

$$\kappa_T = \frac{1}{6(T-1)}. \tag{25.6}$$

Recalling that $T_c = 1$, we observe that this equation states that, at the critical volume, the compressibility diverges as a power law in $(T - T_c)$. We express this,

$$\kappa_T \propto (T - T_c)^{-\gamma}, \tag{25.7}$$

where $\gamma = 1$. So the van der Waals theory predicts this exponent to be equal to 1.

For the exponent associated with pressure, write the equation of state in terms of pressure, temperature, and number density n, and expand about $n = n_c = 1$ at $T = 1$:

$$p = \frac{8nT}{3-n} - 3n^2 = 1 + \frac{3}{2}(n-1)^3 + \cdots \tag{25.8}$$

(there is only a small amount of calculation needed, because by definition we already know the first two derivatives are zero at the critical point). Thus, $(p-p_c) \propto (n-n_c)^3$, so $\delta = 3$.

The exponent β is associated with the density difference between liquid and vapour at temperatures just below the critical point:

$$n_l - n_v \propto (T - T_c)^{\beta}. \tag{25.9}$$

This can be found by employing the Maxwell construction, see Exercise 25.2. One obtains $\beta = 1/2$.

These results from van der Waals theory are summarized in Table 25.2. The experimental values are in the region of these predictions, but not closely matched to them. This is because the van der Waals theory is a mean field theory, so makes no attempt to account for the fact that nearby molecules interact more strongly than far away ones. That is what a more thorough theory has to grapple with.

Table 25.2 *Critical exponents for the liquid–vapour phase transition at its critical point.*

Exponent	Value vdW (mean field)	Experimental value
α	0	0.1
β	0.5	0.32
γ	1	1.2
δ	3	4.8

25.3 Landau mean field theory

The main features of a continuous phase transition at a critical point can be captured by an argument whose essential idea is due to various authors, and which was subsequently elaborated and generalized by Landau. Landau worked out the method in great generality, but a simple example is sufficient to capture its essence.

We consider a system which has various external parameters that may be subject to external constraints, and an internal degree of freedom which is unconstrained, so that its value is dictated by the maximum entropy (in an isolated system) or least Helmholtz free energy (in a system at given temperature). For example, in a magnetic system the internal degree of freedom could be the number of small dipoles in one orientation or another. In a crystalline solid it could be the number of atoms at one kind of site or another. In a liquid mixture of two types of particle it could be the average number of like as opposed to unlike nearest neighbours. The idea is that the internal energy and the entropy are both going to depend on this degree of freedom, and we will examine the effects of both of them together as they influence the free energy.

We first arrange to choose the order parameter m in such a way that a large negative or positive value represents a situation of more structure (less entropy) and a low value represents a situation of less structure (more entropy). For example, in the case of a set of spin-1/2 magnetic particles situated on a three-dimensional lattice, m could be the proportion of particles that are in the 'up' state. Values close to ± 1 then represent the spins nearly all aligned; values close to 0 represent equal numbers of spins in either direction.

In the case of copper and zinc atoms in beta brass, if we define x to be the proportion of copper atoms at A sites in the lattice (see Section 9.4), then the order parameter can be taken as $m = 2x - 1$.

In the case of a gas near the critical point, the order parameter could be taken as the difference in densities of the two phases.

In all these cases we expect the entropy to have a maximum value at $m = 0$ and to show the general shape illustrated in Figure 25.2. We don't need the precise details (which might for example be given by the entropy function, equation (9.14)), only the general form. We argue from symmetry (i.e. a symmetry of the system under study) that S is symmetric about $m = 0$, and we shall assume its Taylor expansion about the stationary point has non-zero quadratic and quartic terms:

$$S(m) = S_0 - a_s m^2 - b_s m^4 + \cdots,$$

where the coefficients S_0, a_s, b_s may depend on T, but this is not expected to be a strong dependence, and the signs are supplied such that a_s is positive and we anticipate b_s may be positive too (but we don't require this).

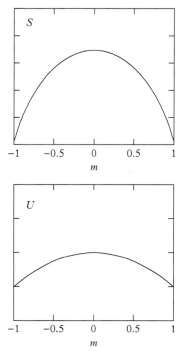

Fig. 25.2 *The generic form of the entropy and energy as a function of order parameter, for a second-order phase transition.*

Next we need the internal energy. Again, we don't need the details, just the general idea that if the system has a symmetry, then the internal energy will be an even function of m, so it goes through either a maximum or a minimum at $m = 0$. We are interested in the case where it goes through a maximum. In the case of the spin system, this could be because neighbouring spins have lower energy when they are pointing in the same direction. In the case of atoms in a bcc crystal structure, it could be because the interaction energy of like nearest neighbours is lower than that of unlike nearest neighbours. In either case, the least energy configuration is the more structured, lower entropy one, and the highest energy configuration is the less structured, high entropy one.

For all these examples we expect the internal energy to have a maximum at $m = 0$, and to be symmetric about $m = 0$. Its gradient at $m = \pm 1$ is not generally expected to be as steep as that of $S(m)$, therefore the two functions are not exactly alike, but they have a maximum at the same place. We expand $U(m)$ about $m = 0$ to obtain

$$U(m) = U_0 - a_u m^2 - b_u m^4 + \cdots,$$

where again the coefficients may depend weakly on T. The energy and entropy as a function of m are shown in Figure 25.2.

If the system were isolated then its energy would be fixed and then there would be a pair of solutions for m, with a unique value of $|m|$, and that in turn would fix the entropy. Such a system can't reach the maximum entropy at $m = 0$ because it can't gain entropy without also gaining energy, and that is not possible if it is isolated. However, this observation hides the interesting physics that takes place inside the system. For, if the system is large enough, then we can always divide it up into many small regions, and then for any such small region, the rest of the system acts like a reservoir. These small regions need not be very small: if there are 10^{24} particles all together, then we can have 10^{12} regions with 10^{12} particles in each. We already know what will happen: the maximum entropy configuration is when each region has the same temperature as the others. Alternatively, just consider the case where the whole system is in equilibrium with a heat bath, which is very often what we have in practice. In either case, we no longer have a constrained internal energy for the system (or part) we are concerned with. It can gain as much energy as it wants from the reservoir, or, indeed, give energy to the reservoir. Instead we have a constrained temperature. In this situation the thermodynamic potential governing the equilibrium is the free energy,

$$F(T; m) = U - TS = U_0 - TS_0 + (Ta_s - a_u)m^2 + (Tb_s - b_u)m^4 + \cdots \quad (25.10)$$

Now comes the beautiful bit.

In equation (25.10) we must keep in mind that this is not an expression for a function of state in terms of equilibrium thermodynamic properties. Rather, m

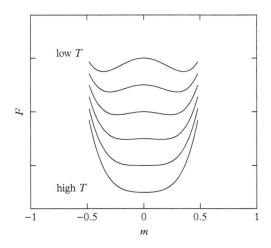

Fig. 25.3 *Free energy $F(T; m)$ given by equation (25.10), as a function of m, for six values of the temperature T.*

is an unconstrained internal parameter and (25.10) gives the free energy for a range of values of m, at any given T. Most of these values of m do not represent thermodynamic equilibrium. The situation is like the study of entropy and free energy for an out of equilibrium system that we presented in Chapter 17. We know what to do. The system will exchange energy with the reservoir until it locates the state of minimum free energy. So we plot the free energy curve, and look for the minimum. This is shown in Figure 25.3.

At high temperature, the entropy 'wins' the trade-off between S and U in the expression for F, and the minimum free energy is at $m = 0$. As the temperature goes down, the contribution to F made by the entropy becomes less significant. The energy is more important, because now the reservoir is more hungry for energy. It has a low temperature and will gobble up any energy it can get in order to raise its entropy. So the energy and entropy contributions come into competition, until they are balanced, and eventually, at the lowest temperatures, the energy wins (or, if you prefer, the reservoir wins) and the least free energy points are at or near $m = \pm 1$.

The balanced case takes place when the m^2 term in $F(T; m)$ changes sign. After this, at lower temperatures, instead of one minimum there are two minima, and on Figure 25.3 they shoot off to left and right as T passes through $T_c = a_u/a_s$. This is a phase transition. After that, at lower temperatures still, the minima continue to move more sedately towards $m = \pm 1$. The behaviour of m is, qualitatively, like the one shown in Figure 25.1 for the magnetization of a magnet at the ferromagnetic transition.

The Landau mean field theory can account for much of the detail of the phase transition. Since it gives the free energy, it gives everything: a phase diagram, the equation of state, heat capacities, susceptibilities, and so on. It gets all this qualitatively right, and quantitatively it achieves a reasonable approximation. It predicts critical exponents, and gets them roughly right.

Let us introduce

$$a = Ta_s - a_u,$$
$$b = Tb_s - b_u, \qquad (25.11)$$

then,

$$F(T; m) = F_0 + a(T)m^2 + b(T)m^4 + \cdots \qquad (25.12)$$

and we no longer need (25.10). It has served its purpose of introducing the ideas, and it gives us reason to expect that the coefficients will have a close to linear dependence on T near the transition. The transition is at the temperature where a changes sign, so $a(T_c) = 0$.

At all T, there is a stationary point of F at $m = 0$; above T_c it is a minimum, below T_c it is a maximum. Below T_c there are two other stationary points, both minima, at

$$m = \pm\sqrt{\frac{-a}{2b}}. \qquad (25.13)$$

Hence the equation of state (i.e. the equation for m as a function of T and whatever parameters determine the coefficients) is

$$m = \begin{cases} 0 & T > T_c \\ \pm\sqrt{-a/2b} & T < T_c. \end{cases} \qquad (25.14)$$

Substituting this into (25.12), we have the free energy of the thermal equilibrium state as a function of the coefficients and T:

$$F(T) = \begin{cases} F_0(T) & T > T_c \\ F_0(T) - a^2/2b & T < T_c. \end{cases} \qquad (25.15)$$

This equilibrium free energy is continuous at $T = T_c$ because $a(T_c) = 0$. Above T_c, the entropy is $S = -\partial F_0/\partial T$. Below T_c the entropy is

$$S = -\frac{\partial F_0}{\partial T} + \frac{a}{b}\left(a' + \frac{a}{2b}b'\right) \qquad (25.16)$$

(N.B. this is obtained by differentiating (25.15) not (25.12)). Therefore the entropy is continuous at $T = T_c$, where a vanishes. However, $\partial S/\partial T$ is discontinuous: just below T_c it has, in addition to $-\partial^2 F_0/\partial T^2$, a term a'^2/b which does not vanish at $T = T_c$. Hence we have a second-order phase transition.

We can use the theory to extract critical exponents. In order to do this, we need to make some assumption about the way the coefficients behave near T_c. We have already suggested they might depend linearly on T, and we know that a changes sign at T_c, so we propose, near to T_c:

$$a \simeq a_0(T - T_c), \qquad b \simeq b_0, \qquad (25.17)$$

where a_0 and b_0 are independent of T. This is a very general and reasonable assumption because all we are assuming is that b does not vanish at the very same temperature as a. Then equation (25.13) tells us that, below T_c,

> **The static scaling hypothesis**
>
> Let $t \equiv |T - T_c|/T_c$, where T_c is the temperature at a critical point. Introduce the *static scaling hypothesis*, which asserts that around the critical point, the free energy is a homogeneous function of its natural variables:
>
> $$F(\lambda^p t, \lambda^q B) = \lambda F(t, B), \qquad (25.21)$$
>
> where λ is a scaling parameter that can take arbitrary values, and the formula holds for all values of p, q.
>
> (i) Defining $M = -\partial F/\partial B$, we obtain
>
> $$\lambda^q M(\lambda^p t, \lambda^q B) = \lambda M(t, B).$$
>
> Now, by picking the value $\lambda = t^{-1/p}$, and evaluating at $B = 0$, we find
>
> $$M(t, 0) = t^{(1-q)/p} M(1, 0).$$
>
> Therefore the critical component β is given by $\beta = (1-q)/p$.
>
> (ii) Now pick $\lambda = B^{-1/p}$, to obtain
>
> $$M(0, B) = B^{-1+1/q} M(0, 1)$$
>
> and hence $\delta = q/(1-q)$.
>
> (iii) Next consider $\chi = \partial M/\partial B$ and choose $\lambda = t^{-1/p}$ again. One obtains $\chi(t, 0) = t^{(1-2q)/p} \chi(1, 0)$, hence $\gamma = (2q-1)/p$.
>
> (iv) For the heat capacity, $C = -T \partial^2 F/\partial t^2$, the same method yields $C(t, 0) = t^{-2+1/p} C(1, 0)$, hence $\alpha = 2 - 1/p$.
>
> (v) By eliminating p and q from the equations for $\alpha, \beta, \gamma, \delta$, we obtain equations (25.19) and (25.20).

$$m \simeq \pm \sqrt{\frac{a_0}{2 b_0}} (T_c - T)^{1/2}, \qquad (25.18)$$

so the critical exponent $\beta = 1/2$. In this approximation, we have, below T_c,

$$S \simeq S_0 + \frac{a_0^2 (T - T_c)}{b_0},$$

so the heat capacity $C = T(\partial S/\partial T) = C_0 + a_0^2/b_0$. This is different from the value just above T_c, so there is a discontinuity, as we have noted previously. The value just below T_c is independent of $(T - T_c)$ so the critical exponent $\alpha = 0$.

In order to get the other two critical exponents γ and δ, we would need to introduce a variable playing the role, for our generic system, that an applied magnetic field plays for a magnetic system. Such a field acts to make the two spin states have different energies, independent of the interactions of the spins with one another, so it adds an asymmetric contribution to the internal energy curve $U(m)$. We can accommodate this by allowing odd powers of m in the expression for free energy. However, we don't need to do this because it can be shown on quite general grounds (see box) that the exponents satisfy

$$\alpha + 2\beta + \gamma = 2 \qquad (25.19)$$

$$\gamma = \beta(\delta - 1), \qquad (25.20)$$

so we find $\gamma = 1$ and $\delta = 3$, just as we did for the van der Waals gas by examining its equation of state (Table 25.2). This is an example of the universality of critical phenomena. It predicts the Curie–Weiss law, which is an example of $\gamma = 1$. This is again a reasonable approximation though not a close match to experimentally measured values, which are nearer to $\gamma \simeq 4/3$.

Further exponents can be defined, related to the length scale on which fluctuations appear in the system, and this characteristic length, called the *correlation length*, is a central idea in a more complete theory. Such a more complete theory, called the Landau–Ginzburg theory, is closely related to the *renormalization group* and the subject of conformal field theory which also arises in particle physics and string theory. It would take us well beyond the remit of the present book.

25.3.1 Application to ferromagnetism

The Landau mean field theory, presented previously, can be applied to a variety of continuous phase transitions, as we have already mentioned. We have focussed on the behaviour near the critical point, in order to keep the discussion general. The same method can also be used to find the behaviour at all temperatures, if we have appropriate equations for $U(m)$ and $S(m)$. In the case of the ferromagnetic transition, we can obtain these from our discussion of the ideal paramagnet in Section 14.4.1. For simplicity, we treat the case of spin half, where the dipoles have just two energy states. For the entropy at any given m, we assume the orientation of the dipoles has no structure except that required to supply the given value of m. That is, we assign probabilities $p_1 = (1-m)/2, p_2 = (1+m)/2$, and use

$$S = -Nk_B \sum p_i \ln p_i$$
$$= \frac{Nk_B}{2} (2\ln 2 - (1+m)\ln(1+m) - (1-m)\ln(1-m)) \qquad (25.22)$$

(note, here we are using m for the order parameter, not the total dipole moment of the sample). For the energy we use

$$U = -(N\mu m)\left(B + \frac{\lambda}{2}N\mu m\right), \qquad (25.23)$$

where we have allowed that the applied magnetic field B may be non-zero. Here, λ is a parameter describing the strength of the inter-particle interactions. N is the number of particles, μ is the magnetic dipole moment of one particle. We can observe that the combination $\lambda N\mu m/2$ acts like an effective magnetic field. In some cases it is indeed a magnetic field caused by the dipoles in the sample, but more common is the case where λ is related to much stronger electrostatic interactions.

The free energy $F = U - TS$ is formed by combining equations (25.22) and (25.23). The stationary points of the free energy are given by

$$\frac{\partial F}{\partial m} = -N\mu B - \lambda(N\mu)^2 m + \frac{Nk_B T}{2}\ln\left(\frac{1+m}{1-m}\right) = 0, \qquad (25.24)$$

which gives
$$m = \tanh x, \qquad (25.25)$$
where
$$x = \frac{\mu(B + \lambda N \mu m)}{k_B T}. \qquad (25.26)$$

This result is just like the paramagnet (equations (14.33) and (14.31)), but with a modified magnetic field. As it stands, (25.25) is an implicit equation for m which cannot be solved by algebraic methods, but it is easily solved by a graphical or numerical method. Define
$$T_c = \frac{\lambda N \mu^2}{k_B}, \qquad (25.27)$$
then we have
$$m = \frac{T}{T_c} x - \frac{\mu B}{k_B T_c}, \qquad (25.28)$$
so equation (25.25) reads
$$\frac{T}{T_c} x - \frac{\mu B}{k_B T_c} = \tanh x. \qquad (25.29)$$

The graphical solution of this equation is illustrated in Figure 25.4. The critical point occurs when the gradient of the straight line on the left-hand side of (25.29) matches the gradient of the tanh function at the origin, i.e. 1. This is when $T = T_c$, which is why we defined T_c as we did. The solution at $B = 0$ is shown in Figure 25.1, and the solution for more general values of B is shown in Figure 25.5.

Now that we have found the equilibrium value of m, we can obtain the equilibrium free energy, and hence all the other thermodynamic information. For example, by substituting the solution for m into equations (25.22), we can find the entropy and hence the heat capacity. This result at $B = 0$ is shown in Figure 25.6, and compared with the observed behaviour. The observed heat capacity C_{exp} follows the characteristic shape of a lambda point, which differs from the mean field result in three respects. First, if the parameter λ of the mean field theory is chosen to fit the high temperature behaviour, where the mean field approach is accurate, then the theory overestimates the critical temperature. Next, C_{exp} is more sharply peaked as T approaches the critical point from below, and it does not fall immediately to zero on the other side of the point where m reaches 0. This indicates that the magnetic entropy does not become immediately independent of T just above the transition: there must be some remaining structure in the system, even though the magnetization is strictly zero. This structure consists of local groups of dipoles which line up with one another, which is not accounted for by using only the average m to describe the spin states. At higher temperatures there are further contributions to the entropy from motion of the particles, which we have not considered.

Fig. 25.4 *Graphical solution of equation (25.29). Solutions occur where one of the straight lines intersects the* tanh *function. The example straight lines shown are all at $B = 0$. A non-zero value of B displaces the lines left or right. This graph shows beautifully how the abrupt behaviour indicated in Figure 25.1 emerges out of all the smooth functions we have employed. At high temperature the straight line is steep. As the temperature falls the line tips over, rotating about its intersection with the x axis. Two new solutions suddenly appear when the gradient reaches 1. The solution at $x = 0$ is stable at high T, unstable at low T.*

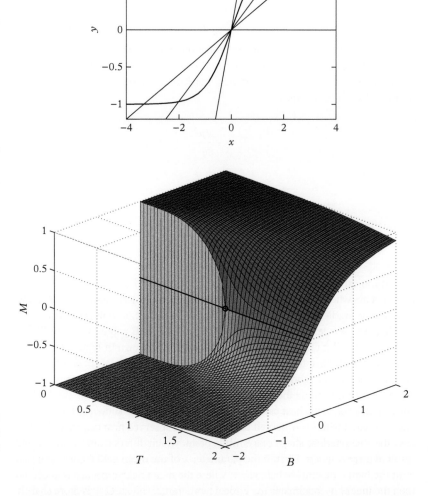

Fig. 25.5 *Magnetization as a function of temperature and applied magnetic field, in the vicinity of the ferromagnetic transition, as given by mean field theory.*

The mean field theory also slightly underestimates the magnetization just below T_c and slightly overestimates it near $T = 0$. In both cases it is because the dipoles are not distributed randomly about their mean orientation, but exhibit further structure such as periodic variations called spin waves. This both contributes to the energy and reduces the entropy.

From Figure 25.5 we learn that when we study the behaviour as a function of B and T we have overall a situation in many respects similar to the liquid–vapour phase transition. Above the critical temperature, there is no phase transition. Along the line $B = 0$ there is the second-order phase transition that we studied

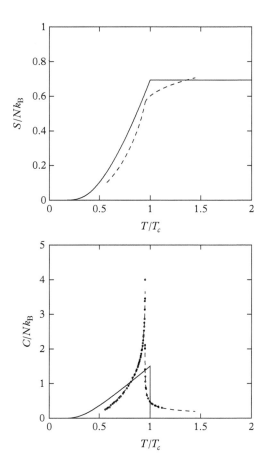

Fig. 25.6 *Entropy and heat capacity of a paramagnetic/ferromagnetic system from mean field theory (continuous lines) and experiment (points and dashed lines). Here, T_c is the critical temperature predicted by mean field theory; the observed value is a little below this. The mean field theory underestimates the structure in the system, because it does not allow for spin waves and short-range clustering, and therefore it overstates the entropy. However, it captures the continuity of S and discontinuity of C, and gives reasonable physical predictions over the whole temperature range.*

previously. For $T < T_c$, when $B \neq 0$ there is a first-order phase transition which can be observed by varying B from a negative to a positive value. The two phases are the two orientations of m. When we approach the transition from the negative B side, the equilibrium value of m is initially negative, then as B passes through zero, the equilibrium value of m changes by a large amount, $|2m|$. The step change in m is a discontinuity in a first derivative of the free energy, so this is a first-order phase change. However, it is an unusual one because in the limit $B \to 0$ the two phases have the same entropy and internal energy so there is no latent heat. In practice, however, some small effect will break the symmetry and then there is an entropy change and a latent heat.

Considering now the behaviour in the ferromagnetic regime as the applied field is varied, we can study the free energy as in Figure 25.7. In a real system the magnetization does not change to the more stable state immediately as B changes sign, because the less stable state is metastable and various effects related to the crystalline structure inhibit the change. The presence of more than one local minimum in F, and hence a metastable state, is correctly predicted by mean field theory (Figure 25.7). When B changes sign, the complete reversal of the state of magnetization only gets under way at a non-negligible speed for $B \neq 0$, and then the two phases differ substantially in both energy and entropy. In ferromagnetic materials such as iron and nickel the resulting hysteresis is one of the dominant features of the behaviour.

25.4 Binary mixtures

It is a common observation that 'oil and water do not mix'. This is largely but not completely true at ordinary temperatures, and at high temperature it is not true. In fact, at ordinary temperatures, when oil and water are mixed together, the liquid separates into two parts, one of which is mostly (but not completely) water, and the other of which is mostly (but not completely) oil.

The term 'miscibility' refers to whether or not one substance will dissolve in another, forming a homogeneous solution. When a pair of substances can form a homogeneous mixture at all proportions, they are said to be miscible. When there is some range of proportions where this does not occur, then there is said to be a *miscibility gap*, also called a *solubility gap*. For example, at room temperature, at most 2.5% by weight of methanol can dissolve in carbon disulphide, and at most 50% by weight of carbon disulphide can dissolve in methanol. So below 2.5% concentration of methanol, the mixture will be homogenous, and it will also be homogenous above 50% concentration of methanol. In between 2.5% and 50% overall composition, a combination of these liquids will separate into two phases, one containing 2.5%, the other 50% of methanol.

Similar considerations may apply to solids.

At high temperatures the miscibility gap often vanishes, and, overall, the phenomenon is an example of a phase transition showing the same qualitative features as those we studied previously using Landau mean field theory.

Consider a binary mixture of components A and B. Let x be the concentration of A particles in the mixture, and $(1-x)$ the concentration of B particles. Define $m = 2x - 1$: this is an order parameter which has the value zero for an equal mixture. The entropy associated with the mixing (as opposed to other considerations such as volume changes) is the entropy of mixing given by equation (25.22). The internal energy has three types of contribution associated with chemical bonds: there is an interaction energy V_{AA} between neighbouring particles (atoms or molecules) of type A, V_{BB} between neighbouring particles of type B, and $V_{AB} = V_{BA}$ between neighbouring unlike particles.

In the absence of any mixing, all the A particles have only type A neighbours, and all the B particles have only type B neighbours, so the total chemical bond energy is

$$U_u = \frac{zN}{2}\left[xV_{AA} + (1-x)V_{BB}\right], \tag{25.30}$$

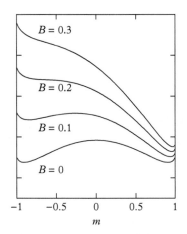

Fig. 25.7 *Effect of a magnetic field on the free energy $F(T;m)$. The free energy is shown for $B = 0$ (lower curve) and three increasing values of B, with the curves separated vertically for clarity (otherwise they would all intersect at $m = 0$). At $B = 0.1$ there is a metastable value of m at -0.9 as well as the stable value near 1. At $B = 0.2$ the metastable solution becomes unstable, and at $B = 0.3$ there is only one stable solution (one stationary point of F). In this way the mean field theory gives a good account (to first approximation) of the metastable solutions and hence the hysteresis observed in ferromagnetic behaviour.*

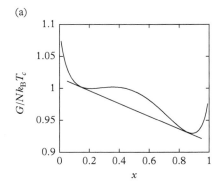 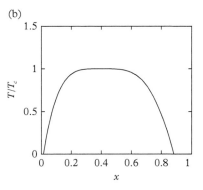

Fig. 25.8 *(a) Gibbs free energy for a binary mixture (equation 25.34) at $T = 0.8T_c$, taking $\mu_A^0 = k_B T_c$, $\mu_B^0 = 1.1 k_B T_c$. (b) Illustrative phase diagram for some generic binary mixture with a solubility gap at low temperatures. The two-phase region is under the curve.*

where z is the number of neighbours per particle, N is the number of particles, and the factor $1/2$ is to avoid double counting.

In the presence of mixing, the total chemical bond energy is calculated similarly, but now, for any given particle, the probability that a neighbour is of type A is x, and the probability that a neighbour is of type B is $(1-x)$. Therefore the number of AA pairs is $zNx^2/2$, the number of BB pairs is $zN(1-x)^2/2$, and the number of AB or BA pairs is $zNx(1-x)$. The total bond energy is now

$$U = \frac{zN}{2}\left[x^2 V_{AA} + (1-x)^2 V_{BB} + 2x(1-x) V_{AB}\right]. \tag{25.31}$$

The change in internal energy owing to mixing is therefore

$$\Delta U = U - U_u = \frac{zN}{2} x(1-x)\left(2V_{AB} - V_{AA} - V_{BB}\right)$$

$$= (1-m^2)\frac{zN}{4}\left(2V_{AB} - V_{AA} - V_{BB}\right). \tag{25.32}$$

Depending on the value of the bracket, this may be of either sign or zero. When $2V_{AB} < V_{AA} + V_{BB}$ then both entropy and energy favour mixing, so the components will mix at all temperatures. When $2V_{AB} > V_{AA} + V_{BB}$ the bracket is positive and then the energy as a function of m is a downward parabola, as in Figure 25.2, so we have a competition between energy and entropy, leading to a phase transition which can be described to first approximation in terms of free energy, just as in Figure 25.3. Therefore the mixture will separate into two phases at low temperature and will be homogeneous at temperatures above a critical temperature.

Mixing is usually discussed in conditions of constant temperature and pressure, so the Gibbs function is the appropriate free energy. At given pressure p, the Gibbs function is

$$G(T;x) = U(x) - TS(x) + pV, \tag{25.33}$$

where we have assumed that the volume depends negligibly on x. Taking the entropy function as in equation (9.14), we find

$$\frac{G}{N} = x\mu_A^0 + (1-x)\mu_B^0 + k_B T \left(x \ln x + (1-x) \ln(1-x) \right) + \alpha x(1-x),$$

(25.34)

where V_{AA} and V_{BB} contribute to $\mu_A^0(T,p)$ and $\mu_B^0(T,p)$, and $\alpha = z(V_{AB} - (V_{AA} + V_{BB})/2)$. More generally, we need not rely on the model of the chemical bonds, but simply remark that (25.34) is the expected behaviour for a simple mixture, with α defined as a parameter that may depend on pressure. The critical point is at $k_B T_c = \alpha/2$. The behaviour of G for $T < T_c$ is shown in Figure 25.8. Below T_c, at any given x the minimum G is obtained for a point on the common tangent line, as shown—this is just as in Figure 17.5 and the proof can be obtained by the reader by suitably modifying the argument leading to equation (17.11). The two phases have compositions corresponding to the two values of x where the common tangent line meets the curve. As x is increased from zero by adding substance A to the mixture, a first-order phase transition occurs between the two tangent points, with a latent heat.

So far we have considered a binary mixture in either fluid or solid form, but not both. When a liquid binary mixture has its temperature reduced, the phase transition to the solid form depends on the concentration and can give rise to a range of types of phase diagram. In the solid phase there can be more than one possible crystalline structure, and as a result, it can happen that the solid shows a miscibility gap while the liquid does not. This leads to behaviour called *eutectic*; Figure 25.9 shows a typical associated phase diagram. This figure shows what is observed when the liquid is miscible but the solid has two crystalline structures, α and β, one favoured at small x, the other at high x, where x is the concentration. As the temperature falls, the chemical potentials of the three molecular arrangements (liquid, solid α, solid β) all rise, the liquid by more than the others. At some temperature one of the solid chemical potential curves $\mu_\alpha(x)$ or $\mu_\beta(x)$ meets the liquid curve $\mu_L(x)$ at an extreme value of the concentration x. This is the melting point of a crystal of the corresponding pure substance. A mixed phase region then opens up as the temperature is lowered further. The other solid phase becomes favourable (i.e. has the lowest μ) at the other extreme of x at some lower temperature, the melting point of a pure crystal of the other substance. Eventually, at a temperature called the *eutectic* temperature T_e, the common tangent line from $\mu_\alpha(x)$ to $\mu_L(x)$ is the same line as the common tangent from $\mu_\beta(x)$ to $\mu_L(x)$. For this case, as x is varied, there is first a pure solid α region, then a mixed phase region of solid α and liquid, and then a mixed phase region of solid β and liquid, and then a pure solid β region. The meeting point of the mixed phase regions at the eutectic temperature is called the *eutectic point*. Below the eutectic temperature the common tangent from $\mu_\alpha(x)$ to $\mu_\beta(x)$ falls below the minimum of $\mu_L(x)$, so the liquid phase is no longer an equilibrium phase at any x.

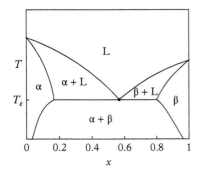

Fig. 25.9 *Eutectic phase diagram. The dot marks the eutectic point.*

If such a mixture is cooled through the eutectic point then it goes directly from a homogenous liquid to an inhomogenous solid. The solid is made of many regions such as layers a few micrometres thick of alternating crystalline structure. Sometimes other shapes are observed, such as rods or globular clusters.

....

EXERCISES

(25.1) *Ehrenfest's equations*

   (i) Show that, for a second-order transition in a pV system, not only are the chemical potentials equal, but so are the volume and entropy per particle.

   (ii) Let v be the volume per particle. Using $dv_1 = dv_2$ for a small change along the phase boundary, obtain

   $$\frac{dp}{dT} = \frac{\beta_2 - \beta_1}{\kappa_2 - \kappa_1} = \frac{\Delta\beta}{\Delta\kappa}. \qquad (25.35)$$

   [Hint: the method is just as for the Clausius–Clapeyron equation]

   (iii) Treating the entropy per particle, obtain

   $$\frac{dp}{dT} = -\frac{1}{vT}\frac{c_{p2} - c_{p1}}{\beta_2 - \beta_1} = \frac{\Delta C_p}{VT\Delta\beta}. \qquad (25.36)$$

   These are called *Ehrenfest's equations*. They apply to second-order phase transitions as long as the quantities involved are all well behaved. (Since for many continuous phase transitions some or all of the response functions diverge, these equations are less useful in practice than the Clausius–Clapeyron equation. However the corresponding results for magnetic properties are useful in the study of superconductivity.)

(25.2) Obtain the critical exponent β for the van der Waals model, as follows. To find the liquid–vapour density difference we need to find the locations of the points A and B indicated in the Maxwell construction shown on Figure 15.6. Let the two volumes be V_A and V_B, then since they are at the same pressure on the same isotherm, we may write

$$p = \frac{8T}{3V_A - 1} - \frac{3}{V_A^2} = \frac{8T}{3V_B - 1} - \frac{3}{V_B^2}.$$

Solve this equation for T, to obtain

$$T = \frac{(3V_A - 1)(3V_B - 1)}{8V_A^2 V_B^2}(V_B + V_A).$$

At the critical point we have $V_A = V_B = 1$ and then the equation gives $T = 1$ as expected. In general we would need the full Maxwell construction to find V_A and V_B, but now argue that near the critical point the two volumes must fall equally either side of 1, i.e. $V_A = 1 - x$ and $V_B = 1 + x$ for some x (this can also be shown by expanding the equation of state to third order in powers of $V - 1$ and looking at the limit when $T \to 1$). Expand the equation for small x and thus find $T \sim 1 - x^2/4$, which implies $(V_v - V_l) \propto (T - T_c)^{1/2}$. Obtain the density difference, and hence the exponent β in equation (25.9).

(25.3) Obtain equation (25.25).

Self-gravitation and negative heat capacity

26

26.1 Negative heat capacity   375
26.2 Black holes and Hawking radiation   378
Exercises   381

In this chapter we examine the influence of gravity on a large gas cloud, such as an interstellar dust cloud, or the interior of a star. Gravitational interactions are unlike the interactions that dominate gases of ordinary size, because they are long range, in the following sense. The gravitational interaction potential between any two particles falls as $1/r$, but the number of particles at any given distance from a given particle in a uniform gas rises as r^2, so the interactions with distant particles can dominate if the gas is large enough. The result is that the distinction between extensive and intensive properties breaks down; this situation is called 'non-extensive' behaviour. Each small region of a large cloud or gas can still be assigned a temperature and a pressure, but different regions are not guaranteed to be in stable equilibrium with one another, and one may have to consider the cloud as a whole in order to discover the overall behaviour. A feature of gravitational phenomena that often arises is that one has a thermodynamically unstable condition, and one of the features of the instability is a negative heat capacity. We will describe both this and some other general ideas.

26.1 Negative heat capacity

Negative heat capacity is the condition $đQ/dT < 0$. It means that, when a system receives some heat (and therefore some entropy when $T > 0$), its temperature falls. This breaks the stability condition (17.6) so it is not possible in stable equilibrium. However, it can happen in unstable cases, such as in the spinodal region of a first-order phase transition, and an important example occurs in a cloud of gas bound together by its own gravitation.

The idea that a higher internal energy might be associated with a lower temperature is counter-intuitive at first, because it never happens for the kinetic energy. However, when potential energy is also involved, either sign is possible. In order to gain familiarity with the ideas, first consider a simple case: a single particle moving in a central potential. We suppose the potential energy is given by a power law:

$$V(r) = \alpha r^n, \tag{26.1}$$

Thermodynamics: A Complete Undergraduate Course. Andrew M. Steane.
© Andrew M. Steane 2017. Published 2017 by Oxford University Press.

where α and n are constants. Two interesting examples are the harmonic potential well, $n = 2$, and gravitational attraction, $n = -1$. Among the possible motions of the particle there exist circular orbits, for which Newton's second law reads

$$\frac{dV}{dr} = m\frac{v^2}{r} \quad \Longrightarrow \quad \frac{1}{2}mv^2 = \frac{n}{2}\alpha r^n = \frac{n}{2}V. \qquad (26.2)$$

Hence, for circular orbits, the kinetic energy K is $(n/2)$ times the potential energy V. In this case the total energy $U = K + V = K(1 + 2/n)$. When $n = 2$, we have $U = 2K$, and when $n = -1$ we have $U = -K$. So in the latter case, when K goes up, U goes down.

Consider now two like particles A and B orbiting around their centre of mass, with a $1/r$ potential energy of attraction. Suppose the orbit is initially circular, at some energy U, and suppose particle A collides with another particle and loses energy. Its orbit will become elliptical, as shown in Figure 26.1. Suppose that it subsequently undergoes another collision, losing more energy, and it happens that its final motion is circular again. In this case the system AB has lost energy overall, and the particles with which A collided have gained energy. If the latter now move off to infinity then AB has emitted energy, its internal energy U is consequently reduced (that is, U has become more negative), and therefore the kinetic energies of both A and B have become *larger*. Their speeds in the final circular orbit are larger than the speeds they had in the initial circular orbit (since $U = -K$).

The argument can be generalized to a cloud of such particles by use of the *virial theorem*. This is a theorem in classical mechanics which states that, for a bound system of N particles, the time-average of the total kinetic energy is given by

$$\langle K \rangle = -\frac{1}{2} \sum_{i=1}^{N} \langle \mathbf{f}_i \cdot \mathbf{r}_i \rangle, \qquad (26.3)$$

where $\mathbf{f}_i$ is the net force on the i'th particle owing to all the others, and $\mathbf{r}_i$ is the location of the i'th particle. If the force between each pair of particles i,j arises from a potential energy of the form (26.1), where $r = |\mathbf{r}_i - \mathbf{r}_j|$, then it may be shown that the virial theorem takes the form

$$n\langle V \rangle = 2\langle K \rangle, \qquad (26.4)$$

where $\langle V \rangle$ is the time-average of the total potential energy. This is the generalization of (26.2). For $n = 2$ (e.g. particles in a solid) one thus find $U = 2\langle V \rangle = 2\langle K \rangle$, and for $n = -1$ (e.g. a self-gravitating cloud) one finds $U = \langle V \rangle/2 = -\langle K \rangle$.

Now consider a gravitating cloud with an initially well-defined temperature. This can be defined in the usual way, (18.9), if we have an understanding of the entropy as well as the energy. In the first instance, let us simply assume that $dS/dU > 0$ so we have positive temperature, and we will discuss this more fully in a moment. Consider a spherical region containing a fixed number of particles N inside such a cloud. Suppose these particles are within a region of radius R. Since

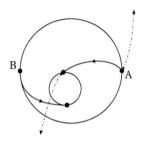

Fig. 26.1 A *and* B *orbit one another initially on the large circular orbit. Then, after two collisions, they orbit one another on the smaller circle. The* AB *system thus loses energy; the energy is carried away by the particles that* A *collided with (dashed trajectories).*

the gravitational potential energy varies as $V \sim -1/R$, we have, for this group of particles,

$$\frac{d\langle V\rangle}{dR} > 0, \qquad \frac{dU}{dR} > 0, \qquad \frac{dS}{dU} > 0, \qquad \frac{dU}{dT} < 0, \qquad (26.5)$$

where the last result holds because the kinetic energy $\langle K \rangle$ is an increasing function of temperature and $U = -\langle K \rangle$. We may write total derivatives because the system state is fixed by a single parameter: as soon as the temperature is fixed, so is the mean kinetic energy and hence the mean potential energy and hence the radius. We are here assuming that the initial state is bound, so that the virial theorem applies. If the initial temperature were high enough, the system would not be bound and the argument would not apply.

The situation can be close to equilibrium, at least temporarily, but it is not a stable equilibrium. For low enough initial temperature, the self-gravitating cloud cannot be stable because when any given part of the whole cloud loses energy, that part gets hotter and shrinks, while another part gains energy, gets colder, and expands. The temperature difference is now enhanced so the process continues. The net result is that the fluctuations in density or temperature in the cloud grow, and the whole process is called *condensation*. The parts that shrink lose entropy, while the parts that expand gain entropy, and there is a net entropy increase because, as usual, the direction of heat flow at any time is such as to guarantee this. There is no violation of the second law. On the contrary: the second law is fully obeyed, as it is in all of physics.

In order to justify the claim that each part of the cloud has a positive temperature, one may argue as follows. We have that $\langle V \rangle \propto -1/R$, so the virial theorem gives

$$\langle K \rangle \propto 1/R. \qquad (26.6)$$

Using $\langle K \rangle = (1/2)m\langle v^2 \rangle$, this implies that the rms speed of the particles varies as $v \propto R^{-1/2}$ and hence the volume of momentum space explored in the motion is $\Delta p_x \Delta p_y \Delta p_z \propto R^{-3/2}$. The volume of position space is $\propto R^3$ so the volume of phase space (position–momentum space) is $\propto R^{3/2}$. Now, according to Boltzmann's statistical definition of entropy, the entropy of the system increases when the accessible phase space volume increases. Therefore we deduce that the entropy is an *increasing* function of R, i.e. $dS/dR > 0$. When R grows, the larger volume 'wins' against the lower temperature, as far as entropy is concerned. We already know that $dU/dR > 0$, so we find $dS/dU > 0$, i.e. a positive temperature, as claimed. In fact, the ideal gas result $\langle K \rangle = (3/2)k_B T$ applies to good approximation within each parcel of gas in the cloud.

26.1.1 Jeans length

So far we have considered a cloud that was initially close to uniform, and we have examined the process of gravitational condensation by making use of the virial theorem. When the initial temperature of a cloud is high, however, it does not condense but disperses. The fate of the cloud is determined by the combination of its temperature and density, as follows.

The gravitational potential energy of a region of gas of total mass M and radius R is

$$\langle V \rangle \simeq -\frac{GM^2}{R}, \tag{26.7}$$

where we have omitted a constant of order 1, which will depend on the density distribution. If the particles have mass m then the number in a region of mass M is $N = M/m$ and the thermal kinetic energy is

$$\langle K \rangle \simeq N k_B T, \tag{26.8}$$

where again we will not be concerned with factors of order 1.

The region will collapse under its gravity if $|\langle V \rangle| > \langle K \rangle$, which gives

$$M > \frac{k_B T}{Gm} R. \tag{26.9}$$

The quantity on the right-hand side of this inequality is called the *Jeans mass*. Putting $M \simeq R^3 \rho$, the condition can be written $R > L_J$, where

$$L_J \simeq \sqrt{\frac{k_B T}{Gm\rho}} \tag{26.10}$$

is the *Jeans length*. We have deliberately omitted the factors of order 1 because they lend a spurious sense of precision unless they are calculated taking full account of the density distribution. This distribution is not uniform in a star, and in a nearly uniform interstellar cloud a full calculation has to study the effect of a ripple on the background density. The essential insight is that there is a competition between gravity and pressure. When pressure wins, a ripple will oscillate (sound waves). When gravity wins, a ripple will grow (gravitational collapse, also called condensation). Gravity wins for $L > L_J$, pressure wins for $L < L_J$. Hence large regions collapse and small regions oscillate.

A more precise treatment gives for the Jeans length,

$$L_J = v_s \sqrt{\frac{\pi}{G\rho}}, \tag{26.11}$$

where v_s is the speed of sound in the cloud.

26.2 Black holes and Hawking radiation

A *black hole* is a region of space where the mass density has become high enough that gravity prevents anything, including light, from escaping. The main properties of the hole are its mass, angular momentum, and electric charge.

Ordinarily any large black hole carrying an electric charge will become neutralized by attraction of matter of the opposite charge. In the following we will only consider neutral black holes.

A black hole with zero angular momentum is called a *Schwarzschild* black hole, one with angular momentum is called a *Kerr* black hole. In practice Schwarzschild black holes are rare, but they are simpler to describe so we will start with them.

The spacetime around a Schwarzschild black hole is characterized by a parameter called the *Schwarzschild radius*, given by[1]

$$r_s = \frac{2GM}{c^2}, \qquad (26.12)$$

where M is the mass of the hole, $G \simeq 6.67428 \times 10^{-11}$ m^{3}kgs^{-2} is the gravitational constant, and c is the speed of light. The region from which nothing escapes is the interior of a spherical surface called the *event horizon*. The area of this spherical surface is $4\pi r_s^2$, so, loosely speaking, one may say that r_s is the radius of the event horizon.

The Kerr black hole also has a spherical event horizon, whose radius is given by

$$r_K = \frac{1}{2}\left(r_s + \sqrt{r_s^2 - 4a^2}\right), \qquad (26.13)$$

where $a = J/Mc$ is a distance scale associated with angular momentum. A knowledge of general relativity is needed in order to get a precise understanding of these statements, but we have included them here mainly in order to point out the simple fact that when the angular momentum is reduced, the radius, and hence the surface area, of the Kerr event horizon grows.

When the properties of black holes were first established from the field equations of general relativity, it was assumed that black hole dynamics could be described by general relativistic mechanics without regard to thermal concepts such as temperature and entropy. This amounts to assuming that black holes are bodies or regions of spacetime at zero temperature. However, a deep connection to thermal physics was later made, chiefly by Stephen Hawking, with significant contributions from Roger Penrose, Yakov Zeldovich, Alexei Starobinsky, Jacob Bekenstein, and various other workers.

The first hint of a connection to thermal physics is the *area theorem*, which states that the area of the event horizon of a black hole cannot decrease, under the dynamics allowed by classical general relativity. For a Schwarzschild black hole, this might appear obvious (since matter can only fall in, causing M and hence r_s to grow), but the theorem also applies to Kerr black holes, and more generally to event horizons of any shape. This is much more interesting, because the mass of a Kerr black hole can decrease in a type of process called the *Penrose process*. The rotating hole interacts with objects in its vicinity, and in the right circumstances, the mass of the hole can decrease while the angular momentum also decreases,

---

[1] A useful mnemonic for remembering this formula is $(1/2)mc^2 = GMm/r_s$, so r_s is the radius at which the escape velocity would be c if Newtonian physics applied.

but the combined effect is always such that r_K increases. In short, the mass can increase or decrease under allowed physical processes, but the horizon area can only increase.

The one-way nature of this result prompted Bekenstein to propose arguments to suggest that an entropy could be associated with an event horizon. His order-of-magnitude arguments were based on considering quantum uncertainty limits on the position and momentum of objects lowered into a black hole.

Subsequently, Hawking placed the idea on a much firmer footing by a tour de force calculation of the quantum electromagnetic field in the vicinity of an event horizon. He showed that one expects electromagnetic radiation to be produced at the horizon, propagating away from the hole, with a distribution over frequency closely related to the Planck form, equation (19.12), and with all the other characteristics of thermal radiation.[2] The temperature of this emitted radiation is given by

$$T = \frac{\hbar \kappa / c}{2\pi k_B}, \qquad (26.14)$$

where κ is a measure of the gravity at the event horizon, called the surface gravity. For a Schwarzschild black hole, this is given by $\kappa = c^4/4GM$, so one has

$$T = \frac{\hbar c^3}{8\pi G M k_B}. \qquad (26.15)$$

Following standard practice, we will refer to this temperature as the temperature of the black hole. The same calculation also results in an entropy associated with the event horizon, given by

$$S = \frac{A}{4\ell_P^2} k_B, \qquad (26.16)$$

where A is the area of the event horizon and $\ell_P = \sqrt{G\hbar/c^3}$ is a length scale called the *Planck length*. For the Kerr black hole, $A = 4\pi r_K^2$.

Various observations can be made from these ideas. First, the black hole temperature is typically low, $T \simeq 61$ nK for a solar mass black hole. On the other hand, the entropy is huge, $S \simeq 4.2 \times 10^{77} k_B$ for a solar mass black hole (for comparison, the number of protons and electrons in the Sun is of order 10^{57} so the entropy of the Sun is of order $10^{57} k_B$). Also, it is interesting that the black hole entropy is proportional to the surface area, not to the volume nor the mass.

The emission of thermal radiation implies that, once quantum physics is taken into account, the area of a black hole event horizon can decrease with time after all, but only by emission of entropy to the rest of the universe. A number of calculations have shown that the overall result of adding the horizon entropy to the entropy of other systems gives a net quantity which is strictly increasing with time. This is called the *generalized second law*. It is reasonably well established for many situations, but entropy in quantum field theory is a subtle concept and there remain open questions when the dynamics are non-trivial.

---

[2] Hawking's original calculation suggested that the emitted radiation is precisely of the Planck spectrum with no small fluctuations around this value. There are arguments to suggest that in fact it may also have the same fluctuations that one expects in ordinary thermal radiation, but this is an open question.

In practice, astronomical black holes are bathed in the cosmic microwave background radiation, so all but the smallest holes will absorb more radiation than they produce. As they do this, the mass grows and hence the temperature falls: the heat capacity is negative, just as it is for other gravitationally bound systems. Consequently a black hole cannot come to a stable thermal equilibrium with other bodies.

In the absence of incoming matter or radiation, or more generally if a black hole is small and hot enough to emit more energy than it absorbs, then a black hole will eventually emit all of its energy and disappear. This is called black hole evaporation. It is a runaway process which would finish in an explosion. It has never been observed, but it is included in astronomical searches since it is possible that the Big Bang may have left behind some black holes of such a mass that they would now be exploding.

The most significant implications of the thermal properties of black holes are that they suggest instructive new ways to think about general relativity, and they give hints at what a more general theory—some sort of quantum treatment of spacetime—might involve. The description of gravity that general relativity provides can be seen as a type of guiding principle for the symmetry structure of a more detailed theory. It thus functions in a similar way to thermodynamics in relation to micro-physics.

EXERCISES

(26.1) A rigid object having negative heat capacity and positive temperature T is placed inside a cavity whose walls are maintained at positive temperature T_R. Explain what happens when $T > T_R$ and when $T < T_R$. Modelling the object as a rigid box containing a self-gravitating gas which initially only fills part of the box, explain what prevents T from either reaching absolute zero or increasing indefinitely.

(26.2) Estimate the heat capacity of the Sun (including the sign!)

(26.3) Estimate the number of moles of argon gas at STP that would be required for the gas to be gravitationally unstable.

(26.4) Find the mass of a black hole whose temperature is equal to that of the cosmic microwave background radiation.

(26.5) Show that the total power in the Hawking radiation from a Schwarzschild black hole of mass M is

$$P = \frac{\hbar c^6}{15360\pi G^2 M^2}.$$

27 Fluctuations

27.1 Probability of a departure from the maximum entropy point 383
27.2 Calculating the fluctuations 385
27.3 Internal flows 393
27.4 Fluctuation as a function of time 395
27.5 Johnson noise 398
Exercises 401

Fluctuations are an inescapable part of the nature of thermal systems. When a system is maintained at constant temperature, for example, it is in continuous interaction with a reservoir, exchanging energy with the reservoir. The system and reservoir together find an equilibrium which maximizes their combined entropy, but this is a dynamic equilibrium. The system's energy has a most likely value U^* (where the entropy is maximized), and an average value $\langle U \rangle$ (which to good approximation is equal to U^*), but it fluctuates as a function of time above and below this value. The standard language of thermodynamic reasoning does not allow for this fluctuation. One simply says the system's internal energy 'is' $U = U^*$, and one does not worry about the fact that it may depart slightly from U^*. Arguments which make that assumption are often useful in practice, because in the thermodynamic limit (the limit of large systems, large energies, etc.) the fluctuations are negligible compared to the most likely values, except in special cases such as around critical points. When fluctuations are not negligible, thermodynamics does not fail completely, but it has to be applied more carefully, taking the fluctuations into account.

Fluctuations have some features which can lead to confusion, so we begin with a brief discussion of those.

We characterize the size of fluctuations in any given parameter such as U or p by giving the variance ΔU^2, Δp^2 or the standard deviation ΔU, Δp. The first thing to note is that these quantities are not functions of state. For example, for a system in a given thermodynamic state (say, with given T, V, N) the value of ΔU will be depend on the constraints the system is under. If the system is isolated, then $\Delta U = 0$ (apart from a small quantum effect discussed later), whereas if it is in thermal contact with a reservoir at a fixed temperature, then $\Delta U \neq 0$. If, on the other hand, one first specifies the constraints, then ΔU can be written in terms of state functions, so then it becomes itself a function of state as long as the constraints don't change.

An intensive property such as pressure has further subtleties, because the pressure fluctuations on a given wall depend to some extent on the character of the wall. If the wall is soft then it 'gives' a little during each collision, which tends to smooth the fluctuations. One consequence of this is that *the equilibrium relations between functions of state do not hold during fluctuations*. For example, for an ideal gas we have that $p = (\gamma-1)U/V$ (Eq. 7.28), but in an isolated system U and V are constant but p is not, so $p + \Delta p \neq (\gamma-1)(U+\Delta U)/(V+\Delta V)$ and in particular one cannot claim, as one might be tempted to, that $\Delta p = (\gamma-1)\Delta(U/V)$, unless some further argument is given to justify the conclusion or explain what is meant by it.

Properties such as T, μ, and S introduce another type of subtlety. In a microscopic model, these are not properties of any one quantum state such as an energy eigenstate, but of the *distribution* over energy eigenstates. If the distribution changes, S will change, and so, in general, will its derivatives such as $(\partial S/\partial U)$, but the temperature obtained that way is not an equilibrium temperature and therefore will not in general satisfy the equation of state. Also, whereas one can constrain properties such as U, V, N, it is not possible to constrain properties such as T in the same way. When a system is in thermal contact with a large reservoir at temperature T_0, the latter stays constant but the system temperature fluctuates, because the system energy does. We will study this case more fully later.

Quantum fluctuations. For an isolated system the total internal energy and volume are fixed and do not fluctuate. According to classical physics their values can in principle be arbitrarily well defined, and according to quantum physics their values can be fixed to very high precision. For example, the statement

$$\Delta E \Delta t \geq \hbar,$$

sometimes called the 'energy–time uncertainty relation' asserts that as time goes on, the energy of an isolated system can be established with greater and greater precision. The principle of conservation of energy asserts that the energy will not fluctuate outside its range of uncertainly, of order $\hbar/\Delta t$. Taking $\Delta t = 1$ second for illustrative purposes, we obtain the range $\Delta E \simeq 10^{-34}$ J. Therefore the relative uncertainty, owing to quantum effects, of an energy of one joule measured for one second is one part in 10^{34}. Although such 'quantum fluctuations' can be important in some phenomena, they are very much smaller than the effects we will be studying in this chapter and we will ignore them from now on. A similar conclusion applies to volume and amount of matter. Therefore we will simply claim that U, V, and N can be regarded as precisely defined, fixed quantities for a large isolated system.

In this chapter we shall explore the size and nature of thermal fluctuations. One cannot deduce this information from the basic laws of thermodynamics alone, but, as with other aspects of thermal physics, once a small amount of further information is provided, thermodynamic reasoning reveals connections between one thing and another. As we have stressed throughout the book, the theme of thermodynamics is to do a lot with a little. In the case of fluctuations, all we need is one reasonably simple idea from statistical theory, and then we can deduce much. This idea is introduced in Section 27.1, and applied in the rest of the chapter.

27.1 Probability of a departure from the maximum entropy point

Let the *maximum entropy point* of an isolated system be defined as the combination of internal properties that maximizes its entropy. In order to reason about fluctuations, we need to reason about an isolated system which is not at its maximum entropy point. We have already learned in Chapter 17 how to do this: we divide the system up into smaller parts, each of which is assumed to be in internal equilibrium. Also, in Section 17.3.1 we introduced the idea of studying a range of non-equilibrium states by introducing a further parameter (e.g. r in Eq. 17.24) such that each non-equilibrium state under consideration would be an equilibrium state if the further parameter were constrained to one particular value. This is the method also adopted in the study of nucleation in Chapter 24 and in Landau mean field theory (Section 25.3). We will adopt it again here.

Consider an isolated system with given U, V, N, and also some internal parameter x. For example, x could be the radius of the balloon in Figure 17.8, or it could be an amount of energy that moves from one part of the system to another, etc. We will be considering the fluctuations of x, and we restrict discussion to cases where the state at each value of U, V, N, x would itself be an equilibrium state if x were constrained to have that value. Let $S(x; U, V, N)$ be the equilibrium entropy which the system has *when x is constrained at the given value*. Then, when x is *not* constrained, we shall only need to assume the following basic idea:

> The probability that an isolated system is found, at any time, in a state of entropy $S(x; U, V, N)$, is given by
> $$P \propto e^{S(x;U,V,N)/k_B}. \tag{27.1}$$

The proportionality constant can be found by normalization, such that $\sum P = 1$ when we consider all possible values of x. The formula (27.1) follows directly from Boltzmann's formula (9.8) and the general idea that a system explores the state space available to it, spending equal times in equal volumes of *phase space* (i.e. the position–momentum space of the microscopic variables that describe the system in detail). However we shall not discuss it further, merely make use of it.

27.1.1 Is there a violation of the second law?

There immediately arises another possible confusion here. The second law asserts that the entropy is maximized, under the given constraints, for a system in thermal equilibrium. Does equation (27.1) say that the system can in fact fluctuate to other states and thus lower its entropy? The answer to this is that the question has confused two different senses of the word 'entropy'. The quantity $S(x; U, V, N)$ is not 'the entropy of the system' except when x has its maximum entropy value x_0. Rather, $S(x_0; U, V, N)$ is 'the entropy of the system' and $S(x; U, V, N)$ is 'the entropy which the system *would* have *if x* were also constrained'.

But someone might insist that, no matter how the words and symbols are defined, the fact remains that a system in equilibrium does not remain fixed in one state, but explores a range of states. This is true. However, if one tries to exploit thermal fluctuations to convert heat into work, then one has scenarios such as those already explored in Chapter 9—things like the Feynman ratchet and Szilard's engine. We have already shown that such examples do not allow one to build a machine that violates the Clausius or Kelvin statements of the second law. Therefore small random changes in $S(x; U, V, N)$ do not constitute the type of change which the second law rules out, as long as they are consistent with (27.1). The correct way to understand the situation is to say that fluctuations are part and parcel of the nature of thermal equilibrium. The nature of thermal equilibrium is that the system is not always precisely at the maximum entropy point x_0, but fluctuates about it with a probability distribution given by (27.1). The maximum entropy principle discussed in Chapter 18 should be understood in this sense.

Note that it is the *randomness* of the fluctuations that prevents their being exploited to convert heat to work. If one could predict the fluctuations then one could construct a Szilard heat engine in which rather than *observing* the location of a particle, one could *predict* it. In this way work could be obtained from Szilard's engine, just as work was obtained by the process shown in Figure 9.12. However in order to make the prediction, information would first have to be gathered. The eventual erasure of that gathered information would be needed to complete a thermodynamic cycle. This erasure step would involve the emission of just as much heat as would be emitted by a Carnot engine operating in the same environment. The second law always wins in the end!

This is not the only area of physics where 'built-in' randomness acts to make impossible what would otherwise be possible. Another example occurs in the physics of quantum entangled particles, where faster than light signalling would be possible if the measurement outcomes for quantum superpositions were not inherently unpredictable.

27.2 Calculating the fluctuations

We now turn to the use of equation (27.1) to calculate the variance of various fluctuating quantities. We will continue to use the simple compressible system as our standard thermodynamic system; the arguments readily generalize to other types of system.

Equation (27.1) says that the most likely value of x is when $S(x; U, V, N)$ is at a maximum. Let this most likely value be x_0. Then, we can Taylor expand $S(x)$ about x_0 (we suppress U, V, N for clarity), giving

$$S(x) \simeq S(x_0) + \frac{1}{2} \left.\frac{\partial^2 S}{\partial x^2}\right|_{U,V,N} (x-x_0)^2 + \cdots \qquad (27.2)$$

The first-order terms vanish because we are at a stationary point, and the higher-order terms are normally negligible for large systems. By substituting this into (27.1) we find that the probability distribution of x is Gaussian to good approximation:

$$P(x) \propto e^{-(x-x_0)^2/2\Delta x^2}, \qquad (27.3)$$

with the variance given by

$$\Delta x^2 = \frac{k_B}{-\frac{\partial^2 S}{\partial x^2}} \qquad (27.4)$$

(the $S(x_0)/k_B$ term has been absorbed into the normalization constant). The sign is introduced in (27.4) because we are at a maximum of S, so the second derivative is negative.

Consider now the application of this result to the case of an isolated system in which energy moves between two parts 1 and 2. Here, U is the total internal energy, and x is the internal energy of the first part. Since we can imagine separating the parts by a thermally insulating barrier, the method invoked in equation (27.1) applies, because we can define $S(x; U, V, N)$.

When energy dU_1 moves from 2 to 1, then we find

$$dS = dS_1 + dS_2 = \left(\frac{1}{T_1} - \frac{1}{T_2}\right) dx,$$

since $dx = dU_1 = -dU_2$ (cf. Eq. 17.2). Therefore,

$$\frac{\partial^2 S}{\partial x^2} = -\left(\frac{1}{C_1} + \frac{1}{C_2}\right)\frac{1}{T^2},$$

where C_1, C_2 are the constant-volume heat capacities of the two parts. Using (27.4) we deduce that

$$\Delta U_1^2 = k_B T^2 \frac{C_1 C_2}{C_1 + C_2}. \qquad (27.5)$$

Now suppose that part 2 is very much larger than part 1. Then it is serving as a thermal reservoir (cf. Figure 27.1) and $C_2 \gg C_1$. In this case we can regard part 1 as our system and hence obtain

---

Energy fluctuations under fixed T, V, N

$$\Delta U^2 = k_B T^2 C_V. \qquad (27.6)$$

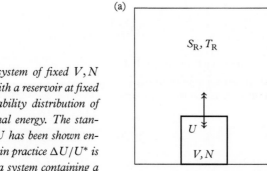
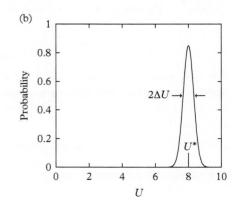

Fig. 27.1 (a): A system of fixed V, N exchanging heat with a reservoir at fixed T_R. (b) The probability distribution of the system's internal energy. The standard deviation ΔU has been shown enhanced for clarity; in practice $\Delta U/U^*$ is of order 10^{-12} for a system containing a mole of particles.

For a monatomic ideal gas, (27.6) gives

$$\frac{\Delta U}{U^*} = \sqrt{\frac{2}{3N}}, \qquad (27.7)$$

where $U^* = (3/2)Nk_B T$ is the most likely energy, i.e. the energy at the maximum entropy point, the quantity we usually call simply U. Therefore the fluctuations in internal energy, for a system in contact with a thermal reservoir at given temperature, are negligible compared to the most likely internal energy when the number of particles is large. For a mole of gas, $\Delta U/U^* \simeq 10^{-12}$, so to assert $U = U^*$, which is what we have asserted throughout this book until now, it is accurate to about one part in 10^{12}.

More generally the *equipartition theorem* asserts that, at high temperature, the thermal contribution to internal energy is equal to $(1/2)k_B T$ for each microscopic degree of freedom of the system. Therefore $U = N_{df}k_B T/2$, where N_{df} is the number of degrees of freedom (roughly speaking, the number of particles multiplied by the number of independent ways each particle can store energy). Hence,

$$\frac{\Delta U}{U} = \sqrt{\frac{2}{N_{df}}}. \qquad (27.8)$$

When the system energy fluctuates without a change in V and N, then so does its temperature. This is an example of Le Chatelier's principle: if U increases then T must also increase, so that a temperature gradient results that causes the energy to flow back again, suppressing the fluctuation—otherwise the equilibrium could not be stable. For the system of fixed volume and particle number, in contact with a thermal reservoir, one would expect the temperature fluctuation to be given by

$$\Delta T = \frac{\Delta U}{C_V} = T\sqrt{\frac{k_B}{C_V}}. \qquad (27.9)$$

We will show that this is in fact the case in Section 27.2.1.

27.2.1 More general constraints

So far we have discussed the internal energy of a rigid closed system in thermal constant with a reservoir. We have applied the probability equation (27.1), to the total entropy of system plus reservoir, and obtained a statement about the fluctuations of the system. In order to generalize to systems under other constraints, we adopt this same approach of applying (27.1) to the total assembly of system plus reservoir. Then,

$$dS_{tot} = dS_R + dS,$$

where dS_R is the change in entropy of the reservoir and dS is the change in entropy of the system, and the probability is proportional to $\exp(S_{tot}/k_B)$. We have

$$dS_{\text{tot}} = (dU_R + p_R dV_R - \mu_R dN_R)/T_R + dS$$
$$= (-dU - p_R dV + \mu_R dN)/T_R + dS$$
$$= -\frac{dA}{T_R}, \qquad (27.10)$$

where $A = U - T_R S + p_R V - \mu_R N$ is the availability (Eq. 17.16). Therefore, for whichever system variable x is under consideration,

$$\frac{\partial^2 S_{\text{tot}}}{\partial x^2} = -\frac{1}{T_R}\frac{\partial^2 A}{\partial x^2} \qquad (27.11)$$

and therefore,

$$\overline{\Delta x^2} = \frac{k_B T_R}{\frac{\partial^2 A}{\partial x^2}}, \qquad (27.12)$$

where the quantities held constant in the partial derivatives depend on the direction in parameter space of the fluctuation we wish to calculate. This is best illustrated by some examples.

In general we have

$$dA = dU - T_R dS + p_R dV - \mu_R dN$$
$$= (T - T_R)\, dS + (p_R - p)\, dV + (\mu - \mu_R)\, dN. \qquad (27.13)$$

27.2.1.1 Constrained T, V, N

First consider the case we have already treated: a system of fixed V, N in contact with a reservoir at temperature T_R. With V and N fixed (and all the reservoir properties fixed), we have $dA = (T - T_R)dS$, so

$$\left.\frac{\partial A}{\partial U}\right|_{V,N} = 1 - \frac{T_R}{T} \quad \Rightarrow \quad \left.\frac{\partial^2 A}{\partial U^2}\right|_{V,N} = \frac{T_R}{T^2 C_V}.$$

Substituting this into (27.12) with $T = T_R$, we obtain (27.6) as before. Also,

$$\left.\frac{\partial A}{\partial T}\right|_{V,N} = C_V\left(1 - \frac{T_R}{T}\right) \quad \Rightarrow \quad \left.\frac{\partial^2 A}{\partial T^2}\right|_{V,N} = \frac{\partial C_V}{\partial T}\left(1 - \frac{T_R}{T}\right) + C_V \frac{T_R}{T^2},$$

which, when evaluated at $T = T_R$, gives (27.9). Note that the temperature constraint that the reservoir provides does not prevent the system temperature from fluctuating, it merely ensures that the system temperature averages to T_R.

27.2.1.2 Constrained T, p, N

Next consider a closed system with unconstrained volume and internal energy, in contact with a reservoir at fixed temperature and pressure. Now we have

$$dA = (T - T_R)\,dS + (p_R - p)\,dV. \tag{27.14}$$

Before considering the energy, it will be useful to first obtain the volume fluctuations. To this end, consider

$$\left.\frac{\partial A}{\partial V}\right|_{T,N} = (T - T_R)\left.\frac{\partial S}{\partial V}\right|_{T,N} + p_R - p.$$

Hence,

$$\left.\frac{\partial^2 A}{\partial V^2}\right|_{T,N} = (T - T_R)\left.\frac{\partial^2 S}{\partial V^2}\right|_{T,N} - \left.\frac{\partial p}{\partial V}\right|_{T,N}.$$

Evaluating this at $T = T_R$, and substituting into (27.12) gives

$$\Delta V^2 = -k_B T \left.\frac{\partial V}{\partial p}\right|_{T,N}. \tag{27.15}$$

In order to discuss the energy fluctuations in a closed flexible system in contact with an environment at constant pressure and temperature, it will be useful to consider the case where (p/T) is constant. In this case we will obtain the variance of the energy fluctuations along a line in the p, T plane that passes through the given state (p, T) and intercepts the axes at the origin. The calculation is most conveniently done by introducing a variable α, defined by

$$\alpha \equiv \frac{p}{T},$$

and then we replace p by αT in the basic expressions. The fundamental relation becomes

$$dU = T\,dS - \alpha T\,dV + \mu\,dN$$

and

$$dA = (T - T_R)\,dS + (\alpha_R T_R - \alpha T)\,dV + (\mu - \mu_R T_R)\,dN.$$

At $T = T_R$ and $\mu = \mu_R$, one finds

$$\left.\frac{\partial^2 A}{\partial U^2}\right|_{\alpha,N} = \left(\left.\frac{\partial S}{\partial U}\right|_{\alpha,N} - \alpha \left.\frac{\partial V}{\partial U}\right|_{\alpha,N}\right)\left.\frac{\partial T}{\partial U}\right|_{\alpha,N} = \frac{1}{T}\left.\frac{\partial T}{\partial U}\right|_{\alpha,N}.$$

Therefore,

$$\Delta U^2 = k_B T^2 \left.\frac{\partial U}{\partial T}\right|_{(p/T),N}. \tag{27.16}$$

This result can also be written (Exercise 27.4)

$$\Delta U^2 = k_B T^2 C_V + \Delta V^2 \left.\frac{\partial U}{\partial V}\right|_{T,N}^2, \tag{27.17}$$

where ΔV^2 is given by (27.15). By comparing this expression with (27.6), we can make a natural physical interpretation. When the closed system can exchange volume as well as heat energy with its environment, then the energy fluctuations are given by the sum of two terms. The first can be interpreted as the contribution from heat, the second as the contribution to the energy fluctuations arising from the volume fluctuations.

To consider the energy fluctuations for more general changes (not just those where p/T stays constant) in this system, one may extend the method to cover simultaneous fluctuations of two or more variables. For example, if there are simultaneous changes in energy and volume, then we perform a double Taylor expansion of $A(U, V)$ about the stationary point. Since the first-order terms vanish at the stationary point, one finds, to second order:

$$A = A_0 + \frac{1}{2}A_{UU}(U - U_0)^2 + A_{UV}(U - U_0)(V - V_0) + \frac{1}{2}A_{VV}(V - V_0)^2 \quad (27.18)$$

where $A_{ij} \equiv (\partial^2 A/\partial i \partial j)$. This implies that the probability distribution is a double Gaussian distribution. One finds

$$\left.\frac{\partial^2 A}{\partial U^2}\right|_V = \frac{T_R}{T^2}\left.\frac{\partial T}{\partial U}\right|_V = \frac{T_R}{C_V T^2} \quad (27.19)$$

$$\left.\frac{\partial^2 A}{\partial U \partial V}\right. = \frac{T_R}{T^2}\left.\frac{\partial T}{\partial V}\right|_U = -T_R \left.\frac{\partial (p/T)}{\partial U}\right|_V \quad (27.20)$$

$$\left.\frac{\partial^2 A}{\partial V^2}\right|_U = -T_R \left.\frac{\partial (p/T)}{\partial V}\right|_U. \quad (27.21)$$

In general, $A_{UV} \neq 0$ and therefore the fluctuations in U and V, are correlated, although the ideal gas is an exception to this.

If, for the same system, we consider the fluctuations of T and V instead of U and V, then a similar method of calculation leads to

$$\frac{\partial^2 A}{\partial T \partial V} = 0, \quad (27.22)$$

so the fluctuations in T and V are uncorrelated. The other terms in the expansion lead to (27.9) and (27.15) as before (Exercise 27.2).

27.2.1.3 Constrained T, V, μ

Finally, let us consider a system of fixed volume and temperature, which exchanges material with its environment. Now the constrained variables are T, V, μ and the free energy is the grand potential (Landau free energy), $\Omega = F - \mu N$. From (27.13) one obtains

$$\left.\frac{\partial^2 A}{\partial N^2}\right|_{V,T} = (T - T_R) \left.\frac{\partial^2 S}{\partial N^2}\right|_{V,T} + \left.\frac{\partial \mu}{\partial N}\right|_{V,T}$$

and hence, by substituting this into (27.12) at $T = T_R$,

$$\Delta N^2 = k_B T \left.\frac{\partial N}{\partial \mu}\right|_{V,T}. \quad (27.23)$$

This can be expressed in terms of more convenient quantities as follows. From the Gibbs–Duhem relation (13.5), we have

$$\left.\frac{\partial \mu}{\partial N}\right|_{V,T} = \frac{V}{N}\left.\frac{\partial p}{\partial N}\right|_{V,T}.$$

Using first the reciprocity relation and then equation (5.16), this becomes

$$\left.\frac{\partial \mu}{\partial N}\right|_{V,T} = -\frac{V^2}{N^2}\left.\frac{\partial p}{\partial V}\right|_{N,T},$$

hence,

$$\Delta N^2 = k_B T \frac{N^2}{V} \kappa_T, \tag{27.24}$$

where $\kappa_T = (-1/V)(\partial V/\partial p)_T$ is the isothermal compressibility.

To find the energy fluctuations, adopt a method similar to the one we used in order to obtain equations (27.16) and (27.17). One finds, for fluctuations along the line $(\mu/T) = $ constant,

$$\Delta U^2 = k_B T^2 \left.\frac{\partial U}{\partial T}\right|_{(\mu/T),V} \tag{27.25}$$

$$= k_B T^2 C_V + \Delta N^2 \left.\frac{\partial U}{\partial N}\right|_{T,V}^2, \tag{27.26}$$

where ΔN^2 is given by (27.23). The energy fluctuations again have two terms, one from heat exchange and the other from the contribution to the energy fluctuations arising from the number fluctuations.

27.2.2 Some general observations

The results obtained in the previous section are gathered in Table 27.1. From all of these, and from the study of further examples, some common themes emerge.

1. *Fluctuation is related to response function*
   In every case, the variance of the quantity under consideration is found to be proportional to a response function related to that quantity. Energy

Table 27.1 *Fluctuations of extensive quantities for a system under various external constraints. The last two results for ΔU^2 are for the cases of constant (p/T) and constant (μ/T) respectively.*

	U, V, N	T, V, N	T, p, N	T, V, μ
ΔU^2	0	$k_B T^2 C_V$	$k_B T^2 C_V + \Delta V^2 \left.\frac{\partial U}{\partial V}\right|_T^2$	$k_B T^2 C_V + \Delta N^2 \left.\frac{\partial U}{\partial N}\right|_{T,V}^2$
ΔV^2	0	0	$-k_B T \left.\frac{\partial V}{\partial p}\right|_T = k_B T V \kappa_T$	0
ΔN^2	0	0	0	$k_B T \frac{N^2}{V} \kappa_T$

fluctuations are proportional to heat capacity, volume fluctuations are proportional to compressibility, etc. It makes sense that a large heat capacity, for example, is associated with large energy fluctuations, because a large C_V implies that more energy can be absorbed or given out by the system before it 'notices' and adjusts its temperature. The relationship is used, for example, when thermal systems are modelled in computer calculations. The variance of the energy or the volume etc. is often used as the most convenient way to derive the value of the relevant response function.

2. These relationships are connected to the *fluctuation–dissipation theorem*, which is a theorem in mechanics that relates dissipative effects such as friction, heat diffusion, etc. to the fluctuations in the associated forces.

3. *The fluctuations depend on the constraints*
   The fluctuation of any quantity depends on the constraints the system is under (see for example Table 27.1).

4. *Fluctuations are mostly negligible in the thermodynamic limit*
   We have already explored this briefly after equation (27.8). More generally, the variances of extensive properties scale in proportion to system size, so the standard deviations scale in proportion to the square root of system size, and therefore the relative size of the fluctuations tends to zero as the system size tends to infinity. For example, for the ideal gas, one finds for the cases described in Table 27.1,

$$\frac{\Delta V}{V} = \frac{\Delta N}{N} = \sqrt{\frac{1}{N}}. \tag{27.27}$$

5. *Thermal fluctuations are macroscopic*
   Notwithstanding the previous point, the changes associated with thermal fluctuations are macroscopic quantities that can in principle be measured without encountering the limits to measurement associated with quantum theory. They are typically small by everyday standards, but detectable by sensitive instruments. The energy fluctuations for a mole of gas at room temperature are of the order of a few nanojoules. Brownian motion and the Johnson noise, discussed in Section 27.5, are examples of the effect of thermal fluctuations, and the most readily observed example is thermal radiation.

6. *Fluctuations are large near critical points*
   Near a critical point many of the response functions diverge. Therefore the related fluctuations diverge also. In this case the fluctuations are not small compared to the average values and it is not accurate to treat the system in terms of a small number of macroscopic parameters. More advanced methods of calculation are required.

7. *Intensive variables are not well defined at a point*
   When we use intensive variables such as temperature, pressure, density, and so on in physical arguments, it is convenient to suppose that they

take on well-defined values at each point in space, for any given system. However, owing to thermal fluctuations, this must be reconsidered. If the energy in a region of volume V fluctuates with standard deviation ΔU, then the average energy density in the region has a standard deviation,

$$\Delta u = \frac{\Delta U}{V}. \tag{27.28}$$

If the region is part of a system in equilibrium, then ΔU is proportional to $\sqrt{V}$, so $\Delta u \propto V^{-1/2}$. As $V \to 0$, $\Delta u \to \infty$. Similar considerations apply to temperature, pressure, and so on. Whenever we refer to temperature and pressure in thermodynamic arguments, we really mean their values after averaging over some finite-sized region.

27.3 Internal flows

Next we shall apply the probability formula to study the situation indicated in Figure 27.2. We imagine an isolated system divided into cells, and we consider a fluctuation whose net effect is to transfer energy q from cell 2 to cell 1 (not necessarily neighbouring). q is a small but not infinitesimal quantity. We have for the total system entropy,

$$\tilde{S} = S_1 + S_2 + S_{\text{rest}}, \tag{27.29}$$

where S_{rest} refers to the rest of the system, which is unaffected by the fluctuation under consideration.[1] Therefore,

$$\frac{\partial \tilde{S}}{\partial q} = \frac{\partial S_1}{\partial q} + \frac{\partial S_2}{\partial q} = \frac{1}{T_1} - \frac{1}{T_2}. \tag{27.30}$$

We now take an interest in

$$\left\langle \frac{\partial \tilde{S}}{\partial q} q \right\rangle.$$

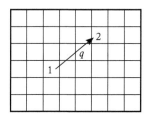

Fig. 27.2 Energy q is transferred from cell 1 to cell 2.

This quantity tells us about the entropy change associated with the internal energy flow. It can also be read as a correlation function, telling us whether fluctuations in $(1/T_1 - 1/T_2)$ are correlated with fluctuations in q, the transferred energy.

We first present an approximate argument, in order to get a feel for the sort of process we are investigating. Then we shall present a more accurate and general argument.

Suppose some energy q moves from region 2 to region 1. The temperature in region 1 changes from T to $T + \delta T$, where

$$\delta T = \frac{q}{C_{V1}} \tag{27.31}$$

---

[1] The notation $\tilde{S}$ here is the same as that adopted in Chapter 17; it draws attention to the fact that we are considering the entropy of a state which is not necessarily the maximum entropy state.

394 *Fluctuations*

and, for two regions having equal volume, the temperature in region 2 changes from T to $T-\delta T$. The average temperatures of the two regions during the energy transport are $T_1 = T + \delta T/2$ and $T_2 = T - \delta T/2$, so, using (27.30),

$$\frac{\partial \tilde{S}}{\partial q} \simeq \frac{1}{T+\delta T/2} - \frac{1}{T-\delta T/2} = -\frac{\delta T}{T^2} = -\frac{q}{T^2 C_{V1}}.$$

Therefore we should expect

$$\left\langle \frac{\partial \tilde{S}}{\partial q} q \right\rangle \simeq -\frac{\langle q^2 \rangle}{T^2 C_{V1}} = -k_B, \tag{27.32}$$

using (27.6). Thus we expect that the size of the entropy fluctuation associated with the internal fluctuation is k_B, independent of the size of regions 1 and 2. This is because a larger region will have larger energy fluctuations, but also larger heat capacity so smaller temperature fluctuations.

Now we present a more careful argument. From (27.1) we have

$$\tilde{S} = k_B \ln P + \text{const} \quad \Rightarrow \quad \frac{\partial \tilde{S}}{\partial x_i} = k_B \frac{\partial \ln P}{\partial x_i},$$

where x_i is any one of a set of independent variables in terms of which $\tilde{S}$ can be expressed. Therefore,

$$\left\langle \frac{\partial \tilde{S}}{\partial x_i} x_j \right\rangle = k_B \left\langle \frac{\partial \ln P}{\partial x_i} x_j \right\rangle$$

$$= k_B \int \frac{\partial \ln P}{\partial x_i} x_j P \, d\mathbf{x}$$

$$= k_B \int \frac{\partial P}{\partial x_i} x_j \, d\mathbf{x}$$

$$= k_B \int [P x_j]_{-\infty}^{\infty} d\mathbf{x}_{i \neq j} - k_B \int \frac{\partial x_j}{\partial x_i} P \, d\mathbf{x}$$

$$= -k_B \delta_{ij}, \tag{27.33}$$

where $d\mathbf{x} \equiv dx_1 dx_2 \ldots dx_n$ and in the penultimate step we perform the i'th integral by parts. We have then used that $P \to 0$ for any large excursion, and (obviously) $\partial x_j / \partial x_i = \delta_{ij}$.

The result (27.33) is interesting in several ways, and we shall make use of it again in Chapter 28. For the case $i \neq j$ it gives zero. This says that the internal fluctuating flow of any given quantity is not correlated with fluctuations in the entropy gradient with respect to other quantities. For example, energy flows are not correlated with pressure fluctuations. For the case $i = j$ it says there is a correlation. Note the overall sign is negative. This is not surprising since one can only go down from a maximum: the fluctuations away from maximum entropy can only ever lower the entropy.

For $i = j$, equation (27.33) predicts that the size of entropy fluctuation associated with the i'th internal fluctuation is k_B, independent of the size of regions 1 and 2, as we already noted, and we see that the initial approximate treatment gave a good insight (Eq. 27.32).

As we have already noted in Chapter 17, for stable equilibrium a given fluctuation tends to create the very gradient which then serves to oppose the fluctuation (Le Chatelier's principle). Thermal fluctuations are a continuous 'attempt' by the system to push against the internal traffic control mechanisms. Since the equilibrium point is a maximum entropy point, the average (27.33) cannot help but be negative. This sometimes leads to confusion about whether or not the second law of thermodynamics is broken in the presence of fluctuations. We have already discussed this in Section 27.1 and Chapter 9 (the Feynman ratchet), and concluded that the second law is upheld.

27.4 Fluctuation as a function of time

So far in this chapter we have examined the *size* of the fluctuations without regard to the timescale on which they occur. Next we will consider the time dependence. There now arises a possible ambiguity in the terminology. The quantities such as q or x_i considered in Section 27.3 were defined as the *amount of departure* (at any given time) from the maximum entropy point:

$$q(t) = U_1(t) - U_1^* = U_2^* - U_2(t),$$

where U_1^* and U_2^* are the values of the energies in regions 1 and 2 when the whole system has its maximum entropy. This is related to, but different from, another idea, namely the *change during a given time interval*. The word 'fluctuation' could refer to either type of quantity. We have been using it to refer to the first type (a synonym for 'departure') and this is the sense we will continue to mean throughout this chapter.

Another area of possible confusion is that the values U_1^* and U_2^* are often called 'equilibrium values'. That term can be misleading, however, because the presence of thermal fluctuations does not mean a system is not in thermal equilibrium. Therefore we refer to U_1^* and U_2^* as 'maximum entropy values'.

The probability formula (27.1) does not make any mention of time, so it does not tell us about the temporal behaviour of fluctuations. It asserts merely that, after averaging over all ways of realizing a system of the given type subject to the given constraints, one will find the system is in one state or another with the probabilities as given. A first step towards predicting the fluctuations as a function of time is to make the

> *Ergodic hypothesis*: A single system evolves as a function of time in such a way that $\tilde{S}(t)$ is described by the same probability distribution as the one describing an ensemble of systems at any given time, i.e. equation (27.1).

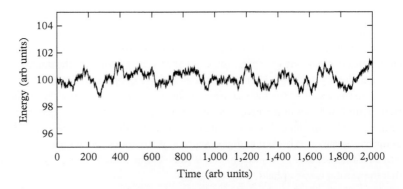

Fig. 27.3 *Typical form of the energy as a function of time, for a system in equilibrium with a reservoir at given temperature. In this example the system has around 2×10^4 degrees of freedom and the relaxation time is approximately 50 units.*

This hypothesis can be motivated by a classical- or quantum-mechanical treatment of the microscopic behaviour of the system. Such a study also yields further information, i.e. not just the probability of any particular configuration, but also the likelihood of each transition from one configuration to another, and thus the full typical time evolution. This is outside our remit but we will make a few general observations.

Figure 27.3 shows an illustrative example of energy as a function of time for a system in thermal equilibrium with a reservoir at non-zero temperature. The energy of the system fluctuates about the value U^*. For short times Δt after any given time t, the energy $U(t+\Delta t)$ is *correlated* with $U(t)$, because the energy cannot abruptly jump from one value to another: there is some 'inertia' in the system. However, for large enough Δt, $U(t+\Delta t)$ is uncorrelated with $U(t)$. There are two timescales that are important to this 'loss of memory' after any given change. They are called the *acceleration time* τ_a and the *relaxation time* τ_r. The acceleration time is the time the system takes to develop the internal currents which represent its response to a given change in conditions. In a gas the acceleration time is of the order of the collision time $\tau_a \simeq \lambda/\bar{v}$, where λ is the mean free path, giving an acceleration time of order 0.5 ns for nitrogen at STP. The relaxation time is the time it takes those internal currents to return the system to the maximum entropy point. This can be estimated by the method discussed in Section 10.2, where we obtained $\tau_r \simeq L^2/D$, where L is the size of the system and D the thermal diffusivity. Using $D \simeq \lambda \bar{v}$ for an ideal gas, one finds

$$\tau_r \simeq \frac{L^2}{\lambda \bar{v}}, \qquad (27.34)$$

which gives, for example, 1 μs for a small (10 micron) region in a sample of nitrogen gas at STP, and much longer times for a large sample of gas. We have

$$\frac{\tau_a}{\tau_r} \simeq \frac{\lambda D}{\bar{v} L^2} \simeq \frac{\lambda^2}{L^2}$$

(cf. Eq. 10.22). Therefore for a gas the acceleration time is small compared to the relaxation time when the dimensions of the system are large compared to the mean free path.

The value of Δt over which the 'loss of memory' occurs (i.e. after which $U(t+\Delta t)$ is uncorrelated with $U(t)$) is called the *correlation time* τ. For many systems the correlation time turns out to be the same as the relaxation time: $\tau \simeq \tau_r$. One typically finds, for $|\Delta t| > \tau_a$,

$$\langle U(t+\Delta t)U(t) \rangle \simeq \langle U^2 \rangle e^{-|\Delta t|/\tau} \quad (27.35)$$

(see Figure 27.4). Note however that equation (27.35) cannot be valid at very short times, i.e. the limit $\Delta t \to 0$, because it predicts the wrong dependence on Δt for very small Δt, as we now show.

Consider

$$\left\langle \frac{d}{dt}U^2 \right\rangle \equiv \lim_{T\to\infty} \frac{1}{2T} \int_{-T}^{T} \frac{d}{dt}U^2 dt = \lim_{T\to\infty} \frac{[U^2]_{-T}^{T}}{2T} = 0,$$

which is valid as long as $U(t)$ does not tend to infinity as $t \to \pm\infty$. It follows that

$$\left\langle \frac{dU}{dt}U \right\rangle = \frac{1}{2}\left\langle \frac{d}{dt}U^2 \right\rangle = 0. \quad (27.36)$$

This equation says that dU/dt is uncorrelated with U. In particular, at any given U, the value of U is as likely to be going up as down. This is because the system is as likely to arrive at U while moving away from the average as it is while journeying back.

Now consider the average value of $U(t+\Delta t)U(t)$, by performing a Taylor expansion:

$$\langle U(t+\Delta t)U(t) \rangle = \left\langle \left(U + \Delta t \frac{dU}{dt} + O(\Delta t^2)\right) U(t) \right\rangle = \langle U^2 \rangle + O(\Delta t^2),$$

by using (27.36). This shows that at short times the correlation function must be quadratic; this rounds off the peak which is typically found at $\Delta t = 0$. It follows that equation (27.35) cannot be correct at very small values of Δt. Equation (27.35) is not completely invalid, it is simply that it does not apply at very small $|\Delta t|$. We can use (27.35) for values of $|\Delta t|$ larger than the acceleration time.

Having thus established the lack of any correlation between U and its instantaneous rate of change, we shall now argue that there *is* a correlation between U and its average rate of change over timescales which are small but not infinitesimal. To see this, notice that the energy does not make a random walk (such as Brownian motion), which would cause it to drift further and further away from U^* as time goes on. This means there must be a correlation between U and the average rate of change of U. For, consider the situation at any given time t. We will find $q(t) = U(t) - U^*$ has a non-zero value (except at instants when it passes through zero). However, the subsequent behaviour must return q to zero, on average. Therefore for given $q(t)$, we must find

$$q(t+\Delta t) \stackrel{\text{average}}{=} 0 \quad \Rightarrow \quad q(t+\Delta t) - q(t) \stackrel{\text{average}}{=} -q(t)$$

for times $\Delta t \gg \tau$. The above statement refers somewhat loosely to the average we have in mind. To make a precise statement, multiply both sides by $q(t)$ and then the precise statement is

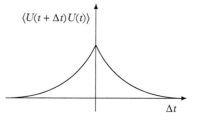

Fig. 27.4 *Typical shape of the distribution function describing energy fluctuations (Eq. 27.35).*

$$\langle (q(t+\Delta t) - q(t))\, q(t)\rangle = -\langle q^2\rangle, \tag{27.37}$$

which is valid for $\Delta t \gg \tau$. We can derive this more generally from (27.35) by applying the fact that averages add, i.e. $\langle A + B\rangle = \langle A\rangle + \langle B\rangle$, so

$$\begin{aligned}\langle (q(t+\Delta t) - q(t))\, q(t)\rangle &= \langle q(t+\Delta t) q(t)\rangle - \langle q(t)q(t)\rangle \\ &= \langle q^2\rangle \left(e^{-|\Delta t|/\tau} - 1\right). \end{aligned} \tag{27.38}$$

This equation gives a general insight into the way thermal fluctuations exhibit the degree of temporal correlation required to ensure that the thermodynamic properties of a system in thermal equilibrium do not drift over long timescales, even though they fluctuate on short timescales.

For $\Delta t \ll \tau$, equation (27.38) gives

$$\frac{\langle (q(t+\Delta t) - q(t))q(t)\rangle}{\Delta t} \simeq -\frac{\langle q^2\rangle}{\tau}. \tag{27.39}$$

This result is valid for $\tau_a < |\Delta t| \ll \tau$. In other words it is valid for Δt small but not infinitesimal. (We have already explained that (27.35) does not apply at $\Delta t \to 0$, where instead one finds

$$\left\langle \frac{\mathrm{d}q}{\mathrm{d}t} q \right\rangle = 0, \tag{27.40}$$

cf. (27.36).)

27.5 Johnson noise

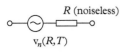

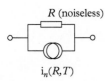

Fig. 27.5 *A resistor at finite temperature, and two equivalent circuit models.*

Johnson noise, also called Johnson–Nyquist noise, is a noise appearing in electrical circuits in thermal equilibrium, owing to thermal fluctuations. It consists of current fluctuations, and associated voltage fluctuations. The main idea is illustrated in Figure 27.5: a resistor at temperature T is electrically equivalent to a noiseless resistor in parallel with a current source which produces a fluctuating current. This is also electrically equivalent to a noiseless resistor in series with a voltage source which produces a fluctuating voltage. The term 'electrically equivalent' does not mean equivalent in all respects, it means merely that the currents and voltages appearing at the terminals, when the resistor is part of a larger circuit, are the same as if one had a noiseless resistor combined with the noisy current or voltage source.

The main ideas can be introduced by analysing the simple circuit shown in Figure 27.6. If $V_C = Q/C$ is the voltage across the capacitor at any time, then the energy stored on the capacitor is

$$\frac{1}{2} C V_C^2. \tag{27.41}$$

The thermal fluctuations in the resistor cause the time average of this energy to be equal to the thermal energy associated with one degree of freedom, according to the equipartition theorem:

$$\frac{1}{2}C\langle V_C^2\rangle = \frac{1}{2}k_B T \quad\Longrightarrow\quad \langle V_C^2\rangle = \frac{k_B T}{C}. \tag{27.42}$$

The voltage fluctuates such that its mean value is zero, therefore $\langle V_C^2\rangle$ is the variance.

Note that the resistance does not appear in this result. However it does appear in the expression for the fluctuations at the source in the equivalent circuit, as we now explain.

In order to gain some familiarity with the relevant ideas, first consider the simple case of an a.c. source oscillating at a single frequency, $V(t) = V_0 \cos\omega t$. The power delivered to a load R_L is $P(t) = (V_0^2/R_L)\cos^2\omega t$ and the time-averaged power is

$$\langle P\rangle = \frac{V_0^2}{2R_L}. \tag{27.43}$$

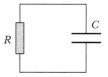

Fig. 27.6 *A simple circuit.*

It will be more convenient in the following to use the complex notation, in which we introduce the mathematical quantity $V(t) = V_0 \exp(i\omega t)$, and call it 'the voltage', on the understanding that the physical voltage will be the real part of this.

Next, consider the case of a source that oscillates at two frequencies, such that the voltage at the source is

$$V(t) = V_1 e^{i\omega_1 t} + V_2 e^{i\omega_2 t}. \tag{27.44}$$

The time-averaged power delivered to a load R_L is now

$$\langle P\rangle = \frac{|V_1|^2 + |V_2|^2}{2R_L}, \tag{27.45}$$

where a term involving the average value of $\mathrm{Re}[V_1 V_2^* \exp i(\omega_1-\omega_2)t]$ has been dropped because it is negligible when the average is taken over long enough times.

Generalizing this idea to a source emitting a continuum of frequencies, we shall have a source voltage,

$$V(t) = \frac{1}{2\pi}\int_{-\infty}^{\infty} \tilde{V}(\omega) e^{i\omega t} d\omega, \tag{27.46}$$

where $\tilde{V}(\omega)$ characterizes the strength at each frequency. The factor 2π is introduced so that $\tilde{V}(\omega)$ is the Fourier transform of $V(t)$. This leads to a total delivered energy,

$$E = \frac{1}{R_L}\int_{-\infty}^{\infty}|V(t')|^2 dt' = \frac{1}{2\pi R_L}\int_{-\infty}^{\infty}|\tilde{V}(\omega)|^2 d\omega. \tag{27.47}$$

This result, including the factor 2π, is proved in Appendix C.2; see equation (C.8). The delivered energy involves an integral over all times from $-\infty$ to $+\infty$, which may be infinite, but in the present context we are interested in the average power. The main role of (27.47) for the present discussion is simply to show that it is the integral of the squared voltage spectrum that is the important quantity.

In order to characterize fluctuating voltages in general, we now introduce a quantity S whose dimensions are voltage-squared per unit frequency range (SI unit V^2s or JΩ). For a source with given $S(\omega)$, the voltage variance is

$$\langle V^2 \rangle = \int_0^\infty S(\omega) d\omega. \tag{27.48}$$

Note that this integral is, by convention, over positive frequencies only. Since $S(\omega)$ is an even function, one could also use the integral over all frequencies of $S/2$.

We are now ready to analyse the circuit of Figure 27.6 again, only this time we treat it using the equivalent circuit shown in Figure 27.7. Using standard a.c. circuit theory, the impedance of the resistor and capacitor in series is

$$Z = R + \frac{1}{i\omega C}. \tag{27.49}$$

Fig. 27.7 *Equivalent circuit model of the circuit in Figure 27.6.*

Therefore, for a single-frequency source voltage, $V = V_0 \exp(i\omega t)$, the current is V/Z and the voltage across the capacitor is

$$V_C = V - IR = \frac{V}{1 + i\omega RC}. \tag{27.50}$$

Hence,

$$|V_C|^2 = \frac{|V|^2}{1 + (\omega RC)^2}. \tag{27.51}$$

For a general noisy source, we replace $|V|^2$ on the right by $Sd\omega$ and $|V_C|^2$ on the left by $S_C d\omega$, and then integrate. Hence we find that the variance of the voltage across the capacitor is

$$\langle V_C^2 \rangle = \int_0^\infty S_C d\omega = \int_0^\infty \frac{S(\omega)}{1 + (\omega RC)^2} d\omega = \frac{k_B T}{C}, \tag{27.52}$$

where in the final step we have used (27.42). This expression is useful because it gives us information about $S(\omega)$:

$$\int_0^\infty \frac{S(\omega) RC d\omega}{1 + (\omega RC)^2} = R k_B T. \tag{27.53}$$

In a classical model of Johnson noise, one finds that $S(\omega)$ is independent of ω, and this is valid in practice at low frequencies. In this approximation, we have, therefore,

$$R k_B T = S \int_0^\infty \frac{RC d\omega}{1 + (\omega RC)^2} = S \frac{\pi}{2}. \tag{27.54}$$

Hence,
$$S(\omega) = \frac{2Rk_B T}{\pi}. \tag{27.55}$$

This is the central result for Johnson noise. It is usually quoted in the following way. If we integrate the voltage noise at the source in the equivalent circuit over a range of frequency $\Delta\omega$, then the variance is

$$\langle V^2 \rangle = S(\omega)\Delta\omega = \frac{2Rk_B T}{\pi}\Delta\omega = 4Rk_B T \Delta f. \tag{27.56}$$

This expression gives the variance of the voltage fluctuations in the resistor when they are integrated over a frequency range Δf. The reason that the value of the resistance does not appear in equation (27.42) is that the resistor plays a dual role in the simple RC circuit. First it causes the source in the equivalent circuit to have the given noise spectrum, and then it acts, with the capacitor, to create a low-pass filter with bandwidth $1/RC$. When the noise is integrated over this bandwidth, the variance of the fluctuations on the capacitor is finite and independent of R.

The above argument is due to van der Ziel. It does not give any reason for the assumption that S is independent of ω. That can be argued by classical statistical mechanics, but it cannot be valid for arbitrarily high frequencies or one will have a problem very much like the 'ultraviolet catastrophe' mentioned in Section 19.2.3.2 in connection with thermal radiation. Indeed, Johnson noise and thermal radiation are closely linked. In a quantum treatment one finds that $k_B T$ in the above formula is replaced by $\hbar\omega/(e^{\beta\hbar\omega} - 1)$, leading to:

$$\langle V^2 \rangle = \int_0^\infty \frac{2R}{\pi} \frac{\hbar\omega}{e^{\beta\hbar\omega} - 1} d\omega, \tag{27.57}$$

where $\beta = 1/k_B T$.

. . . .

EXERCISES

(27.1) Show that, for a rigid system in thermal equilibrium with a reservoir,

$$e^{S_{\text{tot}}/k_B} \propto e^{-F/k_B T}$$

and for a flexible system in equilibrium with a pressure and temperature reservoir,

$$e^{S_{\text{tot}}/k_B} \propto e^{-G/k_B T},$$

where, as usual, S_{tot} is the total entropy of reservoir and system, and F, G, T are properties of the system alone. Comment on the physical implications.

(27.2) Consider joint fluctuations in temperature and volume for a system in equilibrium with a pressure and temperature reservoir. Show that $\partial^2 A/\partial V \partial T = 0$ and obtain equations (27.9) and (27.15).

(27.3) Discuss physically why the temperature and volume fluctuations considered in question 27.2 are uncorrelated, whereas in general the internal energy and volume fluctuations are correlated.

(27.4) Obtain (27.17), as follows. Throughout this exercise, we take N to be constant. By considering $U = U(T, V)$, show that

$$\left.\frac{\partial U}{\partial T}\right|_\alpha = C_V + \left.\frac{\partial U}{\partial V}\right|_T \left.\frac{\partial V}{\partial T}\right|_\alpha, \qquad (27.58)$$

where $\alpha = p/T$. By considering $V = V(T, p)$, show that

$$\left.\frac{\partial V}{\partial T}\right|_\alpha = \left.\frac{\partial V}{\partial T}\right|_p + \left.\frac{\partial V}{\partial p}\right|_T \left.\frac{\partial p}{\partial T}\right|_\alpha.$$

From this, obtain

$$\left.\frac{\partial V}{\partial T}\right|_\alpha = -\frac{1}{T} \left.\frac{\partial V}{\partial p}\right|_T \left.\frac{\partial U}{\partial V}\right|_T.$$

[Hint: use reciprocity on $(\partial V/\partial T)_p$, realize that $(\partial p/\partial T)_\alpha = \alpha = p/T$, and use equation (13.24).] By substituting this into (27.58), equation (27.17) is obtained.

(27.5) Obtain equations (27.25) and (27.26).

(27.6) Show that equation (27.6) can be written

$$\Delta U^2 = \left.\frac{\partial^2 \ln Z}{\partial \beta^2}\right|_{V,N},$$

where $\beta = 1/k_B T$ and $Z = e^{-\beta F}$. This result is useful in statistical mechanics, because for many systems the partition function Z can be obtained from (14.95).

Thermoelectricity and entropy flow

28

28.1 Thermoelectric effects 403
28.2 Entropy gradients and Onsager's reciprocal relations 409
Exercises 418

In this chapter we venture into the area called *non-equilibrium thermodynamics*. It is so called because we shall be dealing with processes in which entropy is continuously produced. Examples are electric current flow through a resistive wire, and heat conduction through an ordinary material, and movement of particles by diffusion. Such processes may be steady—independent of time—but they are always out of equilibrium processes, in the thermodynamic sense, because they involve entropy production. There are also plenty of processes further removed from equilibrium, but these simple flow processes already offer a wealth of physics and chemistry and we shall restrict the discussion to them.

A common theme will be that of *coupled flows*. A temperature difference can cause not just the flow of heat, but also a net movement of particles against an opposing force. Similarly, a pressure difference can cause heat flow. We begin by considering such effects in the context of electric circuits.

28.1 Thermoelectric effects

When an ordinary electrical conductor such as a metal wire is placed in contact with thermal reservoirs so that its ends are at different temperatures (see Figure 28.1), an electric potential difference appears between the ends. This is called the *Seebeck effect*. This is the effect exploited by thermocouple temperature sensors. The typical order of magnitude of the effect is microvolts per Kelvin temperature difference. The origin of the effect is the thermal diffusion of electrons in the conductor: electrons will tend to move away from a region at higher temperature. This tendency is quantified precisely by the chemical potential μ. Chemical potential accounts for all contributions to the energy per particle that influence the average movement of particles from one place to another. In the presence of a gradient of chemical potential, the average force per particle is

$$\mathbf{f} = -\nabla \mu.$$

Therefore, if q is the charge on the carriers, the force per unit charge is

$$\mathcal{E} = -\frac{1}{q}\nabla \mu = \frac{1}{|q|}\nabla \mu, \qquad (28.1)$$

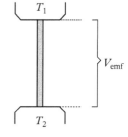

Fig. 28.1 *The Seebeck effect. When an electrical conductor has its ends maintained at different temperatures, and no electrical current flows, an electric potential difference appears between the ends.*

Thermodynamics: A Complete Undergraduate Course. Andrew M. Steane.
© Andrew M. Steane 2017. Published 2017 by Oxford University Press.

where the second version applies in the case of negatively charged carriers such as electrons. This is sometimes called an 'electrochemical field'. An electrochemical field is not an electric field, but by dividing $\nabla \mu$ by the charge per particle we get a quantity whose physical dimensions are the same as those of an electric field, and which concerns forces on particles carrying charge, so in some respects it mimics an electric field. When there is also an ordinary electric field $\mathbf{E}$ in the conductor, then it contributes $q\phi$ to μ, where ϕ is the electric potential (recall the discussion on page 160). Equation (28.1) still applies; μ then includes both this and other terms.

The Seebeck effect is quantified by the observation that when $\mathbf{E} = 0$ and no current flows, for small temperature gradients $\boldsymbol{\mathcal{E}}$ is found to be proportional to the temperature gradient:

$$\boldsymbol{\mathcal{E}} = \epsilon \nabla T \qquad \text{when } \mathbf{j} = 0.$$

The proportionality constant ϵ is called the Seebeck coefficient or the 'thermopower'.[1] (The letter S is often used for the Seebeck coefficient, but we adopt ϵ to avoid confusion with entropy.)

The Seebeck effect is not easy to measure directly on a single piece of wire, because temperature gradients will normally exist in the measuring apparatus also. However, the difference between the effects in two metals can easily be measured by constructing a circuit as shown in Figure 28.2. The e.m.f. developed around such a circuit is

$$V_{\text{emf}} = \int \boldsymbol{\mathcal{E}} \cdot d\mathbf{r} = \int \epsilon \nabla T \cdot d\mathbf{r} = \int_{T_0}^{T_1} \epsilon_B dT + \int_{T_1}^{T_2} \epsilon_A dT + \int_{T_2}^{T_0} \epsilon_B dT$$

$$= \int_{T_1}^{T_2} \epsilon_A - \epsilon_B \, dT. \tag{28.2}$$

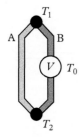

Fig. 28.2 *A thermocouple circuit. Wires of two different materials (A and B) are used. The voltmeter records a voltage which is a function of T_1 and T_2. If one junction is held at fixed temperature while the temperature T of the other is varied, then $dV_{\text{emf}}/dT = \epsilon_A - \epsilon_B$. The voltage is proportional to $T_1 - T_2$ when $T_1 - T_2$ is small.*

A pair of conductors used this way is called a *thermocouple* and is useful in thermometry, as noted in Chapter 6. If one junction is held at constant temperature while the temperature of the other is varied, we find

$$\frac{dV_{\text{emf}}}{dT} = \epsilon_{AB}, \tag{28.3}$$

where $\epsilon_{AB} \equiv \epsilon_A - \epsilon_B$.

Now suppose we construct a thermocouple circuit, but instead of measuring the e.m.f. we use it to drive a device such as an electric motor, allowing a current to flow around the circuit. The energy delivered to the motor, allowing it to perform work, must have come from somewhere, and it is clear that the only available source of energy is heat flow from the reservoirs. In fact this is a simple type of heat engine. Heat flows out of the hotter reservoir, and a smaller amount flows into the colder reservoir. If we now focus our attention purely on what is going on at either one of the reservoirs, we deduce that when an electric current flows across

---

[1] The name 'thermopower' here signifies a propensity to produce electric current; ϵ is not an energy per unit time.

Table 28.1 *Basic processes involving heat and electricity.* **J** *is the heat current density (also called heat flux),* **j** *is the electric current density, H is a power per unit volume (rate of generation or liberation of heat),* **ℰ** *is the electromotive field (= force per unit charge) in the material. The three reversible processes are called 'thermoelectric'.*

Irreversible:	
Joule heating	$H = \mathbf{j} \cdot \boldsymbol{\mathcal{E}}$
thermal conduction	$\mathbf{J} = -\kappa \nabla T$
Reversible:	
Seebeck effect	$\boldsymbol{\mathcal{E}} = \epsilon \nabla T$
Peltier effect	$\mathbf{J} = \Pi \mathbf{j}$
Thomson heat	$H = \mu \mathbf{j} \cdot \nabla T$

a junction between a pair of conductors having different Seebeck coefficients, heat must flow either into or out of the junction (the direction depending on the coefficients and the direction of the electric current). This is called the *Peltier effect*. We have shown that it is a necessary partner to the Seebeck effect.

The underlying mechanism of the Peltier effect is the fact that the electrons carry some thermal energy with them, so electric current flow at non-zero temperature is necessarily accompanied by convective heat flow. At small current densities, the two types of flow are proportional. The proportionality constant (Peltier coefficient) can differ from one material to another, therefore at an interface, whereas the electric current flows across without loss or gain, there can be a net transport of heat either into or out of the interface—see Figure 28.3.

We define the Peltier coefficient Π through

$$\mathbf{J} = \Pi \mathbf{j}, \tag{28.4}$$

where **J** is the heat current density and **j** is the electric current density and the equation applies in the absence of a temperature gradient along the conductor. In the presence of a temperature gradient we must also account for further effects such as ordinary thermal conductivity, which we shall consider shortly. When electric current per unit area **j** flows from a material A to a material B at the same constant temperature, a heat $\Pi_A \mathbf{j}$ is brought up to the interface and a heat $\Pi_B \mathbf{j}$ is carried away into material B, per unit area and time. Clearly the heat $(\Pi_A - \Pi_B)\mathbf{j}$ must be liberated and flows into whatever system is acting as a heat reservoir at the interface.

By applying the laws of thermodynamics to a thermoelectric circuit such as the one shown in Figure 28.2 we shall next deduce a relationship between the Seebeck and Peltier coefficients. However, in order to do this correctly we must take into account three further processes. Two are the familiar Joule heating and thermal

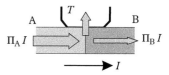

Fig. 28.3 *The Peltier effect. Electric current I flows from A to B. The electric current is accompanied by a heat current ΠI where the coefficient Π depends on the material. The example shows a case where $\Pi_A > \Pi_B$. To conserve energy, some heat must flow into the reservoir even in the absence of any temperature gradient.*

conduction—the relevant formulae are listed in Table 28.1, where κ is the thermal conductivity (as measured in the absence of electric current). The final effect is called the *Thomson heat*. This seems at first to be a partner to Joule heating, because it involves the liberation of heat inside the wire, but in fact it is a quite different type of process because it is reversible.

In the presence of a temperature gradient, as any given group of electrons (or other current carriers) moves along the wire, carrying its thermal energy with it, it arrives at a place whose temperature is slightly different from the one it (the group of carriers) has. The slight temperature difference between the current carriers and the surrounding material causes heat to flow reversibly between them. Obviously this could be in either direction, depending on whether the carriers are moving towards places at higher or lower temperature. The rate of liberation of heat by this process at any given place is proportional to both the temperature gradient and the electric current:

$$H = \mu \mathbf{j} \cdot \nabla T. \tag{28.5}$$

Here H is a power per unit volume and μ is the Thomson coefficient.[2] By integrating over the cross-sectional area of the conductor, one may also write the rate of liberation of heat per unit length of conductor, which is given by $\mu \mathbf{I} \cdot \nabla T$.

28.1.1 Thomson's treatment

We are now in a position to analyse the heat engine shown in Figure 28.4. We consider the energy entering or leaving the circuit, per unit time, in a small amount of time. If the current around the circuit is I in the direction shown, then the heat leaving the hot reservoir owing to the Peltier effect is

$$Q_1 = \Pi_{AB}(T_1)I,$$

where $\Pi_{AB} \equiv \Pi_A - \Pi_B$ and this quantity is to be evaluated at temperature T_1. The heat entering the cold reservoir is

$$Q_2 = \Pi_{AB}(T_2)I.$$

In addition to these contributions, further heat leaves the junction at T_1 and enters the junction at T_2 owing to normal heat conduction. The heat transferred from one reservoir to the other by conduction is

$$dQ_\kappa = \frac{A(\kappa_A + \kappa_B)dT}{L},$$

where $dT = T_1 - T_2$ and we assume both wires have cross-sectional area A and length L.

Joule heating in the circuit leads to heat leaving the wires at the total rate

$$dQ_J = I dV_R,$$

where dV_R is the voltage difference developed around the circuit owing to the electrical resistance R of the wires.

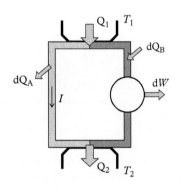

Fig. 28.4 *Heat engine based on a thermocouple circuit.*

---

[2] In older treatments the coefficient was defined with the opposite sign.

Finally, the Thomson heat is

$$dQ_A = \mu_A I dT, \qquad dQ_B = \mu_B I dT.$$

Let dW be the extracted work per unit time. By conservation of energy, it is given by

$$dW + I dV_R = Q_1 - Q_2 - dQ_A + dQ_B = \left(\frac{d\Pi_{AB}}{dT} - \mu_{AB}\right) I dT, \qquad (28.6)$$

where we have chosen on to place the Joule energy on the left along with the work made available, since it represents an amount of electrical work done by the current against the resistance of the wires. On the right we have defined $\mu_{AB} \equiv \mu_A - \mu_B$ and we have used $\Pi_{AB}(T_1) - \Pi_{AB}(T_2) = d\Pi_{AB}$. The conducted heat does not enter into this equation, since we assume all of it simply flows from one reservoir to the other.

After dividing equation (28.6) by I, we shall have on the left-hand side the total e.m.f. developed around the circuit:

$$dV_{\text{emf}} = \frac{dW}{I} + dV_R = \left(\frac{d\Pi_{AB}}{dT} - \mu_{AB}\right) dT. \qquad (28.7)$$

Now we use (28.3) and we find

$$\epsilon_{AB} = \frac{d\Pi_{AB}}{dT} - \mu_{AB}. \qquad (28.8)$$

This is called the *first Thomson relation*. It follows directly from conservation of energy.

Next we consider entropy and we assume, following Thomson, that there is no net entropy production associated with the reversible processes alone. Although this is a reasonable assumption, it is not self-evident and requires careful justification which we present shortly. Under this assumption, the entropy leaving the hot reservoir by the Peltier effect is equal to the entropy leaving the wires via Thomson heat, plus the entropy entering the cold reservoir by the Peltier effect:

$$\frac{Q_1}{T_1} - \frac{Q_2}{T_2} = \int_{T_2}^{T_1} \frac{\mu_A - \mu_B}{T} I dT$$

$$\Rightarrow \quad \frac{d}{dT}\left(\frac{\Pi_{AB}}{T}\right) = \frac{\mu_{AB}}{T}. \qquad (28.9)$$

Combining this result with (28.8), we obtain

$$\Pi_{AB} = T\epsilon_{AB}, \qquad (28.10)$$

which is called the *second Thomson relation* (or Kelvin relation). Using this in (28.8) we obtain

$$\mu_{AB} = T\frac{d\epsilon_{AB}}{dT}. \qquad (28.11)$$

Equations (28.8), (28.10), and (28.11) are together called the *thermoelectric equations*. They show that one can obtain the Peltier and Thomson coefficients once the temperature dependence of the Seebeck (thermocouple) e.m.f. is known. They are amply confirmed by experiments.

Now we return to the assumption involved in the entropy argument used previously. This was the assumption that the net entropy change *associated with the reversible processes alone* is zero. This is not self-evident since there is no general thermodynamic principle that allows us to separate the reversible from the irreversible processes when both are present. It can be derived from the Onsager reciprocal relations, as shown in Exercise 28.1. Here we show how it can be justified to some extent by arguing that there can be circumstances in which the contributions from Joule heating and thermal conduction are negligible.

We have already exhibited in equation (9.7) the rate of generation of entropy by irreversible processes in a wire carrying electric and thermal currents. If we now add the reversible process (Thomson heat), we find the total

$$\frac{dS}{dt} = \frac{\mathcal{J}\Delta T}{T^2} + \frac{I\Delta V_R}{T} + \frac{\mu I}{T}\frac{dT}{dx}\Delta x$$

$$\Rightarrow \frac{1}{\Delta x}\frac{dS}{dt} = \underbrace{A\kappa\left(\frac{1}{T}\frac{dT}{dx}\right)^2}_{\text{conduction}} + \underbrace{\frac{I^2\rho}{AT}}_{\text{Joule}} + \underbrace{\frac{\mu I}{T}\frac{dT}{dx}}_{\text{Thomson}},$$

where Δx is the length of the wire, assumed small enough that the gradient of temperature along it is uniform, and ρ is the resistivity. The conduction and Joule contributions are small compared to the Thomson contribution if

$$\frac{A\kappa}{T}\frac{dT}{dx} \ll \mu I, \qquad I\rho \ll \mu A\frac{dT}{dx}.$$

By making dT/dx the subject of both conditions, one finds that they can be simultaneously satisfied if

$$T \gg \frac{\kappa\rho}{\mu^2}. \tag{28.12}$$

This temperature is very high ($\sim 10^6$ K) for ordinary conductors such as copper, well above the melting point. Therefore for such materials one cannot argue that the irreversible processes are negligible. Thomson was himself aware of this difficulty. However, there is nothing fundamentally wrong with the idea that a material could have high enough Thomson coefficient to make the irreversible processes negligible in comparison with the reversible ones.

Another question that may be raised when treating the entropy is whether or not the process may be regarded as quasi-equilibrium (and hence subject to the laws of equilibrium thermodynamics). In a modern treatment, the methods of non-equilibrium thermodynamics are applied, which are the subject of Section 28.2.

28.2 Entropy gradients and Onsager's reciprocal relations

Thermoelectricity can be regarded as an example of the general concept of *coupled flows*. Onsager proved an important general property of such flows, by analysing the entropy changes associated with thermal fluctuations. Like the discussion of fluctuations in Chapter 27, this subject is on the boundary between macroscopic thermodynamics and microscopic statistical mechanics. However, since the results are of great generality and can be treated mostly in terms of macroscopic concepts, we think it appropriate to present the ideas in this book. In order to do this it is necessary first to understand the nature of the problem.

A standard problem in classical mechanics is to consider the deformation of a solid object subject to tension or compression. For small applied forces, one finds that the strain (the extension in a given direction, relative to the size of the object) is proportional to the stress (the force per unit area). A simple example is Hooke's law for a spring. In a three-dimensional crystal, the internal structure of the crystal comes into play, so that the extension in any one direction is related to all three components of the applied force. The general relationship is

$$\mathbf{x} = M\mathbf{f},$$

where $\mathbf{f}$ is the stress, $\mathbf{x}$ is the strain, and M is a three by three matrix, or, to be precise, a second rank tensor.[3] Although M does not have to be a diagonal matrix, it is not hard to show, from Newton's third law, that M must be symmetric. That is,

$$M_{xy} = M_{yx}, \qquad M_{xz} = M_{zx}, \qquad M_{yz} = M_{zy}.$$

Relations of this kind are sometimes called *reciprocal relations*.

In thermal physics we encounter various transport processes where the flow of some quantity is proportional to a gradient which may be regarded as a type of generalized force. For example, for electric current in the absence of other flows, we have

$$\mathbf{j} = \sigma \mathbf{E}$$

and for a heat current in the absence of other flows we have

$$\mathbf{J} = -\kappa \nabla T.$$

In the simple problems we have discussed so far in this book, we have treated isotropic materials in which σ and κ are scalars. More generally, for a non-isotropic material such as a crystal they can be tensors. It is found empirically that these tensors are also symmetric. However, to prove this property from fundamental considerations is much more difficult than the mechanical problem of stress and strain, because now we have to treat a flow in a system which is not in thermal equilibrium (if it were in equilibrium, the flow would vanish!).

[3] A *tensor* is a mathematical entity that, when multiplying a vector, gives another vector.

Now consider a problem involving two different types of flow simultaneously present in the same region of space. A simple case is that of charged particles subject to gravity and a static electric field. We assume the particles are moving through a material such as a gas that provides a viscous damping force in proportion to the particle velocity **v**, so that in steady state the flow is described by

$$\alpha \mathbf{v} = m\mathbf{g} + q\mathbf{E},$$

where α is a coefficient describing the viscous force, and m and q are the mass and charge of a group of particles moving at velocity **v**. In order to make a point we are working towards, we write this in the form

$$\alpha \mathbf{v} = m\mathbf{g} + qm\mathbf{F},$$

where $\mathbf{F} = \mathbf{E}/m$.

Since the particles carry both mass and electric charge, there are two simultaneous flows. The current densities of mass and charge are given by

$$\mathbf{j}_m = \rho_m \mathbf{v} = \frac{\rho_m m}{\alpha}\mathbf{g} + \frac{\rho_m q m}{\alpha}\mathbf{F},$$
$$\mathbf{j}_q = \rho_q \mathbf{v} = \frac{\rho_q m}{\alpha}\mathbf{g} + \frac{\rho_q q m}{\alpha}\mathbf{F}.$$

This pair of equations has the form of a matrix equation,

$$\mathbf{j}_m = N_{11}\mathbf{g} + N_{12}\mathbf{F}$$
$$\mathbf{j}_q = N_{21}\mathbf{g} + N_{22}\mathbf{F},$$

with

$$N_{11} = \frac{\rho_m m}{\alpha}, \quad N_{12} = \frac{\rho_m q m}{\alpha}, \quad N_{21} = \frac{\rho_q m}{\alpha}, \quad N_{22} = \frac{\rho_q q m}{\alpha}.$$

This matrix is not symmetric. This is easy to show using $\rho_m = (m/q)\rho_q$, giving

$$N_{12} = \frac{\rho_q m^2}{\alpha} \neq N_{21}.$$

However, if we revert to the field **E** instead of **F**, we shall find a symmetric matrix (as the reader can verify). The lesson is that different but connected flows can lead to symmetric matrices, but only when a suitable description of the force is adopted.

If we now turn to transport processes in general, there are two issues. First, how to prove that conductivity tensors (describing a single type of flow in more than one dimension) are symmetric, and second, how to formulate the connection between different but related flows (e.g. heat and electric current in thermoelectricity) so that the matrix describing the coupled flows is symmetric—thus allowing a reciprocal relation to be discovered. We shall first exhibit the method to *formulate the problem* so as to obtain a symmetric matrix L. We shall then present Onsager's argument to show that the matrix is indeed symmetric.

We shall use thermoelectricity as an example.

The thermoelectric phenomena discussed in the previous section give rise to the electric and heat currents described by the equations

$$\mathbf{j} = \sigma(\mathcal{E} - \epsilon \nabla T),$$
$$\mathbf{J} = \Pi \mathbf{j} - \kappa \nabla T. \qquad (28.13)$$

(The sign of the Seebeck term in the first equation is correct since it yields $\mathcal{E} = \epsilon \nabla T$ when $\mathbf{j}=0$, which is the definition of ϵ.) We can write these in the matrix form,

$$\mathbf{j} = N_{11}\mathcal{E} + N_{12}\nabla T$$
$$\mathbf{J} = N_{21}\mathcal{E} + N_{22}\nabla T, \qquad (28.14)$$

where

$$N_{11} = \sigma, \ N_{12} = -\sigma\epsilon, \ N_{21} = \sigma\Pi, \ N_{22} = -\Pi\sigma\epsilon - \kappa.$$

This matrix is *not* symmetric.

To obtain a symmetric matrix, we need to learn how to compare flows of different types of quantity. The crucial insight is that recognized by Thomson (Lord Kelvin) and subsequently generalized by Onsager: *different processes have to be compared in terms of the entropy changes involved.* To be precise, the force terms which 'drive' the currents must be expressed in the form

$$\frac{\partial S}{\partial \alpha},$$

where S is entropy and α is a displacement of the relevant type of quantity. We will explain this idea in the following.

28.2.1 Derivation of Onsager's reciprocal relation

Consider two nearby regions in a conductor, one at temperature T_2 and chemical potential μ_2, the other at T_1, μ_1, as indicated in Figure 28.5. If energy dU_1 and particle number dN_1 move from location 2 to location 1, then the change in entropy S of the system as a whole is governed by

$$\left.\frac{\partial S}{\partial U_1}\right|_{N_1} = \frac{1}{T_1} - \frac{1}{T_2} \equiv f_U \qquad (28.15)$$

$$\left.\frac{\partial S}{\partial N_1}\right|_{U_1} = \frac{\mu_2}{T_2} - \frac{\mu_1}{T_1} \equiv f_N, \qquad (28.16)$$

where we introduce the symbols f_U, f_N for convenience in the following. These quantities are a sort of thermodynamic force, in that one expects the rate of flow to increase with the sizes of these 'forces', though of course the physical dimensions of f_U and f_N are not the same (nor are they measured in newtons). To treat the flows of energy and particles, we now make a *linear approximation*. We assume that, when the driving 'forces' are small (that is, for small differences in temperature and chemical potential), the associated flows are linearly proportional to the entropy gradients that drive them:

$$\frac{dU_1}{dt} = L_{UU}f_U + L_{UN}f_N \qquad (28.17)$$

$$\frac{dN_1}{dt} = L_{NU}f_U + L_{NN}f_N, \qquad (28.18)$$

which can also be written

$$\frac{d}{dt}\begin{pmatrix}U_1\\N_1\end{pmatrix} = \begin{pmatrix}L_{UU} & L_{UN}\\L_{NU} & L_{NN}\end{pmatrix}\begin{pmatrix}f_U\\f_N\end{pmatrix}. \qquad (28.19)$$

Onsager's reciprocal relation is the assertion that when the relationship is expressed this way (i.e. using entropy gradients, as in Eq. 28.15), the matrix will be symmetric:

$$L_{UN} = L_{NU}.$$

We will prove this first for the case under discussion, and then generalize to a very wide class of entropy-driven flow processes.

As it stands, equation (28.19) describes flows that are taking place when a system is out of equilibrium, and consequently entropy is being generated. The essential concept in Onsager's argument is that these same flows may also come about in the course of the thermal fluctuations that a system undergoes when it is in thermal equilibrium. Suppose the regions 1 and 2 under discussion are nearby regions in a system in thermal equilibrium. Then U_1 and U_2 will fluctuate, and so will N_1 and N_2, owing to a fluctuating movement of energy and particles between the regions. But these thermal fluctuations do not cause U_1 or N_1 to wander further and further from their equilibrium values (as would be the case for a random walk, for example). Rather, U_1 and N_1 continually return to their equilibrium values. Why is this? It is because (argues Onsager) of the same processes that are described by (28.19). The idea is that if a system is out of equilibrium, it will relax back to equilibrium in the same way, irrespective of how the out of equilibrium state was produced—including if it was caused by an unusually large fluctuation in the course of the fluctuations that accompany thermal equilibrium! Of course this will only be true if the state in question is not too outlandish, but we are not here discussing outlandish states, we are discussing states in which the regions 1 and 2 can be assigned a temperature and a chemical potential. In the vast majority of cases, and on average, the assumption will be valid. But since it is an assumption that may need to be justified more fully in particular cases, we underline that it is an assumption by giving it a name:

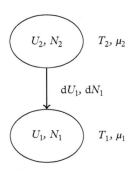

Fig. 28.5 *Coupled flows of heat and particles.*

Onsagers's regression hypothesis:[4] 'The regression to the mean of microscopic thermal fluctuations at equilibrium follows the macroscopic law of relaxation of small non-equilibrium disturbances,'

where the word *regression* refers to the average decay of the fluctuation following any given starting condition. The Onsager regression hypothesis can be supported by statistical arguments for a wide class of systems. It can be shown to follow from the fluctuation–dissipation theorem. Overall, then, it can be regarded as very secure. Nevertheless, it is not quite as secure as more basic results such as Carnot's theorem and Clausius's theorem.

Equation (28.19) expresses dU_1/dt in terms of L_{ij} and f_i. With the regression hypothesis in mind, we now argue that this same quantity is given by taking the appropriate average over fluctuations in thermal equilibrium:

$$\frac{dU_1}{dt} \stackrel{\text{average}}{=} \frac{U_1(t+\Delta t) - U_1(t)}{\Delta t}, \tag{28.20}$$

where Δt is large compared to the acceleration time but small compared to the relaxation time (recall the discussion in Section 27.4), and the word 'average' is here employed as a place-holder. What we really wish to consider is the correlation between dU_1/dt and N_1, which is expressed by the following time average:

$$\left\langle \frac{dU_1}{dt} N_1 \right\rangle = \frac{\langle U_1(t+\Delta t)N_1(t)\rangle - \langle U_1(t)N_1(t)\rangle}{\Delta t} \tag{28.21}$$

$$= L_{UU} \left\langle \frac{\partial S}{\partial U_1} N_1 \right\rangle + L_{UN} \left\langle \frac{\partial S}{\partial N_1} N_1 \right\rangle, \tag{28.22}$$

where for the second version we have used equation (28.17). A similar result holds for dN_1/dT, and we can express both results together by

$$\left\langle \frac{dq_i}{dt} q_j \right\rangle = \sum_k L_{ik} M_{kj}, \tag{28.23}$$

where q_i is one of the quantities under consideration and

$$M_{kj} \equiv \left\langle \frac{\partial S}{\partial q_k} q_j \right\rangle. \tag{28.24}$$

Now, the right-hand side of (28.23) is a product of two matrices. We can solve the equation for L_{ij} as long as M_{kj} is non-singular. But M_{kj} is an example of the quantity we have already considered in equation (27.33). We showed in (27.33) that this matrix is diagonal and non-singular, therefore it is symmetric and invertible. Therefore L_{ik} is symmetric if and only if $\langle q'_i q_j \rangle$ (the left-hand side of (28.23)) is symmetric.

Now examine equation (28.21), which can be written more generally:

$$\left\langle \frac{dq_i}{dt} q_j \right\rangle = \frac{\langle q_i(t+\Delta t)q_j(t)\rangle - \langle q_i(t)q_j(t)\rangle}{\Delta t}. \tag{28.25}$$

[4] L. Onsager, *Phys. Rev.* 37, 405 (1931); 38, 2265 (1931)).

The second term on the right is obviously symmetric in i and j. For the first term we argue (for reasons expounded in the next paragraph)

$$\langle q_i(t+\Delta t)q_j(t)\rangle = \langle q_i(t-\Delta t)q_j(t)\rangle = \langle q_i(t)q_j(t+\Delta t)\rangle \quad (28.26)$$
$$= \langle q_j(t+\Delta t)q_i(t)\rangle,$$

therefore this is symmetric also. Hence all the matrices in (28.23) are symmetric, including L_{ik}. QED

The second equality in (28.26) is straightforward: it is simply a shift in the origin of time which obviously makes no difference. The first equality in (28.26) follows from an idea called the *principle of microscopic reversibility*. Roughly speaking, this is the idea that a fluctuation is as likely to go in one direction as another. At the microscopic level, the system state is being carried from one location in state-space to another at a rate given by a set of generalized velocities. If the microscopic equations of motion are time-symmetric, then another solution exists having the same state but reversed velocities. The idea we need is that this second solution is as likely to occur as the first, in the average we are examining. If one accepts this assertion, then the result follows.

Note that a statement about microscopic behaviour is here being applied to learn something about the macroscopic behaviour, i.e. about $q(t)$ which are macroscopic (thermodynamic) variables, not properties of individual particles (as are all the quantities in (28.23)). However, in Onsager's argument it is only necessary to use the phrase 'microscopic reversibility' to mean the first equation of (28.26). This is the sense in which Onsager originally used it. Any argument which justifies this equation is sufficient to prove the reciprocal relation, whether or not the discussion involves the microscopic processes and the laws that describe them. The minimum we require for Onsager's argument to hold is that the correlation function has the same symmetry with respect to Δt as is observed in autocorrelation functions associated with thermal fluctuations, such as the one illustrated in Figure 27.4.

28.2.1.1 Generalization to many variables

In order to generalize the Onsager reciprocal relation to any number of variables, the main requirement is to develop a suitable notation. In the case of flow through a continuous system, the situation illustrated by Figure 28.5 takes place between neighbouring spatial regions of the system. If the distance between the regions is δx then the flow in question involves a movement of internal energy δU through a distance δx. We define the *energy displacement* as $\delta \alpha = \delta x \delta U$. In terms of this quantity, equation (28.15) becomes

$$\frac{\delta S}{\delta U} = \delta\left(\frac{1}{T}\right) \quad (28.27)$$

$$\Rightarrow \frac{\partial S}{\partial \alpha} = \frac{\partial}{\partial x}\left(\frac{1}{T}\right). \quad (28.28)$$

For a displacement in a general direction in three dimensions this becomes

$$\frac{\partial S}{\partial \boldsymbol{\alpha}} = \nabla\left(\frac{1}{T}\right) \tag{28.29}$$

(cf. Eq. 27.30). To be clear, the notation $\delta\boldsymbol{\alpha} = \delta\mathbf{x}\delta U$ refers to an amount of energy *moving through* a displacement $\delta\mathbf{x}$ (not contained in a distance $\delta\mathbf{x}$, which in any case makes no sense in three dimensions). This is not a second-order effect (it is not a correction to a first-order approximation, for example); it is simply the correct way to quantify an energy displacement at lowest-order approximation. This is why it is best thought of as a first-order not second-order small quantity, as the notation $\delta\boldsymbol{\alpha}$ implies. δS is the resulting change in the entropy of the system (by contrast, δU here is *not* a change in the total system energy (which stays constant), but an amount of energy moving from one place to another within the system). It should be understood that in the notation employed in (28.29), the quantity on the left-hand side is a type of gradient operator acting on S. The equation tells us the relationship between a temperature gradient and a movement of energy and entropy.

For a displacement of matter in the x direction the corresponding result is

$$\frac{\partial S}{\partial \boldsymbol{\alpha}} = \nabla\left(\frac{-\mu}{T}\right). \tag{28.30}$$

More generally, we can always write the fundamental relation in the form

$$\mathrm{d}s = \sum_i \phi_i \mathrm{d}\rho_i, \tag{28.31}$$

where the ρ_i are the various types of density (such as u, v, n), and the 'potentials' ϕ_i are

$$\phi_i \equiv \frac{\partial s}{\partial \rho_i}.$$

Then the entropy gradients are

$$\frac{\partial S}{\partial \boldsymbol{\alpha}_i} = \nabla\phi_i.$$

The ϕ_i can be regarded as 'generalized potentials'. These 'potentials' have the physical dimensions of *entropy* per unit 'charge', not energy per unit 'charge'.

The set of current densities is given, in the linear approximation, by

$$\mathbf{j}_i = \sum_k L_{ik} \nabla\phi_k, \tag{28.32}$$

where $\mathbf{j}_i$ is the flux of the quantity whose density is ρ_i in (28.31). We now have a general notation such that the Onsager reciprocal relation is the statement that the matrix L_{ik} in (28.32) is symmetric. This reveals a very general property of dissipative and hence irreversible entropy-generating processes. It asserts that when the flow $\mathbf{j}_i$ associated with process i is influenced by the generalized force $\nabla \phi_k$ arising from another process k, then the flow $\mathbf{j}_k$ is also influenced by the force $\nabla \phi_i$ through the same coefficient.

Ilya Prigogine, in his Nobel prize lecture, commented:

> The importance of the Onsager relations resides in their generality. They have been submitted to many experimental tests. Their validity has for the first time shown that nonequilibrium thermodynamics leads, as does equilibrium thermodynamics, to general results independent of any specific molecular model. The discovery of the reciprocity relations corresponds really to a turning point in the history of thermodynamics.

The generalization of the proof given previously is now straightforward. One simply has to note that each q_i in equation (28.23) can be understood to be that part of a given quantity (such as internal energy, charge, particle number, etc.) in a given region that is carried by the flow in a given direction. That is, the index i runs over not just the types of physical quantity, but also over some large number of spatial regions into which the system is divided (e.g. in a cubic lattice) for the purpose of analysis, and over the three directions of movement (in three dimensions) from any one region to neighbouring regions. In the case of a cubic lattice, a given spatial component of a given current density is related to one of the q_i by

$$j_i = \frac{1}{l^2} \frac{\mathrm{d}q_i}{\mathrm{d}t},$$

where l is the side of each cube, and the corresponding component of the generalized gradient is

$$\nabla_\sigma \phi_i \equiv \frac{\partial \phi_i}{\partial x_\sigma} = \frac{1}{l} \frac{\partial S}{\partial q_i},$$

where the index σ indicates the spatial direction of the i'th part of the flow. By the same argument as before, one obtains (28.23) again, apart from a factor l, but since l is a constant it does not affect the conclusion.

28.2.2 Application

Applied to thermoelectricity, the currents are $\mathbf{j}$ and $\mathbf{J}$ (adopting again the notation we used in Section 28.1) and the gradients are $\mathcal{E}/T$ and $\nabla(1/T)$, so the relation between them is written

$$\mathbf{j} = L_{11}\frac{\mathcal{E}}{T} + L_{12}\nabla\left(\frac{1}{T}\right),$$
$$\mathbf{J} = L_{21}\frac{\mathcal{E}}{T} + L_{22}\nabla\left(\frac{1}{T}\right).$$

These equations should be contrasted with (28.14). The equations have been constructed such that Onsager's analysis applies to L, and therefore $L_{12} = L_{21}$. Recalling the definitions of the Seebeck and Peltier coefficients ϵ and Π, and using $\nabla(1/T) = (-1/T^2)\nabla T$, we find

$$L_{11} = \sigma T, \; L_{12} = \sigma\epsilon T^2, \; L_{21} = \sigma\Pi T, \; L_{22} = (\Pi\sigma\epsilon + \kappa)T^2.$$

The reciprocal relation allows us to deduce that

$$\sigma\epsilon T^2 = \sigma\Pi T \quad\Rightarrow\quad \Pi = \epsilon T. \tag{28.33}$$

This is the second Thomson relation (cf. equation (28.10)). The derivation by Onsager's method is free of the questionable assumptions associated with Kelvin's argument and is generally accepted as sound.

In the approach by reciprocal relations, we only needed to examine the *currents* and therefore it was not necessary to bring the Thomson heat into the argument. With Thomson's second relation in hand, we can now derive the first Thomson relation from energy conservation. Since $\int \nabla \cdot \mathbf{J} dV$ is the net flux out through a surface, the rate of liberation of heat per unit volume, not including the heat conduction term, must be (using (28.4))

$$\nabla \cdot (\Pi\mathbf{j}) = \mathbf{j} \cdot \nabla\Pi,$$

since $\mathbf{j}$ is divergenceless when there is no electric charge build-up. Now use the Onsager reciprocal relation (= second Thomson relation), (28.33):

$$\nabla \cdot (\Pi\mathbf{j}) = \mathbf{j} \cdot \nabla(\epsilon T) = \mathbf{j} \cdot \epsilon\nabla T + T\mathbf{j} \cdot \nabla\epsilon$$
$$= \mathbf{j} \cdot \mathcal{E} + T\mathbf{j} \cdot \frac{d\epsilon}{dT}\nabla T,$$

where $\mathcal{E}$ refers to the part of the total e.m.f. due to the Seebeck effect. The first term can be interpreted as a contribution to the Joule heating. The second term is proportional to $\mathbf{j} \cdot \nabla T$ so its coefficient must be the Thomson coefficient μ. Therefore we find

$$\mu = T\frac{d\epsilon}{dT}, \tag{28.34}$$

which is equation (28.11).

28.2.3 Entropy current, entropy production rate

A further property of the conduction matrix L, and a useful general statement about entropy production, can be obtained by considering the flow of entropy. This is described by an entropy current density $\mathbf{J}_s$, given by

$$\mathbf{J}_s = \sum_k \phi_k \mathbf{j}_k,$$

which follows from (28.31). The rate of production of entropy per unit volume is

$$\Sigma = \nabla \cdot \mathbf{J}_s + \frac{\partial s}{\partial t}$$

$$= \sum_k (\phi_k \nabla \cdot \mathbf{j}_k + \mathbf{j}_k \cdot \nabla \phi_k) + \sum_k \phi_k \left.\frac{\partial \rho_k}{\partial t}\right|$$

$$= \sum_k \mathbf{j}_k \cdot \nabla \phi_k,$$

where the last step uses the continuity equations $\nabla \cdot \mathbf{j}_k + (\partial \rho_k / \partial t) = 0$ which express the conservation of each of the quantities (energy, matter, etc.).

Now we make the linear approximation equation—(28.32)—and obtain

---

Entropy production rate per unit volume, in the linear approximation

$$\Sigma = \sum_{i,k} \nabla \phi_i \cdot L_{ik} \nabla \phi_k. \tag{28.35}$$

Assuming we can apply thermodynamic, i.e. macroscopic, reasoning to a small region, the second law of thermodynamics implies that $\Sigma \geq 0$. That is, the second law forbids the entropy in any region to increase by less than the net entropy flowing into that region from elsewhere. This is a stronger statement than a restriction only on the global entropy. It implies for example that L is a positive-definite matrix, i.e. all its eigenvalues must be positive.

Equation (28.35) can be regarded as a summary of the themes of this chapter. It emphasizes that what the different types of flow (whether of energy, electric charge, mass, etc.) have in common is that they are all ways of transporting entropy. Either equation (28.35) or equation (28.32) can be regarded as a definition of the conduction matrix L (once the linear assumption has been made). The symmetry of the form of equation (28.35) is a strong hint, or else a reminder, that L is itself symmetric.

....

EXERCISES

(28.1) *Another look at Thomson's argument.*

(i) Starting from equations (28.15)–(28.18), obtain

$$\frac{dS}{dt} = f_U \frac{dU_1}{dt} + f_N \frac{dN_1}{dt}$$

and hence

$$\frac{dS}{dt} = L_{UU}f_U^2 + L_{NN}f_N^2 + (L_{UN} + L_{NU})f_U f_N$$

(making no assumption about the symmetry of L_{ij}).

(ii) From (28.18), obtain $f_N = (\dot{N}_1 - L_{NU}f_U)/L_{NN}$ and hence

$$\frac{dS}{dt} = \left(L_{UU} - \frac{L_{UN}L_{NU}}{L_{NN}}\right)f_U^2$$
$$+ \left(\frac{L_{UN} - L_{NU}}{L_{NN}}\right)f_U \dot{N}_1 + \frac{\dot{N}_1^2}{L_{NN}}.$$

Applying this to thermoelectricity, we may note that when no current flows, $\dot{N}_1 = 0$ and in this case only the first term on the right-hand side survives. It follows that this term must describe entropy production associated with ordinary thermal conduction owing to a temperature gradient (the temperature gradient is expressed by f_U). Similarly, when $f_U = 0$ only the third term survives, so this must be the term describing Ohmic or resistive heating. To derive the second Thomson relation using equation (28.9), Thomson assumed that these two processes account for all the entropy production—that is, the reversible processes do not themselves generate entropy, and this is still true even in the presence of the irreversible processes. We can now see that Thomson's assumption is valid if and only if $L_{UN} = L_{NU}$.

(28.2) *Entrained entropy.* Show that the rate at which entropy enters region 1 in Figure 28.5 is

$$\frac{dS_1}{dt} = \frac{1}{T}\left[\left(L_{UU} - \frac{L_{UN}L_{NU}}{L_{NN}}\right)f_U + \left(\frac{L_{UN}}{L_{NN}} - \mu\right)\dot{N}_1\right].$$

Interpret the first term, and hence show that the second term is the *entrained entropy*, that is, the extra entropy carried along by the particle current. Note, this is not the same as (and has little relation to) the equilibrium entropy per particle in the fluid.

(28.3) *A thermodynamic argument for the reciprocal relation.* Consider a situation of steady-state flow between two reservoirs separated by a thin membrane of thickness δx. The membrane is not perfectly conducting but does allow heat and particles, etc. to pass. We assume that the flow is slow enough for each reservoir to be characterized by a well-defined temperature, pressure etc., even though it is slow in releasing or absorbing entropy. The membrane is in steady state, so any entropy generated in it is passed to one or other of the reservoirs.

(i) Show that the rate of entropy generation in the whole system, per unit volume of the membrane, is

$$\frac{ds}{dt} = \sum_i (\nabla \phi_i) \cdot \mathbf{j}_i, \qquad (28.36)$$

where the potentials ϕ_i and currents $\mathbf{j}_i$ are as described in (28.31) and (28.32). This implies that $\dot{s}$ can be regarded as a function of the gradients and the currents:

$$\dot{s} = \dot{s}(\boldsymbol{\phi}'_1, \boldsymbol{\phi}'_2, \ldots, \mathbf{j}_1, \mathbf{j}_2, \ldots),$$

where $\boldsymbol{\phi}'_i \equiv \nabla \phi_i$.

(ii) From (28.36), show that $\mathbf{j}_i = (\partial \dot{s}/\partial \boldsymbol{\phi}'_i)$ and hence (using (28.32)) obtain

$$L_{ik} = \frac{\partial^2 \dot{s}}{\partial \boldsymbol{\phi}'_k \partial \boldsymbol{\phi}'_i}.$$

Similarly,

$$L_{ki} = \frac{\partial^2 \dot{s}}{\partial \boldsymbol{\phi}'_i \partial \boldsymbol{\phi}'_k}.$$

Since $\dot{s}$ is a well-behaved analytic function, it follows that $L_{ki} = L_{ik}$. Thus we appear to have obtained Onsager's reciprocal relation without the need to consider thermal fluctuations, and without the need to invoke the regression hypothesis or the principle of microscopic reversibility.

(iii) The above argument is plausible. However, when Thomson and others first discussed this subject, they were careful to raise questions over whether this type of argument is valid, because equilibrium concepts are being applied to out of equilibrium processes.[5] Does the presence of the flow invalidate the use of the concepts being used to describe it? Consider the case of not one thin layer, but many adjacent thin layers. Then temperature, and other intensive properties, are being ascribed to *a system with a non-negligible temperature gradient, through which energy and other quantities are flowing, and within which entropy is being generated.* Is (28.36) correct, and does it involve any hidden assumptions?

---

[5] Thomson described the required assumption in these terms: 'The electromotive forces produced by inequalities of temperature in a circuit of different metals, and the thermal effects of electric current circulating in it, are subject to the laws which would follow from the general principles of the thermodynamic theory of heat if there were no conduction of heat from one part of the circuit to another.'

Electric and magnetic work

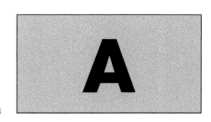

We present some basic results of classical electromagnetism that are used in Section 14.5.

| A.1 Quasistatic electric fields with polarization | 421 |
| A.2 Quasistatic magnetic fields with magnetization | 422 |

A.1 Quasistatic electric fields with polarization

First consider a system that consists of two conductors and, nearby, some dielectric material, all of arbitrary but fixed shape. Conductor A carries a charge q, and conductor B carries a charge $-q$. The electric potential difference between the conductors is defined by the line integral

$$\phi_A = \int_B^A -\mathbf{E} \cdot \mathbf{dr}, \tag{A.1}$$

where the path is from the surface of B to the surface of A. Ignoring the atomic structure of the conductors, we may take it that the electric potential is constant all around the surface of either conductor, and we can choose to assign the value 0 to the potential at B. Then the electric potential at conductor A is ϕ_A.

We consider first the quantity $q\phi_A$. This can be written

$$\phi_A q = \phi_A \int_A \rho \, dV = \int \phi \rho \, dV, \tag{A.2}$$

where $\rho(x, y, z)$ is the density of free charge (not polarization charge) at any point, and $\phi(x, y, z)$ is the electric potential at any point. The first integral is over the volume of conductor A, but the second integral is over all space. This is correct because we are assuming the case where the free charge is only located on conductors A and B, and we have assigned $\phi = 0$ at B so the charge there does not contribute to the integral. Using now the Maxwell equation (Gauss's law) $\nabla \cdot \mathbf{D} = \rho$, we have

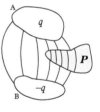

Fig. A.1

$$\phi_A q = \int \phi \nabla \cdot \mathbf{D} \, dV. \tag{A.3}$$

Now note the general vector identity,

$$\nabla \cdot (\phi \mathbf{D}) = (\nabla \phi) \cdot \mathbf{D} + \phi \nabla \cdot \mathbf{D}. \tag{A.4}$$

Using this to replace $\phi \nabla \cdot \mathbf{D}$ in the previous equation, we have

$$\phi_A q = \int \nabla \cdot (\phi \mathbf{D}) - (\nabla \phi) \cdot \mathbf{D} \, dV = \oint \phi \mathbf{D} \cdot \mathbf{dS} + \int \mathbf{E} \cdot \mathbf{D} \, dV, \tag{A.5}$$

where d**S** is a surface element for a surface enclosing the region over which the volume integral is evaluated. We said earlier that the volume integral was over all space, but we do not need to include the interior of conductor B in the volume integral, because in any case both ϕ and **D** are zero there. Therefore we can take as the closed surface for the surface integral the surface of B. Since $\phi = 0$ everywhere on this surface, that integral is zero. Hence,

$$\phi_A q = \int \mathbf{E} \cdot \mathbf{D} \, dV. \qquad (A.6)$$

Now suppose a small amount of charge δq is moved from B to A. By essentially the same argument, one finds

$$\phi_A \delta q = \int \mathbf{E} \cdot \delta \mathbf{D} \, dV. \qquad (A.7)$$

A.2 Quasistatic magnetic fields with magnetization

We consider a conducting loop with negligible electrical resistance. The loop could, for example, be wound into a helical shape and connected to a source of e.m.f., which we shall call a 'current source'. As long as the current in this circuit is constant the source does no work. In the vicinity of the loop there is a magnetizable body of arbitrary shape. Let I be the current and Φ the flux through the loop. When a process happens that causes Φ to change, there will be a back e.m.f. $d\Phi/dt$ which opposes the change. The rate of doing work by the source is then $I d\Phi/dt$ and the net work done after time δt is

$$\delta W = I \delta \Phi.$$

Now consider

$$I\Phi = I \int \mathbf{B} \cdot d\mathbf{S}, \qquad (A.8)$$

where the integral is over the surface bounded by the conducting loop. Hence,

$$I\Phi = I \int (\nabla \wedge \mathbf{A}) \cdot d\mathbf{S} = I \oint \mathbf{A} \cdot d\mathbf{l}, \qquad (A.9)$$

where **A** is the vector potential and d**l** is a line element around the conductor, in the direction of the current. The product of the current I with this line element can be written

$$I d\mathbf{l} = \mathbf{j} \, dV,$$

where **j** is the current density and dV is a volume element. Hence,

$$I\Phi = \int \mathbf{j} \cdot \mathbf{A} \, dV, \qquad (A.10)$$

where the integral is over the volume of the conductor in the first instance, but we can immediately extend this to all space since we assume there are no conduction currents elsewhere. Now use the Maxwell equation $\mathbf{V} \wedge \mathbf{H} = \mathbf{j} + \mathrm{d}\mathbf{D}/\mathrm{d}t$, where the second term is negligible in a quasistatic process. Hence,

$$I\Phi = \int (\mathbf{V} \wedge \mathbf{H}) \cdot \mathbf{A} \, \mathrm{d}V$$
$$= \int \mathbf{V} \cdot (\mathbf{H} \times \mathbf{A}) \, \mathrm{d}V + \int \mathbf{H} \cdot \mathbf{V} \wedge \mathbf{A} \, \mathrm{d}V, \quad (\mathrm{A}.11)$$

where the integrals are over all space. The first can be converted into a surface integral which will vanish as the surface tends to infinity, because the product of the fields falls faster than r^{-2}. Hence we obtain

$$I\Phi = \int \mathbf{H} \cdot \mathbf{B} \, \mathrm{d}V. \quad (\mathrm{A}.12)$$

By essentially the same argument, one also finds

$$I\delta\Phi = \int \mathbf{H} \cdot \delta\mathbf{B} \, \mathrm{d}V \quad (\mathrm{A}.13)$$

and

$$\Phi\delta I = \int \mathbf{B} \cdot \delta\mathbf{H} \, \mathrm{d}V. \quad (\mathrm{A}.14)$$

More on natural variables and free energy

B.1 Minimum energy principle   425
B.2 The general concept of natural variables   426

Natural variables were introduced in Chapter 13, in connection with thermodynamic potentials such as U, F, H, and G. In Chapter 17 we went on to consider the concept of *free energy* (also called availability), and argued that F and G are examples: for a closed system, F is the free energy in conditions of fixed T and V, and G is the free energy in conditions of fixed T and p. The free energy in each case is the function which is minimized in equilibrium.

Enthalpy and internal energy also have minimum principles associated with them. First we consider the case of an isolated system, and the internal energy.

We suppose that the entropy of an isolated system is a function of internal energy U and some internal parameter X. For example, the system could consist of two chambers with a movable partition between them, or a chemical reaction in an isolated vessel. There is only one internal equilibrium state under the external constraints of fixed U, V, N, but there are other nearby states that the system could adopt if it had a different U or if the internal parameter X were constrained to some particular value. Therefore we can define a function $S(U, X)$ which is the entropy the system would have if it were in the state at U, X (cf. Figure 17.3).

The maximum entropy principle for the isolated system tells us that, in equilibrium,

$$\left.\frac{\partial S}{\partial X}\right|_U = 0 \quad \text{and} \quad \left.\frac{\partial^2 S}{\partial X^2}\right|_U < 0. \tag{B.1}$$

Now we take an interest in the behaviour of the internal energy when S is fixed. That is, we are interested in $(\partial U/\partial X)_S$. Let us name it Y:

$$Y \equiv \left.\frac{\partial U}{\partial X}\right|_S = -\left.\frac{\partial U}{\partial S}\right|_X \left.\frac{\partial S}{\partial X}\right|_U = -T\left.\frac{\partial S}{\partial X}\right|_U \tag{B.2}$$

and the latter quantity is zero at the equilibrium point. So we find that at the equilibrium determined by the maximum entropy principle, $Y = 0$, i.e. $U(S, X)$ is at a stationary value for fixed S. It remains to discover whether this is a maximum or a minimum. We have

$$\left.\frac{\partial^2 U}{\partial X^2}\right|_S = \left.\frac{\partial Y}{\partial X}\right|_S = \left.\frac{\partial Y}{\partial U}\right|_X \left.\frac{\partial U}{\partial X}\right|_S + \left.\frac{\partial Y}{\partial X}\right|_U, \tag{B.3}$$

where the last step follows from considering $Y = Y(U, X)$. Now, using $(\partial U/\partial X)_S = Y = 0$, which we have just discovered, we have

Table B.1

Variables constrained		Equilibrium condition
Often used	U, V	S *max*imum
	T, V	F minimum
	T, p	G minimum
	S, V	U minimum
Rarely used	U, S	V minimum
	S, p	H minimum
	G, T	p *max*imum
	G, p	T minimum
	F, T	V minimum
	F, V	T minimum
	H, S	p *max*imum
	H, p	S *max*imum

$$\left.\frac{\partial^2 U}{\partial X^2}\right|_S = \left.\frac{\partial Y}{\partial X}\right|_U = -\left.\frac{\partial T}{\partial X}\right|_U \left.\frac{\partial S}{\partial X}\right|_U - T \left.\frac{\partial^2 S}{\partial X^2}\right|_U \qquad (B.4)$$

$$= -T \left.\frac{\partial^2 S}{\partial X^2}\right|_U > 0. \qquad (B.5)$$

Thus $U(S, X)$ is minimized in the closed system at fixed S, V.

A similar argument can be made about the enthalpy of a closed system that can exchange volume but not heat with a reservoir. Such a system does not receive entropy from its environment, so its equilibrium is governed by a maximum of its own entropy, subject to a constraint of constant pressure. One can also consider other combinations, and one thus finds the various equilibrium conditions listed in Table B.1.

B.1 Minimum energy principle

The case of a system undergoing reversible behaviour (so S is constant) is sufficiently important that its equilibrium condition, obtained in equations (B.2) and (B.5), has earned the name of 'minimum energy principle', as we discussed in Section 17.1.2. This is often stated as the principle that the thermal equilibrium state, for a system undergoing reversible changes in its internal configuration, is the one that minimizes the internal energy. However, that statement may lead to confusion because energy is strictly conserved, so if a system is not changing its entropy nor exchanging energy with outside bodies, then it cannot come to equilibrium. In practice, if such a system is already, for some reason, in the state of minimum internal energy, then it stays there. If it is not, then it never comes to equilibrium unless it can exchange energy with other systems.

For example, consider a cylinder of gas divided into two compartments by a movable partition that can slide without friction. We assume this partition is a thermally insulating piston, so no heat passes between the two compartments 1 and 2. The internal energy of the system is

$$U = U_1 + U_2. \tag{B.6}$$

When the piston moves by a small amount, the internal energy changes by

$$dU = -p_1 dV_1 - p_2 dV_2, \tag{B.7}$$

where dV_1 and dV_2 are the changes in volume of the two chambers. Since the total volume of the system is constant, we have $dV_2 = -dV_1$. The condition for a stationary value of U is

$$p_1 dV_1 + p_2 dV_2 = 0, \tag{B.8}$$

and therefore $p_1 = p_2$. You can further confirm that this is a minimum not a maximum value of U as a function of V_1.

If the pressures are not equal at the outset, then the piston in such a system will oscillate, and in the absence of friction or other damping the oscillation will continue indefinitely. In this case the equilibrium state is never reached.

The influence of damping can be modelled by leaving the internal configuration of the system unchanged, so that everything within the system remains reversible, but allowing the moving piston to interact with a second system which offers viscous damping. For example, one could imagine a rod connected to the piston and extended into a bowl of syrup or attached to a vane in a further gas container. Then we consider the complete composite system, and suppose that this combined system is isolated. Its total entropy is

$$S_{\text{tot}} = S + S_d, \tag{B.9}$$

where S is the entropy of the reversible system under study, and S_d is the entropy of the system offering damping. Equilibrium of the composite isolated system is attained at a maximum of S_{tot}, and since S is constant, this implies a maximum of S_d. But the nature of viscous damping is that the internal energy of the damping system, U_d, can only increase under the action of the moving vane or rod (the viscous forces always oppose the motion). In this configuration, U_d is therefore a monotonic increasing function of S_d. It follows that when S_d is maximized, U_d is also maximized, and therefore U is minimized since the total $U + U_d$ is constant.

B.2 The general concept of natural variables

The general concept of natural variables is a mathematical idea that arises whenever we consider functions of more than one variable. The issue is essentially to do with the difference between differentiation and integration.

Suppose we would like to know some function $f(x)$ which only depends on one variable, but we have only been able to discover a formula relating f to its first derivative df/dx. The situation is not too bad: we have a first-order differential equation to solve, but we can tackle it either analytically or if necessary by numerical methods, and we shall soon learn f up to some constant of integration.

Suppose however that f is a function of two variables, say x and y, and we have in our possession a formula such as

$$f = \frac{y}{3} \left.\frac{\partial f}{\partial x}\right|_y.$$

We can try to solve the equation

$$\frac{1}{f}\left.\frac{\partial f}{\partial x}\right|_y = \frac{3}{y} \quad\Rightarrow\quad \ln f = \frac{3x}{y} + g(y) \quad\Rightarrow\quad f = e^{g(y)} e^{3x/y},$$

where g is some unknown function. This is of some use, but our knowledge of f remains severely limited because g here can be any function whatsoever. We know almost nothing about $(\partial f/\partial y)_x$, for example.

This situation regularly arises in thermodynamics in a slightly hidden form. A quantity such as $(\partial f/\partial x)_y$ might itself be an interesting function of state, and have earned itself a name and a symbol, such as temperature T.

Suppose

$$T \equiv (\partial f/\partial x)_y,$$

then the formula above reads

$$f(T, y) = \frac{Ty}{3}. \tag{B.10}$$

This looks like a nice straightforward formula, telling us that f can be considered to be a function of T and y, and we appear to have complete information about f for all states of the system. The trouble is, this appearance is misleading: we still can't deduce $(\partial f/\partial y)_x$.

If instead we had information about f in terms of x and y, say

$$f = y e^{3x/y}, \tag{B.11}$$

then we *can* discover everything else. $(\partial f/\partial y)_x$ is available immediately, and so is T.

In a situation like this, we say that x and y are the *natural variables* or *proper variables* of f, whereas T and y are not. Note that it is easy to derive (B.10) from (B.11), but impossible to derive (B.11) from (B.10). We have already tried to do the latter, and we ended up with an unknown function g in the formula.

C Some mathematical results

C.1 The Riemann zeta function

For real $s > 1$, the Riemann zeta function can be defined

$$\zeta(s) \equiv \frac{1}{\Gamma(s)} \int_0^\infty \frac{x^{z-1}}{e^x - 1} dx,$$

where $\Gamma(s)$ is the gamma function.

For integer values of s, one may show that

$$\zeta(s) = \sum_{n=1}^\infty \frac{1}{n^s}.$$

This converges for $\Re[s] > 1$ and may be taken as a more general definition of $\zeta(s)$.

Values for small integer s:

s	$\zeta(s)$	$\Gamma(s)$	$\Gamma(s)\zeta(s)$
1	∞	1	∞
2	$\pi^2/6$	1	$\pi^2/6$
3	$\simeq 1.2020569$	2	2.4041138
4	$\pi^4/90$	6	$\pi^4/15$
5	$\simeq 1.036927755$	24	24.88626612344
6	$\pi^6/945$	120	$8\pi^6/63$

There is a beautiful connection between the Riemann zeta function and the prime numbers. The following identity was proved by Leonhard Euler:

$$\sum_{n=1}^\infty \frac{1}{n^s} = \prod_{p \text{ prime}} \frac{1}{1 - p^{-s}}.$$

To prove this, use

$$\frac{1}{1 - p^{-s}} = 1 + \frac{1}{p^s} + \frac{1}{p^{2s}} + \frac{1}{p^{3s}} + \cdots$$

in each term in the product on the right-hand side of Euler's formula. Upon multiplying out the brackets, you discover that the product is a sum of terms of the form

$$\frac{1}{p_1^{k_1 s} \cdots p_n^{k_n s}} = \frac{1}{(p_1^{k_1} \cdots p_n^{k_n})^s},$$

where $p_1 \cdots p_n$ are different primes and $k_1 \cdots k_n$ are positive integers, and each such combination occurs exactly once. But, by the fundamental theorem of arithmetic, each such product is a unique integer, and all the integers will appear in the sum. This establishes Euler's formula.

C.2 The Wiener–Khinchin theorem

Define the Fourier transform

$$\tilde{f}(\omega) \equiv \int_{-\infty}^{\infty} f(t) e^{-i\omega t} dt. \tag{C.1}$$

It may be shown that one then has

$$f(t) = \frac{1}{2\pi} \int_{-\infty}^{\infty} \tilde{f}(\omega) e^{i\omega t} d\omega. \tag{C.2}$$

This second result is called the inverse Fourier transform.

If $f(t) = \delta(t - t')$ for some constant t', then

$$\tilde{f}(\omega) = \int_{-\infty}^{\infty} \delta(t - t') e^{-i\omega t} dt = e^{-i\omega t'}. \tag{C.3}$$

Substituting this into the inverse Fourier transform, we find

$$\delta(t - t') = \frac{1}{2\pi} \int_{-\infty}^{\infty} e^{i\omega(t - t')} d\omega, \tag{C.4}$$

or

$$\int_{-\infty}^{\infty} e^{i\omega(t - t')} d\omega = 2\pi \delta(t - t'). \tag{C.5}$$

Now let

$$V(t) = \int_{-\infty}^{\infty} V_\omega e^{i\omega t} d\omega. \tag{C.6}$$

Consider the auto-correlation function,

$$C(t) \equiv \langle V^*(0) V(t) \rangle = \int_{-\infty}^{\infty} V^*(t') V(t' + t) dt' \tag{C.7}$$

$$= \int_{-\infty}^{\infty} dt' \left[\int_{-\infty}^{\infty} V_\omega^* e^{-i\omega t'} d\omega \right] \left[\int_{-\infty}^{\infty} V_{\omega'} e^{i\omega'(t' + t)} d\omega' \right]$$

$$= \int_{-\infty}^{\infty} \int_{-\infty}^{\infty} d\omega d\omega' \left[\int_{-\infty}^{\infty} V_\omega^* V_{\omega'} e^{i\omega' t} e^{i(\omega' - \omega)t'} dt' \right]$$

$$= \int_{-\infty}^{\infty} \int_{-\infty}^{\infty} d\omega d\omega' V_\omega^* V_{\omega'} e^{i\omega' t} 2\pi \delta(\omega' - \omega)$$

$$= 2\pi \int_{-\infty}^{\infty} |V_\omega|^2 e^{i\omega t} d\omega. \tag{C.8}$$

Now take the inverse Fourier transform of this equation. One obtains

$$(2\pi)^2 |V_\omega|^2 = \int_{-\infty}^{\infty} \langle V^*(0) V(t) \rangle e^{-i\omega t} dt. \tag{C.9}$$

This is the *Wiener–Khinchin theorem*. It states that the power spectral density is equal to the Fourier transform of the auto-correlation function.

Bibliography

Introductory

C. J. Adkins, Equilibrium Thermodynamics, 3rd ed (Cambridge University Press, 1984).
C. Kittel and H. Kroemer, Thermal Physics, 2nd ed (Freeman 1980).
S. J. Blundell and K. M. Blundell, Concepts in Thermal Physics (Oxford University Press, 2nd ed, 2009).
G. Carrington, Basic Thermodynamics (Oxford University Press, 1994).
R. P. Feynman, R. B. Leighton, and M. Sands, The Feynman Lectures on Physics, Volume I (Addison-Wesley, 1963).
E. Fermi, Thermodynamics (Dover Publications; New Ed edition, 1956).
F. Mandl, Statistical Physics (John Wiley, London, 1971).
H. C. Van Ness, Understanding Thermodynamics (Dover Publications, 1983).
A. B. Pippard, The Elements of Classical Thermodynamics (Cambridge University Press, Cambridge, 1957).
J. A. Ramsay, A Guide to Thermodynamics (Chapman and Hall, 1971).
H. Rosenburg, The Solid State (Oxford, 3rd ed, 2008).
D. V. Schroeder, An Introduction to Thermal Physics (Addison-Wesley, 2000).
R. E. Sonntag, C. Borgnakke, and G. J. Van Wylen, Fundamentals of Thermodynamicis (John Wiley & Sons, 1998).
J. R. Waldram, The Theory of Thermodynamics (Cambridge University Press, 1985).
M. W. Zemansky, Heat and Thermodynamics, 7th ed (McGraw-Hill, 1996).
R. Endem, Nature 141, 908 (1938).
E. A. Guggenheim, J. Chem. Phys. 13, 253 (1945).

Intermediate or advanced

H. B. Callen, Thermodynamics and an Introduction to Thermostatistics (John Wiley & Sons, 1985).
H. Gould and J. Tobochnik, Statistical and Thermal Physics: With Computer Applications (Princeton University Press, 2010).
E. A. Guggenheim, Thermodynamics: An Advanced Treatment for Chemists and Physicists (North Holland, 8th ed, 1986).
L. D. Landau and E.M. Lifshitz, Statistical Physics (Course of Theoretical Physics Volume 5) (Butterworth-Heinemann; 3rd ed, 1980).
J. C. Lee, Thermal Physics: Entropy and Free Energies (World Scientific, New Jersey, 2002).
R. H. Swendsen, An Introduction to Statistical Mechanics and Thermodynamics (Oxford University Press, 2012).
A. L. Wasserman, Thermal Physics: Concepts and Practice (Cambridge University Press, 2012).
J. Wilks, The Third Law of Thermodynamics (Oxford University Press, 1961).

Data

Much of the data in this book has been drawn from Kaye and Laby online http://www.kayelaby.npl.co.uk/ provided by the National Physical Laboratory of Great Britain.

Index

A
absolute
  entropy 111–112
  temperature 53, **97–101**, 230
  zero 32, 102, 111, 179, 184, 266, 331–335
activation energy 301
activity 306
adiabatic
  atmosphere 87
  compressibility 28
  expansion 86, 89–91, 107, 187, 228, 237, 283
  index 78, 83–85, 181, 280, 290
  lapse rate 88
  process 26, 93, 182, 333
  surface 128–131, 135, 180, 263–265, 335
adiathermal **25–27**, 88, 130, 155, 263–265
  work, principle of 70–72
Adkins 90
aerofoil 232
affinity 306
albedo 273, 294
allotrope 315, 334, 356
anomaly, specific heat 120
argon 162, 213, 225, 239
arrow of time 131
aspect 267, 331
atmosphere
  adiabatic 87
  atm 37
  Earth's 273, 294–297
  isothermal 163
atmospheric engine 156
availability **250–253**, 342, 424
Avogadro's number 36, 119, 300

B
balloon 25, 254
bar 37
bath (reservoir) 29
beer 158, 346
Bekenstein, J. 379
Bernoulli equation 231
binary mixture 358, 370–373
binodal line 339
black body 269, 292
  radiation 270–291
black hole 4, 378–381
Bohr magneton 38, 79
boiling 38, 311, 345
  point 90, 219, 318, 321, 348, 351
Boltzmann
  constant 36
  entropy formula 114, 384
  factor 36, 126, 197, 212
  Ludwig E. 114
Boomeria 153
Bose-Einstein
  condensation 356, 358
Boyle's law **59–61**, 95, 116, 184, 229
Boyle temperature 218, 223, 226
brass 121, 271, 356, 361
Brayton cycle 240–242
brick 271
bubble 188, 191, 345
bulk modulus 28, 33, 62, 76
Bureau International 99

C
calibration 53, 66–68
calorimetry 309
capacitor 134, 168, 200, 389
Cape Grim 296
Caratheodory 31, 263
carbon 36, 271, 278, 311, 315
  dioxide 296, 299, 346
Carnot
  cycle 93–97, 334
  engine 110, 155
  theorem 97–100, 262
Carrington, G. 69, 200
cavity radiation, see black body
Celsius scale 32
cerium magnesium nitrate 199
chemical potential 19, 121, **157–162**, 169–173, 244, 260, 301, 335
  of ideal gas 161, 185
  molar 306
  and phase change 321, 326, 337–348, 372
  of some compounds 300
  of thermal radiation 281–282, 290
Clapeyron, see Clausius-Clapeyron
Clausius
  statement 30, 96, 123, 244, 262, 276, 385
  theorem 104, 130
  -Clapeyron equation 315, 320–324, 328–330, 334, 350
climate change 295–297
cloud chamber 345
CMB radiation 288, 381
coexistence curve 233, **312–325**, 328, 330, 341, 348
colligative 347–353
component, chemical 158, 299, 325, 358, 371
compressibility 28, 33, 179, 392
compression factor 218
compressor 90, 236–242
concave function 248, 258
condensation 311
  Bose-Einstein 356
  gravitational 377
  nuclei 345
conduction, thermal **136–145**, 403–418
conservation law 259, 418
constraint 21, 114, 252–254, 302, 325
  and fluctuations 384, 387–392, 424
contact potential 160
continuity equation 137, 418
control surface 12
convection 136, 149, 297, 405
convex envelope 293
coordinate,
  generalized **15–17**, 128, 263–264, 415
corona, solar 279
corresponding states, law of 222–224
cosmic, see CMB
critical
  exponent 356, 359–361, 365
  opalescence 358
  point **312–314**, 324, 328, 339, 341, 356
    ferromagnetic 367, see also Curie point
  radius 344–347
  temperature, see critical point
cryoscopic 352
crystalline structure 114, 122, 181, 190, 314, 334, 361
Curie
  law **64–66**, 194–196, 332
  point 199, 356
  temperature 199
  -Weiss law 199, 213, 215, 365

434   Index

cycle
  Brayton 240
  Carnot 93
  Ericsson 155
  Otto 153
  Rankine 241
  Stirling 155
  see also engine

D
daemon 122–125
Daniel cell 307
demagnetization 195, 197–200, 213, 215
  -zing factor 214
diamagnetism 63, 327
diamond 90, 112, 315
diatomic gas 28, 80, 87, 181, 284
  see also nitrogen
dielectric 202
diesel engine 155
Dieterici gas 217, 225–226, 239
diffusion equation 137
diffusivity 140, 396
dispersion relation 140
dissociates 307
dissociation 307, 310, 351
droplet 341–346

E
Earth 11, 163, 273, 294–297
ebullioscopic 351
efficiency
  Carnot engine 94, 98, 102
  gas turbine 238
  heat pump 96
  Otto cycle 154
effusion 275
Ehrenfest
  classification scheme 355
  equations 373
Einstein, A. 4, 53, 288
electric cell 307
electrochemical field 404
emissivity
  (emittance) 271–272, 277, 297–298
Endem 90
endothermic 306
energy

equation 80, 184, 192, 221
  internal 13, 18, 60, 70–73, 86, 169, 263
  see also free energy
engine
  Carnot 93
  heat 93–96, 406
  internal combustion 153
  Newcomen 156
  Stirling 151, 155
  Szilard 123
  turbojet 242
  see also cycle
entangled state 385
enthalpy 172–177, 186, 231, 237, 256
  of fusion 38
  magnetic 210
  of reaction 305, 309
  of water 241
entropy 1–420
environment 12, 25, 252
equilibrium
  chemical 301–309
  phase 321–326
  state 14
  thermal 22
  thermodynamic 20
  see also stability conditions
equipartition theorem 387, 399
ergodic hypothesis 395
Ericsson cycle 155
eutectic 372
evaporation 316, see also boiling
exact differential, see proper differential
exergetic measure 238
exothermic 306
expansivity 27–28, 76, 179, 189
extensive 18–20

F
Faraday's law 64, 207
Fermi
  energy 159
  temperature 69
ferromagnetism 63, 199, 319, 331, 356–358, 366–370

Feynman
  Richard 100
  -Smoluchowski ratchet 395
first law 30, 70
Flanders, D. 7
flow process 230, 236–242
fluctuation 248, 331, 382–402
  -dissipation theorem 392, 413
  and phase change 318, 338–339, 345, 358–359
flux 136
fountain effect 349
Fourier analysis 143, 399, 429
free energy 173–174, 196, 221, 253–258
  see also Helmholtz function; Gibbs function
free expansion 85–86, 115, 135, 228–230
freezing 74, 351–352
  point 56, 67, 347
  see also phase change
friction 24, 75, 152, 392
fugacity 306
fundamental relation
  dielectric 205
  differential form 105, 169–170, 172, 243, 299, 415
  elastic rod 188
  integral form 170, 186
  magnetic 194, 327
  surface 191
fusion see freezing

G
Gardner, Martin 6
geophysics 62
Gibbs
  -Duhem relation 171, 186, 299, 321, 326, 335, 391
  energy, function 172–176, 221, 254–256, 281
  and chemical reations 299–307
  and phase change 317, 322, 330, 340–343, 352–353, 371
  -Helmholtz equation 177, 187
  Josiah W. 158, 170

paradox 5, 116–118
phase rule 325
Glaser, D. 345
glass 278, 290, 331
graphite 35, 112, 315
gravitation 20, 59, 87, 163, 375–378
greenhouse effect 294–297
Grüneisen parameter 181

H
Hawking radiation 4, 378–381
heat 72–73
  see also bath; capacity; engine; pump
heat capacity 19, 27–28, 34–35, 78–80, 111, 178, 332
  anomaly 120–122, 369
  of black body radiation 281
  difference 179
  ratio 83, 180
  of real gas 226–227
  of superconductor 329
  of water 34–35, 90
helium 35, 38, 90, 223, 233, 239–240, 316, 318, 349
Helmholtz function 172–177, 186–187, 253–256, 361–364
  of cavity (black body) radiation 290
  and Curie-Weiss law 213
  of ideal elastic 213
  of ideal paramagnet 196
  of Peng-Robinson gas 227
  of van der Waals gas 221
Hodgson, Ralph 108
Hooke's law 62, 90, 189
hydrodynamics 147, 232
hydrogen 34–38, 233, 304
hyperfine structure, atomic 118, 161
hypersurface 128
hysteresis 24, 63, 264, 370

I

ice 315, 323, 329, 347
improper, see proper
inaccessible 31, 263–265
indistinguishability 117–118
integration 45–51
intensive 18
ionization 165–168, 289, 309
iron 199
irreversible, see reversible
isentropic 25
isobar, isobaric 27–28, 77
isochoric 27
isotherm, isothermal 27

J

Jeans length 177
jet engine 242
Joule
  expansion, see free expansion
  experiment 71
  James 70

K

Kelvin
  relation 407
  temperature unit 101
  statement 30, 96, 103–107, 130, 262
  William Thomson, Lord Kelvin 123, 420
Kerr black hole 379
Kirchhoff's law 271–278, 290

L

Landau
  free energy (grand potential) 390
  L., mean field theory 361
Landauer, R. 123
latent heat 37–38, 90, 109, 111, 151, 250, 317–319, 372
  relation to colligative properties 351
  temperature dependence 324, 328
lattice 114, 122, 197
Law Dome 296
Lederman 121
Le Chatelier's principle 249, 306, 387, 395
Legendre transformation 172
Linde process 236, 240
liquifier 236

M

macrostate 114–115, 124
magnetic work 207–210
magnetism 39, 63–65, 79, 192–196, 326
  see also demagnetization; ferromagnetism
magnetization 19, 63
magneton, Bohr 39
maximum entropy principle 31, 111, 243–245, 247, 250, 259–263
maximum work theorem 152
Maxwell
  construction 222, 257, 339, 373
  daemon 122, 125
  equations 202, 421–423
  relation 173, 176, 180, 187, 192, 320, 332
mean field theory 221, 360–370
Meissner effect 327
metastability 266, 319, 331, 334, 338–346, 369
methane 218, 223
minimum energy principle 246
miscibility 370–372
molality 351
mole, molar 35
Mollier (hs) chart 238–241
Murnaghan equation of state 62

N

natural variables 173, 175, 183, 186, 290, 365, 424–427
Navier-Stokes equation 147
negative temperature 265–267
neon 185
Newcomen engine 156
nitrogen 35, 38, 79, 87, 223, 225, 233–236, 240, 396
noble gas 67, 223
noise 67, 398
nucleation 341–347
Nyquist, see Johnson noise

O

Onsager's
  reciprocal relation 412, 418
  regression hypothesis 413
opalescence 358
osmosis 347–350, 354
Ozymandias 133

P

paramagnetism 63, 192
  see also magnetism
path (sequence of states) 16, 23, 48
Peltier effect 405
Peng-Robinson gas 225–227
Penrose process 379
permittivity 205–207
perpetual motion (perpetuum mobile) 107
phase
  change (transition) 216, 220–222, 241, 249–250, 311–331, 334
  continuous phase transition 355–374
  diagram 312–315
  metastable 338
  rule (Gibbs) 325
  see also nucleation; colligative
photosphere 165, 168
pinking 155
Planck-Simon statement (third law) 31, 331
spectrum 274
polarization
  electric 19, 202–207, 421
  electromagnetic waves 269, 274
polymer 63, 261
potato 110
pressure 75
Prigogine, I. 416

Index 435

principle
  of adiathermal work 70
  of detailed balance 127
  heat capacities 78
  of microscopic reversibility 414
  of virtual work 247
probability 118, 210, 371, 383–386
proper differential 46–48
pudding 142
pump, heat 11, 95–96, 106, 151
pyrometer 66–67

Q

quasistatic 27, 33, 103, 145, 230

R

radiation (radiant heat), see black body radiation
randomness 125, 132, 384–385
Rankine cycle 241–242
Raoult's law 350–351
ratchet 125–128, 384
reciprocal
  relation, see Onsager
  theorem 43
reciprocity theorem 43
Redlich-Kwong gas 217, 225–226
reflectivity 269
refreshments
  beer 158, 346
  coffee 6, 16, 20–23, 149
  orange juice 16
  whisky 158
  see also water
refrigeration 95, 151–152, 199, 242
regenerator 155
relativistic gas 148
reservoir 29
response function 28, 332, 355–357, 391
reversible 23–25
Riemann zeta function 283, 428
Rüchardt's method 89

S

Sackur-Tetrode equation 170, 182, 282
Sadus, R. 227
Saha equation 165–168, 309
Salamon and Lederman 121
saturated vapour 313, 343
saturation curve 220, 241, 313
scaling hypothesis 365
Schottky anomaly 121
Schwarzschild radius 379
second law 30–31
Seebeck effect 403–405
semi-permeable membrane 121
Shakespeare, W. 188
shape factor 293, 297
Shelley, P. B. 133
Simon statement 31, 331
Smoluchowski ratchet 125–127
solar
  irradiance 294, 297
  power 151
solenoid 193, 207
solubility 370
sound, speed of 67, 146–148
spinodal 339, 375
stability conditions 247–249, 258, 266
standard temperature and pressure, *see* STP
star 270, 377
Starobinsky, A. 379
Stefan-Boltzmann
  constant 69, 270, 335
  law 270–275
Stirling
  approximation 186
  cycle 155
  engine 151, 155
stoichiometric coefficient 301–303
STP 37
stress 188
sublimation 311
Sun 11, 271, 273, 279, 380
superconductivity 319, 326–329, 355–358
supercooling, superheating 319, 339–344
superfluidity 349, 356–358
supersaturated vapour 338
susceptibility 28, 193, 199, 214, 332
symmetry 314, 331, 381
Swann, D. 7
Szilard engine 123–125, 385

T

temperature
  absolute 99, 260
  empirical 54–58
thermal de Broglie wavelength 161, 185
thermal radiation, *see* black body radiation
thermistor 66, 68
thermocouple 68, 404
thermoelectric
  effect 403
  equations 407
thermometer, thermometry 66–69
thermopower (Seebeck coefficient) 404
thermostat 29
thermostatistics 5, 114
Thomson
  heat, coefficient 405–408
  relation 407, 417
  W. 123, 420
  see also Kelvin
Trouton's rule 318, 324
turbine 155, **236–242**
turbojet 242

U

ultraviolet catastrophe 287–288
unattainability theorem 32, 333–334
universality class 359
unlivability 158

V

van der Waals
  critical exponents 359–360, 373
  equation 17, 59, 217, 225, 339, 359
  gas 33, 68, **219–223**, 226, 229, 234, 239, 340
van der Ziel 401
van 't Hoff equation 305
vaporization, *see* boiling; evaporation
vapour 311
virial
  coefficient, expansion 217–219, 226, 233, 239

theorem 376–378

W

water, properties of
  critical and van der Waals parameters 223
  cryoscopic and ebullioscopic constants 351
  density 56
  emissivity 271
  latent heats 38, 317–318, 324, 329
  Mollier (hs) chart 241
  phase diagram 315
  specific heat capacity, 34–35, 90
  surface tension 190–191
  triple point 101
  vapour pressure 324
Weiss
  model 357, 366–369
  see also Curie-Weiss law
Wien, W. 284
  displacement law 270, 287
  distribution law 286
Wiener-Khinchin theorem 399, 429

Y

Young's modulus 90, 188

Z

Zeldovich, Y. 379
zeroth law 29, 32, **52–59**, 68, 265
zeta function 283, 428